Fetal Origin of Diseases

Hui Wang • Liaobin Chen • Dan Xu
Editors

Fetal Origin of Diseases

Springer

Science Press
Beijing

Editors
Hui Wang
Department of Pharmacology
School of Basic Medical Science, Wuhan University
Wuhan, Hubei, China

Liaobin Chen
Division of Joint Surgery and Sports Medicine, Department of Orthopedic Surgery
Zhongnan Hospital of Wuhan University
Wuhan, Hubei, China

Dan Xu
Department of Pharmacology
School of Pharmaceutical Sciences, Wuhan University
Wuhan, Hubei, China

ISBN 978-981-97-5729-9 ISBN 978-981-97-5730-5 (eBook)
https://doi.org/10.1007/978-981-97-5730-5

Jointly published with Science Press

This Springer imprint is published by the registered company Springer Nature Singapore Pte Ltd.
The registered company address is: 152 Beach Road, #21-01/04 Gateway East, Singapore 189721, Singapore

Foreword

"Fetal-originated diseases" refers to non-communicable diseases in offspring caused by abnormal fetal development due to an adverse intrauterine environment. In the early years, research in this field primarily focused on maternal malnutrition during pregnancy. However, there was scant research on the effects of exogenous substances (such as environmental toxins, therapeutic drugs, and unhealthy diet) on fetal development and fetal-originated diseases. As the chief editor of this book, Professor Hui Wang has exceptional forward-thinking and keen research intuition in this domain. She excels in identifying new scientific questions from research and continually broadens innovative approaches. Through over two decades of systematic study, she has successfully extended drug developmental toxicology to xenobiotic developmental toxicology. She adeptly utilizes the multidisciplinary and integrative aspects of research on fetal-originated diseases, involving basic medicine, clinical medicine, pharmacy, preventive medicine, and biology. Internationally, she was the first scholar to organically combine developmental toxicology with fetal-originated diseases, achieving a truly interdisciplinary and cross-disciplinary systematic research approach. This not only opened a new era for developmental toxicology research but also led the new development in international research on fetal-originated diseases.

This book is a comprehensive and detailed monograph in the field of fetal-originated diseases, incorporating the authors' research results, innovative theories, and deep thinking over the past twenty years. It provides valuable references for scholars to further research the international hot topic—the "Developmental Origins of Health and Disease (DOHaD)" theory. At the same time, the systematic research of Professor Hui Wang's team on the intrauterine programming mechanisms and early prevention strategies of fetal-originated diseases also provides important guidance for the etiological analysis and therapeutic intervention of adult diseases. I believe that the publication of this book is a milestone in the field of fetal-originated disease research and plays an important guiding role in directing scholars to deeply research and elucidate complex life science problems.

Harbin, China — Baofeng Yang

Preface

In the 1990s, British scholar David Barker published a groundbreaking study in The *Lancet*. The study found that infants born with low birth weight, shorter body length, and smaller head circumference were more likely to develop diseases such as hypertension, coronary heart disease, type 2 diabetes, and osteoporosis in adulthood. Based on these findings, Barker innovatively proposed that certain phenomena in early life "imprint" on a child's lifelong health. This led to the widely accepted theory of the "Developmental Origins of Health and Disease (DOHaD)" and the concept of "fetal origins of adult diseases."

This book, based on the series of research work by chief editors' team, covers the entire process from physiological development to the onset of diseases. It spans multiple levels from environment to individual, from mother to fetus, from intrauterine to after birth, while integrating elements of basic research and clinical translation. Using the typical examples of caffeine and dexamethasone exposure during pregnancy, the book comprehensively summarizes the characteristics of xenobiotic developmental toxicity—multi-pathway, multi-organ, multi-target effects, as well as its noticeable gender differences, multi-generational genetic effects, and disease susceptibility, *etc*. At the same time, it systematically depicts the phenomena of offspring multi-organ (liver, pancreas, brain, long bones, cartilage, adrenal glands, kidneys, testes, ovaries, etc.) developmental programming and susceptibility to various chronic diseases in adulthood induced by xenobiotic exposure during pregnancy. These include metabolic syndromes, diabetes, non-alcoholic fatty liver disease, hypercholesterolemia, epilepsy, osteoporosis, osteoarthritis, adrenal insufficiency, glomerulosclerosis, testicular dysgenesis syndrome, and premature ovarian failure, etc. Innovatively, it summarizes the common mechanisms of fetal-originated diseases related to exogenous developmental toxicity from three different perspectives—oxidative stress imbalance and multi-cellular function damage, epigenetic modification and developmental programming of multi-organs, endocrine developmental programming, and susceptibility to multiple diseases. It involves cutting-edge topics such as cellular toxicity molecular mechanisms, mechanisms of abnormal epigenetic modification, intrauterine glucocorticoid programming mechanisms, and mechanisms of gut microbiota. In addition, the book also details the methods of establishing animal models in author's team's research on fetal-originated diseases, as well as their study progresses in exploring early warning and intervention targets.

In summary, this book comprehensively and systematically elucidates the intrauterine programming mechanisms and early prevention strategies that lead to offspring's susceptibility to multiple diseases due to exogenous exposure during pregnancy. It proposes innovative theories such as "maternal glucocorticoid programming offspring's susceptibility to multiple diseases", "there are two-programming and two-strike mechanisms in fetal-originated diseases", and "fetal-originated diseases belong to the category of metabolic syndromes" for the first time. We believe these research results from this book will be of important theoretical value and significant practical significance for guiding healthy life during pregnancy, optimizing the reproductive process, and improving the quality of population survival.

Wuhan, China Hui Wang
Wuhan, China Liaobin Chen
Wuhan, China Dan Xu

Brief Introduction

This book focuses on the fetal origin of adult disease. By summarizing the global research progress and the findings of the authors' team for more than 20 years, it reveals the developmental origin, intrauterine programming mechanism, early prevention and treatment strategies of a variety of adult diseases (such as metabolic syndrome, metabolic diseases, neuropsychiatric diseases, bone and joint diseases, reproductive system diseases, etc.). It is the first time to put forward the systematic innovative theories such as "maternal glucocorticoid programming offspring's susceptibility to multiple diseases", "two-programming and two-strike mechanisms in fetal-originated disease", and "fetal-originated diseases belong to the category of metabolic syndrome". This book is the most comprehensive and detailed monograph on fetal-originated disease in the world. It provides important guidance for the etiology analysis and treatment intervention of adult diseases, and provides valuable reference for researchers to further study the international hot issue of "Developmental Origins of Health and Disease (DOHaD)" theory.

Contents

About the Editors

Hui Wang holds the prestigious position of professor of pharmacology, Wuhan University. She is celebrated as the "LUOJIA Distinguished Scholar" at Wuhan University and the leading scientist of the National Key R&D Program. In her current role, she steers the Hubei Provincial Key Laboratory for Developmentally Originated Diseases and shines as a distinguished professor at the Women's and Children's Hospital, affiliated with Zhongnan Hospital of Wuhan University, and an adjunct professor, Peking University Health Science Center. Her leadership extends beyond academia into influential roles in professional societies: she holds vice-chairmanships in the Biochemical and Molecular Pharmacology Committee of the Chinese Pharmacological Society, the Reproductive Toxicology Committee of the Chinese Society of Toxicology, and the Environmental and Developmental Diseases Committee of the Chinese Society of Environmental Mutagens. She primarily engaged in research of developmental toxicology of xenobiotics and prevention strategies for fetal-originated diseases. Her academic achievements mainly include: the creation of various animal models for intrauterine growth retardation, systematically elucidating the programming mechanisms of the thrifty phenotype; a two-step method to establish various animal and cell models of fetal-originated diseases, systematically elucidating their intrauterine programming mechanism; the development of an original, innovative theory on maternal glucocorticoid programming, which explains offspring's susceptibility to multiple diseases; and the construction of early comprehensive prevention strategies for fetal-originated diseases based on the glucocorticoid programming theory. She has led over 50 research projects, including key special projects of the National Key R&D Program, 5 key projects of the Natural Science Foundation of China, and major scientific research projects of the Ministry of Education. She has published over 300 SCI papers in multidisciplinary mainstream journals such as *Cell Reports Medicine*, *Acta Pharmaceutica Sinica B*, *Advanced Science*, *Journal of Advance Research*, *Science of The Total Environment*, and *Metabolism*. The Comparative Toxicogenomics Database (CTD) lists 82 of her research findings on toxic action targets and pathways. She holds 17 authorized national invention patents, of which 3 have been transferred, and has published academic monographs *Fetal-Originated Diseases* and *The Fetal Origin of Bone and Joint Diseases*. Her scientific research achievements have won several awards, including the First Prize in Natural Science of National Maternal and Child Health, the First Prize in Scientific and Technological Progress of Hubei Province, and the Second Prize in Scientific and Technological Achievements from the China Birth Defects Intervention and Rescue Foundation.

Liaobin Chen is a professor and consultant surgeon at Zhongnan Hospital of Wuhan University. Currently, he serves as the head of the Department of Orthopedic Teaching and Research, the director of the Division of Joint Surgery and Sports Medicine, and the director of the Joint Disease Research Center at Wuhan University. In the academic realm, he holds prestigious positions such as the vice president of the International Cartilage Repair Society China Division (ICRS-China), the deputy leader of the Lower Limb Group under the Sports Medicine Branch of the Chinese Medical Association, the standing committee member of the Sports Medicine Branch of the Chinese Medical Doctor Association and the head of the Basic and Translational Medicine Group, the deputy head of the Osteoarthritis Study Group of the Chinese Medical Doctor Association, the director of the Hubei Provincial Base of Arthroscopy Training Institute under the Chinese Medical Doctor Association, and the chairman of the Sports Medicine Branch of the Hubei Medical Association. Beyond his foundational and clinical research work in orthopedics, Prof. Chen primarily focuses on the "mechanisms and prevention of fetal-originated bone and joint diseases." He has led one key project and five general projects of the Natural Science Foundation of China, served as the PI for one major sub-project under the National Key R&D Program, led one major project of the Hubei Provincial Key R&D Program, and one key project of the Hubei Provincial Natural Science Foundation. To date, he has published 155 SCI papers in multidisciplinary mainstream journals, including *Acta Pharmaceutica Sinica B*, *Journal of Advanced Research*, *Stem Cell Research and Therapy*, *Arthroscopy*, *The American Journal of Sports Medicine*, and *The Bone and Joint Journal*; he holds eight national invention patents and one new practical patent. As for research achievements, he won the First Prize of Hubei Province Science and Technology Progress in 2016 and 2018 and the Third Prize of Chinese Medical Science and Technology in 2017, as well as the First Prize of National Maternal and Child Health Natural Science in 2019. He is the chief editor of academic monographs *The Fetal Origin of Bone and Joint Diseases* and *Dislocation After Total Hip Arthroplasty: The Causes and Management*.

Dan Xu is a professor and doctoral advisor at the School of Pharmaceutical Sciences, Wuhan University, where she leads the discipline of pharmacology. She has earned numerous prestigious accolades, including National Excellent Young Scientist, Hubei Province Outstanding Young Scientist, and the "LUOJIA Young Scholar" title at Wuhan University. Xu holds several influential positions in academic circles, serving as the Chair of the Young Committee of the Chinese Pharmacological Society's Division of Analytical Pharmacology, Executive Committee Member of the Youth Committee of the Chinese Society of Toxicology, Executive Committee Member of the Committee on Environmental and Developmental Origin Diseases of the Chinese Society of Environmental Mutagens, and Vice President of the Wuhan Pharmacological Society. Her research primarily explores the developmental toxicology of xenobiotics, with a special focus on intrauterine programming mechanisms and the prevention of fetal-originated neuropsychiatric disorders induced by these xenobiotics. Xu has spearheaded six projects under the National Natural Science Foundation of China, including the Excellent Young Scientists Fund Project, General Program, and Youth Program. She has been instrumental in a major national key R&D program investigating developmental programming and its metabolic regulation. As the first author or corresponding author, Xu has published over 50 research articles, with 90% appearing in JCR Q1/Q2 ranked SCI journals, in leading multidisciplinary publications such as *Microbiome*, *Acta Pharmaceutica Sinica B*, *BMC Medicine*, and *Pharmacological Research.* The Comparative Toxicogenomics Database cites 22 of her studies on toxic action targets and pathways, and two of her theories have been incorporated into academic textbooks. She holds six national invention patents. Xu's contributions to pharmacology have been acknowledged with several awards, including the Young Pharmacologist Award from the Chinese Pharmacological Society, the Outstanding Young Scientist Award from the Chinese Society of Toxicology, and the Outstanding Young Award from the Teratology Committee of the Chinese Society of Environmental Mutagens. Additionally, she has received the First Prize in Scientific and Technological Progress from Hubei Province, the First Prize in Natural Science for National Maternal and Child Health, and the Second Prize in Scientific and Technological Achievement from the China Birth Defects Intervention and Rescue Foundation, among other honors.

Chapter 1
Prenatal Adverse Environment and Susceptibility to Fetal-Originated Disease

Yang Tan and Yawen Chen

Abstract Fetal development is a complex process that is susceptible to physiological and pathological influences. Epidemiological studies have highlighted the impact of adverse prenatal environments, including exogenous and maternal factors, on pregnancy outcomes. These adverse environments can result in developmental toxicity of multiple organs in fetuses and increase the susceptibility of adults to metabolic diseases, neuropsychiatric disorders, reproductive system illnesses, immune-related diseases, and tumors, through the interplay of the mother-placenta-fetus biological unit. This chapter aims to present the adverse pregnancy outcomes resulting from adverse environments during pregnancy, the developmental toxicity of multiple organ systems in offspring, and their long-term consequences. By combining relevant literature and the latest research findings from the authors' team, this chapter provides a theoretical and experimental basis for understanding the occurrence and development of fetal-originated disease.

1 Introduction

In the 1990s, David Barker, a British scholar, proposed the "Developmental Origin of Adult Diseases" theory, which suggests a correlation between low birth weight and the increased incidence of adult metabolic syndrome based on large-scale epidemiological surveys (Barker 1999). Over the past three decades, scholars from various countries have conducted numerous studies on the link between adverse prenatal environments, fetal birth weight, and chronic diseases in adulthood. These studies have led to the development of the "Developmental Origins of Health and

Y. Tan (✉)
Division of Joint Surgery and Sports Medicine, Department of Orthopedic SurgeryZhongnan, Hospital of Wuhan University, Wuhan, Hubei, China

Y. Chen
Department of Pharmacology, School of Basic Medical Science, Wuhan University, Wuhan, Hubei, China

H. Wang et al. (eds.), *Fetal Origin of Diseases*,
https://doi.org/10.1007/978-981-97-5730-5_1

Disease (DOHaD)" concept. The short-term effects of fetal developmental toxicity caused by adverse environmental factors during pregnancy can result in poor pregnancy outcomes and birth defects in offspring. However, the long-term effects can lead to an increased susceptibility to various diseases, including metabolic diseases, neuropsychiatric disorders, reproductive system illnesses, immune-related diseases, and tumors. According to epidemiological surveys, noncommunicable chronic diseases were responsible for approximately 40 million deaths globally in 2015, making them the leading cause of human mortality in the twenty-first century (NCD Countdown 2030 collaborators 2016). Studies have suggested that some of these chronic diseases are linked to adverse early-life environments (York et al. 2014). This chapter provides a theoretical and experimental basis for understanding the short and long-term hazards of fetal developmental toxicity caused by adverse environments during pregnancy. It also offers a theoretical foundation for the prevention of fetal-originated disease based on relevant international literature reports and the research findings of the authors' team.

2 Poor Pregnancy Outcomes Caused by Adverse Intrauterine Environments during Pregnancy

Fetal developmental toxicity is largely attributed to poor intrauterine environments during pregnancy, in addition to congenital genetic factors. Adverse environments during pregnancy can result in many adverse pregnancy outcomes and abnormal multi-organ development and dysfunction in individuals with intrauterine growth retardation (IUGR).

2.1 Adverse Intrauterine Environments during Pregnancy

Exogenous environments during pregnancy mainly refer to the xenobiotics that the mother is exposed to during pregnancy. Xenobiotic exposure during pregnancy is one of the most definite and dangerous causes of harmful intrauterine environments and adverse pregnancy outcomes. The xenobiotic agents that have been proven to cause poor outcomes of pregnancy mainly include the following: ① Environmental toxicants, such as organic volatiles, heavy metals, smoke, and air pollutants. Studies have shown that smoking during pregnancy is associated with a reduction in fetal birth weight and head circumference as well as an increased risk of birth defects (Walfisch et al. 2013; Dessi et al. 2018). Prenatal trichloroethylene exposure has also been linked to adverse pregnancy outcomes, including preterm birth, low birth weight, and birth defects (Forand et al. 2012). ② Drugs, such as glucocorticoids, caffeine, and antibiotics. Prenatal dexamethasone treatment has been shown to reduce neonatal weight (Ristic et al. 2016). Excessive caffeine intake during pregnancy can cause fetal reproductive and developmental toxicity, such as premature delivery, abortion, growth retardation, and congenital malformations (Wang et al. 2018). ③ Poor diets, such as alcohol, coffee, and tea. Alcohol consumption during pregnancy has been associated with lower birth weight or preterm delivery (Nykjaer et al. 2014). Infection is also one of the important factors leading to adverse pregnancy outcomes. For example, human immunodeficiency virus infection during pregnancy has been confirmed to induce low birth weight and a high IUGR rate in the fetus (Lopez et al. 2015).

The maternal environment during pregnancy encompasses various factors such as maternal stress, nutrition, and disease state, which can significantly impact fetal growth and development. Maternal stress, including anxiety and depression, can negatively affect the maternal environment, leading to adverse fetal outcomes (Painter et al. 2008). Inadequate or excessive intake of nutrients during pregnancy can also directly or indirectly interfere with fetal endocrine system and impair fetal growth and development (Song et al. 2020). The three major nutrients, namely, sugars, proteins, and fats, are crucial for fetal development, and their deficiencies during pregnancy can result in unfavorable pregnancy outcomes. Additionally, maternal disease states such as gestational diabetes mellitus and pre-eclampsia can have detrimental effects on fetal development. Pre-eclampsia, for instance, can cause fetal growth restriction (Crispi et al. 2009), while gestational diabetes can result in macrosomia (Veenendaal et al. 2013).

2.2 *Poor Pregnancy Outcomes*

The fetus is highly susceptible to adverse environments during pregnancy, which can lead to impaired growth and multi-organ development. The severity and stage of fetal developmental toxicity can result in various outcomes, such as premature delivery, malformation, growth retardation, and abortion, as outlined in Table 1.1.

Abortion, defined as pregnancy termination before 28 weeks of gestation with a fetus weighing less than 1000 g, can result from exposure to multiple adverse environmental factors during pregnancy. These factors include excessive maternal exposure to harmful chemicals, such as arsenic, lead, benzene, formaldehyde, chlorobutadiene, ethylene oxide, and physical factors such as radiation, noise, and high temperature (Patelarou and Kelly 2014). Studies indicate that prolonged maternal exposure to phenols can also cause spontaneous abortion (Chen and Cheng 2015). A recent investigation on the relationship between urinary levels of various phenols and spontaneous abortion in pregnant women residing in the middle and lower reaches of the Yangtze River reported that exposure to 4-n-octylphenol and alkyl phenols significantly increased the risk of spontaneous abortion compared to the low-exposure group (Chen et al. 2013).

Preterm delivery is defined as a pregnancy lasting more than 28 weeks but less than 37 weeks, resulting in newborns weighing between 1000 and 2499 g. Preterm birth is a significant public health issue globally and accounts for 5% ~ 15% of total

Table 1.1 Fetal developmental toxicity caused by adverse environments during pregnancy

Adverse environments	Classification	Fetal developmental toxicity	Species	Main references
External environments				
Environmental toxicants	Cadmium	Developmental retardation/deformity	Rat	Shi et al. (2020)
	Nicotine	Developmental retardation	Rat	Chen et al. (2007)
	Arsenic	Developmental retardation	Human	Rahman et al. (2017)
	Air pollutant	Developmental retardation	Human	Vinikoor-Imler et al. (2014)
Drug	Dexamethasone	Developmental retardation	Human/rat	Vidaeff and Blackwell (2011)
	Caffeine	Developmental retardation	Rat	Xu et al. (2012b)
	Cocaine and marijuana	Deformity/death	Human	Church et al. (1998)
	Methylamphetamine	Brain developmental toxicity	Human	Nguyen et al. (2010)
Food and drink	Ethanol	Developmental retardation/premature birth	Human/rat	Sabra et al. (2018); Yu et al. (2018)
Biotic environment	Bacteria	Absorbed fetus/stillbirth/malformation/developmental retardation	Rat	Baboonian et al. (2003)
	Virus	Developmental retardation	Human	Di mascio et al. (2020)
	Toxoplasma gondii	Brain developmental toxicity	Human	Anna (2010)
Maternal environments				
Stress state	Stress	Developmental neurotoxicity	Human	Lautarescu et al. (2020)
	Anxiety and depression	Developmental neurotoxicity	Human	De Asis-Cruz et al. (2020)
Malnutrition	High-fat diet	Brain developmental toxicity	Human	Song et al. (2020)
	Low-protein diet	Developmental retardation/skeletal muscle dysplasia	Rat	Vidyadharan et al. (2022)
Pregnancy illness	Gestational diabetes	Macrosomia/shoulder dystocia/premature delivery	Human	Kc et al. (2015)
	Eclampsia	Developmental retardation/stillbirth/premature birth	Human	Crispi et al. (2009)

births in China. It is a leading cause of neonatal death, with approximately 15% of premature babies dying during the neonatal period. Malnutrition during pregnancy and maternal hypertension, heart disease, liver and kidney diseases, diabetes, severe tuberculosis, endocrine disorders, chorionicamnitis, lower reproductive tract, and urinary tract infections during pregnancy can cause premature delivery (Ericson and Laughon 2015). Other studies have found that maternal smoking, exposure to formaldehyde or nitrogen dioxide, and environmental factors such as pyrotechnic chemicals emitted by fireworks factories can also cause premature delivery (Medsker et al. 2015; Li et al. 2018a). A study conducted in Liuyang, China, revealed that the residential density of fireworks factories was negatively correlated with premature birth (Li et al. 2018a). Additionally, exposure to environmental pesticides and heavy metals during pregnancy increases the risk of preterm delivery (Jiang et al. 2018).

Fetal malformation refers to the disruption of fetal development caused by genetic and environmental factors during the embryonic stage, leading to various abnormalities in fetal morphological structures and functions (Corsello and Giuffre 2012). The incidence of birth defects in newborns has been gradually increasing over the past two decades due to environmental pollution, genetic variation, and other factors. Common birth defects include anencephaly, hydrocephalus, open spina bifida, meningocele, cleft lip, cleft palate, congenital heart disease, and trisomy 21 syndrome. Infection during pregnancy or improper use of drugs can result in fetal malformations. Exposure to radiation, chemicals, or infections, hyperglycemia, high fever, and drug abuse during early pregnancy have been linked to fetal developmental malformations (Viteri et al. 2015). Severe deformities can lead to fetal/newborn death or severe disability. The thalidomide event in Europe, which caused several babies to be born with seal-like deformities in some European countries, demonstrated that drugs taken in early pregnancy can significantly impact fetal growth and development, with up to 50%–80% of fetuses of pregnant women taking the drug experiencing teratogenic effects. This event drew attention to the teratogenic hazards of xenobiotics during early pregnancy.

IUGR is characterized by low birth weight and is diagnosed clinically as a birth weight less than 2500 g or less than two standard deviations of the average weight of a fetus over 37 weeks of gestation. During fetal development, IUGR fetuses experience impaired growth potential and normal growth retardation, leading to specific tissue structure and functional development alterations. Numerous epidemiological investigations have demonstrated that IUGR can cause fetal distress, neonatal asphyxia, and perinatal death and can also extend to postnatal life, resulting in increased susceptibility to multiple chronic diseases in adulthood, such as metabolic and neuropsychiatric diseases (Nobile et al. 2022). The global incidence of IUGR ranges from 2.75% to 15.53%, with the incidence in developing countries being six times greater than that in developed countries. Adverse environments during pregnancy are vital factors leading to IUGR. A series of animal experiments conducted by the author's team confirmed that exposure to nicotine (Xu et al. 2013), ethanol (Huang et al. 2015), caffeine (Xu et al. 2012a), dexamethasone (Xu et al. 2011), and azithromycin (Liu et al. 2020) can cause IUGR in offspring.

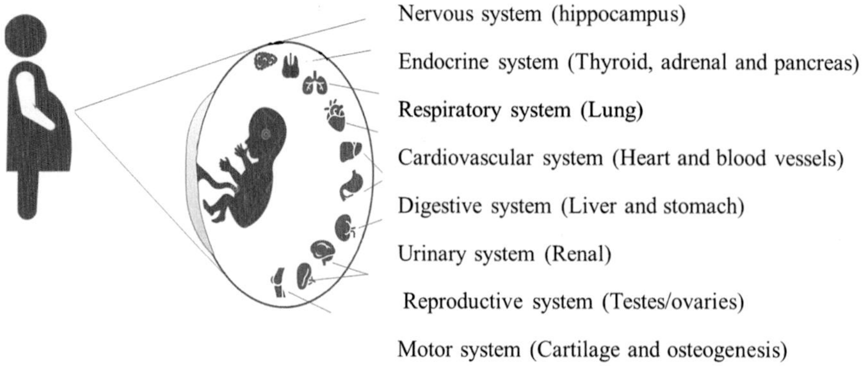

Fig. 1.1 Developmental toxicities of multisystem caused by adverse environments during pregnancy

3 Developmental Toxicities of Fetal Organs Caused by Adverse Environments during Pregnancy

Numerous studies have confirmed that adverse environments during pregnancy can lead to the developmental toxicity of multiple organs in offspring, including the motor, endocrine, digestive, respiratory, urinary, reproductive, cardiovascular, and nervous systems (Fig. 1.1). Among them, in response to adverse environmental exposure during pregnancy, IUGR fetuses will alter their metabolic processes, redistribute their limited energy, and limit the energy consumption of secondary organs (such as bone, kidney, and gonads) to ensure the development of vital organs (such as the brain and liver), resulting in a "thrifty phenotype" development. This change can last for a long time, even permanently, to maintain the survival and development of these IUGR fetuses caused by adverse environmental exposure during pregnancy.

3.1 Motor System

Exposure to adverse environmental factors during pregnancy has a lifelong impact on the structural and functional development of long bone and articular cartilage. It has been reported that maternal exposure to bisphenol A (BPA) during pregnancy can alter the femur geometry in both male and female offspring (Lejonklou et al. 2016). Furthermore, studies have also suggested that changes in maternal estrogen levels during pregnancy can affect the development of fetal bone tissue and result in permanent bone changes (Pienkowski et al. 2015). In a recent study, Tan et al. (2012) found that caffeine exposure during pregnancy can reduce the synthesis function of extracellular matrix in fetal cartilage, delay the growth of growth plates, delay the differentiation of growth plate chondrocytes into calcified cells, and ultimately lead to delayed growth of fetal bone and shortened bone length. Although the properties of articular cartilage and growth plate cartilage differ, they share the same embryological origin. Therefore, the developmental retardation of articular cartilage and growth plate cartilage in fetal rats reflects the multi-temporal effects of prenatal caffeine exposure on chondrogenesis to some extent (Tan et al. 2018). Exposure to various xenobiotics during pregnancy, such as ethanol (Ni et al. 2018), nicotine (Xie et al. 2018), and dexamethasone (Zhang et al. 2016), can also result in articular chondrodysplasia. Studies have shown that caffeine or nicotine exposure during pregnancy delays osteoclast formation and endochondral ossification in the fetal long bones of rats (Shangguan et al. 2017; Hu et al. 2018). Similarly, dexamethasone exposure during pregnancy has been found to delay endochondral ossification by inhibiting chondrocyte maturation and osteoclast differentiation (Zhang et al. 2016). Furthermore, prenatal exposure to dexamethasone significantly inhibits the development of fetal articular cartilage and long bone. This is characterized by decreased body weight and length, roughness of the articular cartilage surface, reduced length of the femur and primary ossification center, decreased number and thickness of bone trabeculae (Chen et al. 2018), and reduced peak bone mass in offspring (Xiao et al. 2018). It is important to note that this phenomenon of dysplasia and the "development programming" of the disease (Moisiadis and Matthews 2014) can be transmitted to the second and third generations, indicating multigenerational genetic effects.

3.2 Endocrine System

Numerous studies have demonstrated that exposure to adverse environments during pregnancy can result in structural and functional damage to the fetal endocrine system. The adrenal glands serve as terminal efficient organs in the hypothalamic-pituitary-adrenocortical (HPA) axis. In the treatment of prenatal synthetic glucocorticoids, Karsli et al. (2017) observed a significant reduction in fetal adrenal volume and delayed fetal growth through two-dimensional ultrasound. In IUGR rat models induced by xenobiotics (such as caffeine and ethanol) and diet restriction during pregnancy, the author's research team discovered that fetal adrenal weight was lower than that of the control group, cell proliferation was reduced while apoptosis remained unchanged, the maximum adrenal cross-sectional area was higher than that of the control group, cell arrangement was disordered, cortical cell swelling occurred, intercellular space significantly increased, and mitochondrial structure was disordered. In adulthood, adrenal cortex cell morphology, cell proliferation, and the maximum cross-sectional area did not differ significantly from the control group (Yan et al. 2014; Wu et al. 2015). The findings indicate that prenatal exposure to xenobiotics can lead to fetal adrenal dysplasia. However, the adrenal cortex demonstrates "catch-up" development in morphology after birth and approaches normalcy in adulthood (He et al. 2017; Chen et al. 2018; Huang et al. 2018). The pancreas is a critical organ for regulating blood glucose homeostasis. Prenatal exposure to dexamethasone does not alter the total number of endocrine precursor cells in offspring, but it reduces the number of differentiated and mature beta cells while increasing the number of acinus cells in the exocrine part. This results in a partial imbalance between the internal and external secretion, especially the endocrine part with insulin synthesis function decreased (Kou et al. 2020). Additionally, a high-fat diet after birth can also cause islet dysfunction in offspring (Guo et al. 2012). Exposure to phthalates during pregnancy has a long-term impact on the hypothalamus-pituitary-thyroid (HPT) axis and further affects thyroid development in offspring (Qian et al. 2020). The consequences of this effect are likely to be permanent. Moreover, dexamethasone exposure during pregnancy can decrease the level of sex hormones and peripheral blood deionization in offspring, thus affecting the differentiation of thyroid cells (Eishi et al. 1983).

3.3 Digestive System

The liver is a crucial metabolic organ in the body. An epidemiological survey (Jaquet et al. 2005) revealed that adults who were IUGR newborns had a 5.75 times higher incidence of metabolic syndrome (MS) than those who were normal fetuses. Various adverse environmental exposure models during pregnancy have suggested that an adverse developmental environment in early fetal life is associated with liver lipid metabolism disorders. The author's research team confirmed that IUGR rats induced by prenatal food restriction displayed significant abnormal lipid metabolism after birth under a high-fat diet. This was characterized not only by increased blood triglyceride levels but also by increased abdominal fat, increased fat cell volume, and liver triglyceride accumulation. However, the accumulation of triglycerides in the liver and lipid metabolism disorders can lead to an increased susceptibility to nonalcoholic fatty liver disease (Zhang et al. 2016). Exposure to pyrrolizidine alkaloids during pregnancy resulted in fetal growth retardation, placental damage, and fetal liver damage (Li et al. 2018b). Methimazole exposure during pregnancy can induce developmental toxicity in the fetal liver, which is characterized by obvious cytoplasmic vacuolation of fetal liver cells, inhibition of liver development-related gene expression, and changes in metabolic function to varying degrees (Wang et al. 2018). The stomach and small intestine are the core organs for nutrient absorption, and almost all nutrients must pass through the gastrointestinal tract, which has become the most critical organ for digestion. Studies have demonstrated that both perinatal hypothermia and decreased metabolic function can result in abnormal gastrointestinal development in IUGR piglets (Amdi et al. 2016). The offspring of mothers with gestational diabetes mellitus exhibit reduced duodenal ghrelin expression, which may result in gastrointestinal tract dysplasia in offspring (Aydın et al. 2017).

3.4 Respiratory System

The respiratory system comprises the respiratory tract (nasal cavity, pharynx, larynx, trachea, bronchus) and lungs. Compared with other organs of the respiratory system, the development of the lungs is more vulnerable to adverse environmental impacts during pregnancy. Clinical research has observed delayed fetal lung development in diabetic mothers (Azad et al. 2017). Experimental studies have shown that exposure to intrauterine hyperglycemia may lead to fetal lung structural changes, such as a reduced mesenchymal ratio, smaller alveolar size, septal thickening, and delayed endothelial basement membrane (Koskinen et al. 2012). Prenatal smoke exposure can alter lung function during the neonatal period, resulting in changes in tidal breathing patterns, reduced passive breathing compliance, and reduced expiratory flow (Carlsen et al. 1997). Cholestasis during pregnancy can cause bile acid accumulation in the fetal lung, accompanied by structural changes in the fetal lung, including bronchial edema, collapse of alveolar space in the alveolar cavity, precipitation of clear substances in the alveolar cavity, and inflammatory cell infiltration (Herraez et al. 2014). The author's research team recently discovered that prenatal dexamethasone exposure can cause susceptibility to pulmonary fibrosis in offspring rats (Zhang et al. 2022).

3.5 Urinary System

In the urinary system, the kidney is one of the most vulnerable organs to damage during various stages of embryonic development (Dorey et al. 2014). Epidemiological data indicate that adverse environmental exposure during pregnancy can affect the kidney development of IUGR offspring (Janot et al. 2014), and IUGR newborns often exhibit a decrease in the number of nephrons and a reduction in kidney volume (Crobe et al. 2014). Fetal mice with IUGR exposed to alcohol (Zhu et al. 2018) and caffeine (Zhu et al. 2019) during pregnancy displayed renal podocyte dysplasia. IUGR fetal mice induced by a low-protein diet exhibited increased renal cell apoptosis and decreased glomerular number 2 months after birth (He et al. 2015). Although IUGR fetuses undergo catch-up growth after birth, restoring kidney volume, nephrons do not reform, and thus the individual suffers permanent nephron loss (Cebrian et al. 2014).

3.6 Reproductive System

Prenatal exposure to dexamethasone, ethanol, and nicotine has been shown to cause abnormal testicular development, as evidenced by both aberrant testicular morphology and steroidogenic function (Liu et al. 2018, 2019, 2021, 2022; Zhang et al.

2020). Moreover, dexamethasone exposure during pregnancy has been found to induce ovarian developmental toxicity, with transgenerational genetic effects, including decreased cytochrome P450 aromatase (P450arom) expression mediated by miR-320a-3p and decreased ovarian reserve mediated by miR-17-5p (Gong et al. 2021a, b). Additionally, maternal caffeine exposure during pregnancy has been shown to have multigenerational genetic effects on offspring ovarian steroidogenesis through high levels of glucocorticoids (Lv et al. 2021). Furthermore, prenatal alcohol and nicotine exposure has been found to affect gene expression related to ovarian function, by directly damaging ovarian granulosa cells or interfering with the level of sex hormones secreted by the ovary through its receptor, resulting in abnormal estrous cycle and follicle development (Fan et al. 2019; Ni et al. 2019).

3.7 Cardiovascular System

Recent epidemiological studies have highlighted the close association between BPA and heart disease. In particular, exposure to BPA during pregnancy has been found to impair mitochondrial function in the hearts of newborn rats, as reported by Jiang et al. (2014). Similarly, perfluorooctane sulfonic acid (PFOS), a widely present cardiotoxic compound in the environment and organisms, has been shown to cause delayed development and reduced perinatal survival after birth when exposed during pregnancy. PFOS is known to accumulate in the heart, second only to the liver, and can cause developmental toxicity by inducing cardiac mitochondrial swelling, vacuolar structure, and loss of cardiac crest, leading to cardiac malformation or delayed cardiac development, ventricular septal defect, and enlargement of the right atrium (Zhang et al. 2011; Zeng et al. 2015). In addition to chemical exposure, physical environmental factors during pregnancy have also been implicated in developmental toxicity to the offspring's heart. For instance, exposure to 900 MHz electromagnetic fields has been reported to cause oxidative stress and histopathological changes in the hearts of fetal rats (Turedi et al. 2015).

3.8 Nervous System

Exposure to endocrine disruptors during early pregnancy, such as dip-chlorophenol trichloromethanes and polychlorinated biphenyls (PCBs), has been found to affect hypothalamic development, leading to nerve cell death and changes in reproductive function and HPA axis activity (Ishola and Adeniyi 2013). Similarly, exposure to alcohol during pregnancy has been associated with significant defects in hippocampal cell loss and neurogenesis in offspring (Gil-Mohapel et al. 2010). Our team's systematic studies in recent years have further demonstrated that exposure to xenobiotics, including caffeine, nicotine, and ethanol, during pregnancy can lead to abnormal development of the fetal hippocampus and changes in functional programming of the HPA axis after birth, characterized by low basal activity and high sensitivity to chronic stress (Zhang et al. 2014). Moreover, alcohol exposure during pregnancy has been shown to increase the potential excitability of the hypothalamus, hypersensitivity of the HPA axis, and mild abnormalities in hippocampal morphology in fetal rats (Lu et al. 2018). Likewise, exposure to caffeine, nicotine, ethanol, and dexamethasone during pregnancy has been found to cause morphological abnormalities in the hypothalamus, including increased or decreased neuronal generation and decreased endplate dendrite structure (Lu et al. 2018, 2020a, b).

4 Susceptibility to Multiple Diseases in Offspring Caused by Adverse Environments during Pregnancy

Extensive epidemiological investigations have consistently demonstrated that IUGR resulting from adverse prenatal environments can increase the susceptibility of affected individuals to chronic diseases in adulthood. Such diseases include metabolic disorders, neuropsychiatric disorders, reproductive system disorders, immune-related disorders, and tumors (Fig. 1.2).

4.1 Metabolic Diseases

MS is a cluster of hypertension, hyperglycemia, dyslipidemia, and obesity in the body, which can directly lead to fatty liver, diabetes, and other metabolic disorders. MS can be diagnosed if any three of the five indicators, including abdominal obesity, hypertriglyceridemia, low high-density lipoprotein-cholesterolemia, hypertension, and fasting hyperglycemia, are present. Studies have consistently shown a close relationship between IUGR and the occurrence of MS in adulthood. For instance, a study involving 600 children aged 5 to 16 years found that IUGR newborns had higher fasting blood glucose, insulin levels, and insulin resistance. Similarly, our team's studies in IUGR rats induced by various adverse prenatal environments, such as xenobiotic exposure and nutrient deficiency, have confirmed their increased susceptibility to MS in adulthood (He et al. 2015; Kou et al. 2017a; Pei

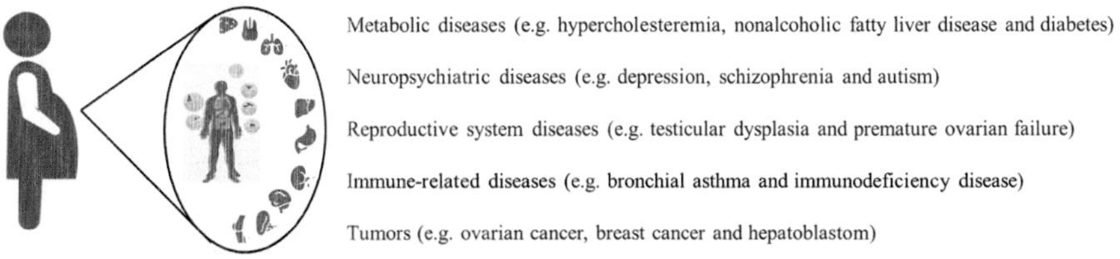

Fig. 1.2 Susceptibility to multiple diseases caused by adverse environments during pregnancy

et al. 2017). Furthermore, our team has shown that prenatal caffeine exposure can increase the blood cholesterol level of IUGR offspring rats after birth, which can be aggravated by a high-fat diet (Xu et al. 2018a). With this model, we also found programming changes in intrauterine liver triglyceride metabolism, which increase the susceptibility to non-alcoholic fatty liver disease in adulthood (Hu et al. 2019) as well as impaired glucose metabolism and susceptibility to diabetes (Kou et al. 2017b). Additionally, prenatal iron exposure has been found to increase the susceptibility to type 1 diabetes in offspring (Stordal et al. 2018).Our team's studies have also demonstrated that prenatal exposure to xenobiotics, such as ethanol and dexamethasone, as well as food restriction, can cause impaired glucose metabolism and increase susceptibility to diabetes in offspring rats after birth (Xiao et al. 2017, 2019; Kou et al. 2020). Moreover, exposure to adverse environments during pregnancy, such as high salt and hypoxia, has been shown to increase the susceptibility to hypertension in adult offspring, possibly due to changes in the fetal renin-angiotensin system (RAS) (Crispi et al. 2010). In conclusion, adverse prenatal environments can increase the susceptibility of offspring to multiple metabolic disorders.

4.2 Neuropsychiatric Disorders

A growing body of evidence suggests that stress-induced changes in brain development during pregnancy or early life can persist into adulthood, exerting lifelong effects on behavior, cognition, and emotional systems, and increasing susceptibility to neuropsychiatric disorders such as depression, schizophrenia, and autism. Epidemiological studies have found that alcohol consumption during pregnancy increases the incidence of depression in offspring, while children of mothers who smoked at least one pack of cigarettes per day during pregnancy were more than twice as likely to exhibit depressive symptoms compared to offspring of nonsmoking mothers (Paintner et al. 2012). Similarly, diethylstilbestrol exposure during pregnancy has been associated with an increased risk of depression in adulthood (Titus et al. 2019), while preterm and hypoxic fetuses have a significantly increased risk of developing schizophrenia after birth (Nosarti et al. 2012). Human brain development begins in early gestation, and adverse environmental conditions during fetal and infant development can affect the expression of key genes related to neurodevelopment. For instance, high-dose mercury exposure during pregnancy has been found to be associated with twice the risk of autism compared to low-dose mercury exposure, while arsenic exposure during pregnancy has been significantly linked to autism or behavioral problems (Golding et al. 2018). Our team's studies have further shown that adverse prenatal environments, including exposure to alcohol, nicotine, caffeine, and food restriction, can lead to abnormal development of the HPA axis in offspring (Zhang et al. 2013; He et al. 2017; Lu et al. 2018; Xu et al. 2018b), thereby increasing susceptibility to neuropsychiatric diseases (Dong et al. 2018; He et al. 2017; Jiang et al. 2021). In conclusion, adverse prenatal environments can increase the susceptibility of offspring to multiple neuropsychiatric disorders.

4.3 Reproductive System Diseases

Adverse environments during pregnancy may increase the risk of reproductive diseases such as testicular dysplasia syndrome, ovarian dysplasia, and polycystic ovary syndrome. A study conducted in European countries found that diethylstilbestrol treatment during early pregnancy significantly increased the incidence of cryptorchidism and hypospadias in offspring (Mitsui 2021). Additionally, both PFOS exposure and smoking during pregnancy were associated with hypospadias and cryptorchidism in offspring (Anand-Ivell et al. 2018). Furthermore, caffeine exposure during pregnancy was found to impair testicular development and reduce sperm count in adulthood (Pei et al. 2019). The pathogenesis of polycystic ovary syndrome (PCOS) involves multiple factors, including environmental, genetic, and intrauterine factors. Exposure to endocrine disruptors, such as BPA, phthalates, and androgens during pregnancy, may alter the endocrine hormone state through fetal programming, leading to the development of PCOS and possible cross-generational genetic effects (Rochester 2013). Exposure to dibutyl phthalate and 2-ethylhexyl phthalate during pregnancy may also induce polycystic changes in the ovaries of offspring and changes in hormone levels similar to those in PCOS (Hewlett et al. 2017). Therefore, it is imperative to recognize that adverse environments during pregnancy can have significant implications for the reproductive health of offspring.

4.4 Immune-Related Diseases

Bronchial asthma is a chronic airway inflammation that involves multiple cells and cellular components. Maternal smoking during pregnancy has been associated with increased fetal lung function and postnatal wheezing susceptibility, as well as a higher incidence of asthma, chronic obstructive airway disease, and lung function impairment in adults (Ji et al. 2016). Additionally, studies have shown that exposure to BPA during pregnancy is associated with asthma and lung function development in children (Spanier et al. 2014). Furthermore, insufficient maternal nutrient intake during pregnancy may directly affect fetal physical development, lung, and respiratory tract development, thus increasing the risk of childhood asthma (Patel et al. 2012). Immunodeficiency disease is a disorder that results from defective development or immune response in the human immune system. Maternal protein malnutrition during pregnancy can lead to an increased incidence of low birth weight, increased incidence of infection in offspring, and permanent immune deficiency in individuals, which cannot be corrected by late nutritional supplementation (Hornsby et al. 2018). Moreover, mutations in n-3 polyunsaturated fatty acids in breast milk from mothers with allergic diseases can increase the incidence of allergic diseases in infants (Mayer and Seeger 2008). Thus, it is crucial to recognize that adverse environments during pregnancy can increase the susceptibility to immune-related diseases.

4.5 Tumor

Ovarian cancer is one of the most common malignant tumors affecting female reproductive organs. Titus-Ernstoff et al. reported in 2011 that overexposure to maternal estrogen during pregnancy was associated with an increased risk of ovarian cancer in adulthood. Additionally, women exposed to diethylstilbestrol during pregnancy have a higher expected incidence of ovarian cancer in their offspring. Maternal obesity and smoking during pregnancy may lead to early menarche, which in turn may lead to prolonged exposure to estrogen, thereby increasing the risk of female reproductive organs developing cancer (Behie and O'Donnell 2015). Breast cancer is a malignant tumor that occurs in the gland epithelial tissue of the breast. Studies have found that there is a positive correlation between birth weight and the risk of breast cancer, and the mechanism is abnormal breast differentiation caused by high estrogen and insulinlike growth factor 1 in the intrauterine microenvironment, thus increasing the sensitivity to environmental carcinogenic factors after birth (Denholm et al. 2016). Hepatoblastoma is a type of malignant embryonic tumor with various differentiation modes. Exposure to traffic-related air pollution during pregnancy may increase the risk of hepatoblastoma (Kumar et al. 2018). Additionally, there is evidence that smoking before or during pregnancy is associated with an increased risk of hepatoblastoma in offspring (Spector and Birch 2012). Therefore, it is essential to recognize that adverse environments during pregnancy can increase the susceptibility to tumors.

5 Summary and Prospect

Adverse environments during pregnancy can result in a range of detrimental pregnancy outcomes and cause developmental toxicity of fetal multi-organs, leading to susceptibility to long-term diseases such as metabolic diseases, neuropsychiatric diseases, reproductive system diseases, immune diseases, tumors, and others (Fig. 1.3). It is imperative to systematically explore the mechanisms underlying the

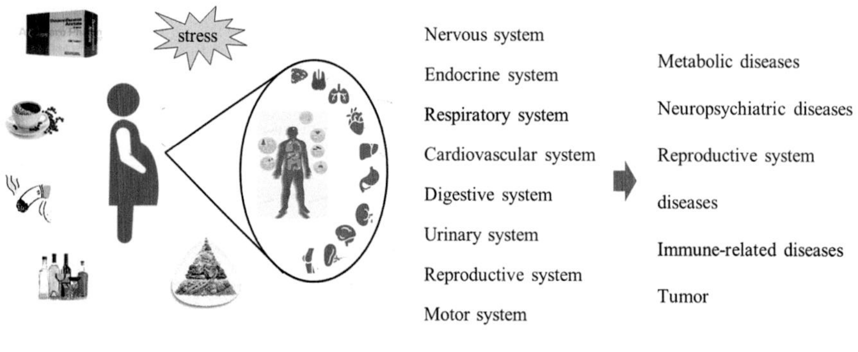

Fig. 1.3 Adverse environments during pregnancy and fetal-originated disease

susceptibility of diseases in offspring caused by adverse environmental conditions during pregnancy in the twenty-first century. This research has significant theoretical and practical implications for the analysis of the origin of fetal development of adult diseases, eugenics, and prenatal health care.

References

Amdi C, Klarlund M V, Hales J, et al. 2016. J Anim Sci, 94(11): 4583–4590.
Anand-Ivell R, Cohen A, Norgaard-Pedersen B, et al. 2018. Front Physiol, 9: 406–416.
Anna A. 2010. Duodecim, 126(2): 159–166.
Aydın H A, Derbent Uysal A, Erol O, et al. 2017, Gynecol Endocrinol, 33(2): 132–135.
Azad M B, Moyce B L, Guillemette L, et al. 2017, Paediatr Respir Rev. 21: 19–26.
Baboonian C, Smith D A, Shapland D, et al. 2003. Cardiovasc Res, 60(1): 165–169.
Behie A M, O'Donnell M H. 2015. Hum Reprod, 30(4): 957–962.
Carlsen K C Lødrup, Jaakkola J J, Nafstad P, et al. 1997. Eur Respir J. 10(8): 1774–1779.
Cebrian C, Asai N, D'Agati V, et al. 2014. Cell Rep, 7(1): 127–137.
Chen C, Cheng Y. 2015. Wei Sheng Yan Jiu, 44(2): 312–316.
Chen M, Wang T, Liao Z X, et al. 2007. Exp Toxicol Pathol, 59(3–4): 245–251.
Chen X, Chen M, Xu B, et al. 2013. Chemosphere, 93(2): 217–222.
Chen Z, Zhao Z, Li Y, et al. 2018. Toxicol Lett, 286: 1–9.
Church MW, Crossland WJ, Holmes PA, et al. 1998. Ann N Y Acad Sci, 846: 12–28.
Corsello G, Giuffre M. 2012. J Matern Fetal Neonatal Med, 25 Suppl 1: 25–29.
Crispi F, Comas M, Hernández-Andrade E, et al. 2009. Ultrasound Obstet Gynecol, 34(6): 660–665.
Crispi F, Bijnens B, Figureueras F, et al. 2010. Circulation, 121(22): 2427–2436.
Crobe A, Desogus M, Sanna A, et al. 2014. Am J Physiol Renal Physiol, 307(9): F1033–F1040.
David J P Barker. 1999. Ann Med, 31(sup1): 3–6.
De Asis-Cruz J, Krishnamurthy D, Zhao L, et al. 2020. JAMA Netw Open, 3(12): e2022349.
Denholm R, De Stavola B, Hipwell J H, et al. 2016. Breast Cancer Res, 18(1): 102–108.
Dessì Angelica, Laura Corona, Roberta Pintus, et al. 2018. Expert Rev Proteomics.15(8): 647–656.
Di Mascio D, Khalil A, Saccone G, et al. 2020. Am J Obstet Gynecol MFM, 2(2): 100107.
Dong W, Xu D, Hu Z, et al. 2018. Toxicol Lett, 283: 1–12.
Dorey E S, Pantaleon M, Weir K A, et al. 2014. Reproduction, 147(6): R189-R198.
Eishi Y, Hirokawa K, Hatakeyama S. 1983. Clin Immunol Immunopathol, 26(3): 335–349.
Ericson J E, Laughon M M. 2015. Clin Perinatol, 42(1): 155–165.
Fan G, Zhang Q, Wan Y, et al. 2019. Food Chem Toxicol, 128: 256–266.
Forand Steven, Elizabeth Lewis-Michl, Marta Gomez. 2012. Environ Health Perspect, 120(4): 616–621.
Gil-Mohapel J, Boehme F, Kainer L, et al. 2010. Brain Res Rev, 64(2): 283–303.
Golding J, Rai D, Gregory S et al. 2018. Mol Autism, 23(9): 30–35.
Gong X, Zhang J, Ge C, et al. 2021a. Pharmacol Res, 105435.
Gong X, Dai S, Wang T, et al. 2021b. Cell Biol Toxicol, 5(21): 645–656.
Guo Y, Zhang L, Shen L, et al. 2012. Drug Metab Rev, 44: 131–137.
He X, Xie Z, Dong Q, et al. 2015. Nephrology (Carlton), 20(1): 34–39.
He Z, Lv F, Ding Y, et al. 2017. Arch Med Res, 48(6): 488–497.
Herraez E, Lozano E, Poli E, et al. 2014. J Mol Med (Berl), 92(4): 359–372.
Hewlett M, Chow E, Aschengrau A, et al. 2017. Reprod Sci, 24(1): 19–27.
Hornsby E, Pfeffer PE, Laranjo N, et al. 2018. J Allergy Clin Immunol, 141(1): 269–278.
Hu H, Zhao X, Ma J, et al. 2018. Toxicol Lett, 295: 249–255.
Hu S, Xia L, Luo H, et al. 2019. Toxicology, 417: 23–34.
Huang H, He Z, Zhu C, et al. 2015. Toxicol Appl Pharmacol, 288(1): 84–94.

Huang H, Liu L, Li J, et al. 2018. Toxicol Appl Pharmacol, 345: 36–47.
Ishola A O, Adeniyi P A. 2013. Niger Med J, 54(6): 411–414.
Janot M, Cortes-Dubly M L, Rodriguez S, et al. 2014. Reprod Biol Endocrin, 12: 62.
Jaquet D, S Deghmoun, D Chevenne, et al. 2005. Diabetologia. 48(5): 849–855.
Ji B, Zhao GZ, Sakurai R, et al. 2016. Lung, 194(4): 535–546.
Jiang T, Hu SW, Dai SY, et al. 2021. Cell Biol Toxicol. 38(1):69–86.
Jiang Y, Liu J, Li Y, et al. 2014. J Appl Toxicol, 34(9): 1012–1022.
Jiang Y, Xia W, Zhang B, et al. 2018. Environ Pollut, 233: 971–976.
Karsli T, Strickland D, Livingston J, et al. 2017. J Matern Fetal Neonatal Med, 32(3): 377–383.
Kc K, Shakya S, Zhang H. 2015. Ann Nutr Metab, 66 Suppl 2: 14–20.
Koskinen Anna, Lukkarinen Heikki, Moritz Niko, et al. 2012. Pediatr Pulmonol, 47(3): 275–282.
Kou H, Shen L, Luo H W, et al. 2017a. Reprod Toxicol, 74: 85–93.
Kou H, Wang G H, Pei L G, et al. 2017b. Naturwissenschaften, 104(11–12): 89.
Kou H, Gui S, Dai Y, et al. 2020. Toxicol Appl Pharmacol, 404: 115187.
Kumar S V, Lupo P J, Pompeii L A, et al. 2018. Int J Environ Res Public Health, 15(3): 505.
Lautarescu A, Craig M C, Glover V. 2020. Int Rev Neurobiol, 150: 17–40.
Lejonklou M H, Christiansen S, Orberg J, et al. 2016. Chemosphere, 164: 339–346.
Li X, Tan H, Luo M, et al. 2018a. J Toxicol Environ Health A, 81(6): 154–159.
Li X, Yang X, Xiang E, et al. 2018b. Drug Metab Dispos, 46(4): 422–428.
Liu K, Wang G, Li L, et al. 2020. Biochem Pharmacol, 180: 114130.
Liu M, Chen B, Pei L, et al. 2018. Toxicology, 408: 1–10.
Liu M, Zhang Q, Pei L, et al. 2019. Epigenetics, 14(3): 245–259.
Liu M, Liu Y, Pei L, et al. 2021. Acta Pharmacol Sin, 43(6): 1461–1472.
Liu Y, Liu M, Liu Y, 2022. Biochem Pharmacol, 199: 115018.
Lopez M, Palacio M, Gonce A, et al. 2015. Eur J Clin Microbiol Infect Dis, 34(2): 223–230.
Lu J, Jiao Z, Yu Y, et al. 2018. Cell Death Dis, 9(6): 659.
Lu J, Li Q, Ma G, et al. 2020a. Food Chem Toxicol, 141: 111419.
Lu J, Li Q, Xu D, et al. 2020b. Toxicol Lett, 331: 33–41.
Lv F, Fan GL, Wan Y, 2021. Sci Total Environ, 789: 147691.
Mayer K, Seeger W. 2008. Curr Opin Clin Nutr Metab Care, 11(2): 121–127.
Medsker B, Forno E, Simhan H, et al. 2015. Obstet Gynecol Surv, 70(12): 773–779.
Mitsui T. 2021. Int J Urol, 28(9): 882–889.
Moisiadis V G, Matthews S G. 2014. Nat Rev Endocrinol, 10(7): 403–411.
NCD Countdown 2030 collaborators. 2016. Lancet, 392(10152):1072–1088.
Nguyen D, Smith L M, Lagasse L L, et al. 2010. J Pediatr, 157(2): 337–339.
Ni Q, Lu K, Li J, et al. 2018. Toxicol Sci.164(1): 179–190.
Ni Y, Xu D, Lv F, et al. 2019. J Endocrinol, 19: 63–74.
Nobile Stefano, Chiara Di Sipio Morgia, Giovanni Vento. 2022. J Pers Med, 25;12(2):157.
Nosarti C, Reichenberg A, Murray R M, et al. 2012. Arch Gen Psychiatry, 69(6): E1-E8.
Nykjaer C, Alwan N A, Greenwood D C, et al. 2014. J Epidemiol Community Health, 68(6): 542–549.
Painter R C, Osmond C, Gluckman P, et al. 2008. BJOG, 115(10): 1243–1249.
Paintner A, Williams A D, Burd L. 2012. J Child Neurol, 27(2): 258–263.
Patel S P, Rodriguez A, Little M P, et al. 2012. J Epidemiol Community Health, 66(9): 809–814.
Patelarou E, Kelly F J. 2014. Int J Environ Res Public Health, 11(6): 5904–5933.
Pei L G, Yuan C, Guo Y T, et al. 2017. Reprod Toxicol, 71: 150–158.
Pei L G, Zhang Q, Yuan C, et al. 2019. J Endocrinol, 242(1): M17-M32.
Pienkowski W, Wolski H, Drews K, et al. 2015. Ginekol Pol, 86(8): 622–625.
Qian Y, Shao H, Ying X, et al. 2020. Front Public Health, 8: 366.
Rahman M L, Valeri L, Kile M L, et al. 2017. Environ Int, 108: 32–40.
Ristic N, Severs W, Nestorovic N, et al. 2016. Cells Tissues Organs, 201(2): 148–158.
Rochester JR. 2013. Reprod Toxicol, 42: 132–155.
Sabra S, Malmqvist E, Almeida L, et al. 2018. Alcohol, 70: 43–49.

Shangguan Y, Jiang H, Pan Z, et al. 2017. Cell Death Dis, 8(10): e3157.
Shi X T, Zhu H L, Xiong Y W, et al. 2020. Chemosphere, 254: 126785.
Song L, Yan J, Wang N, et al. 2020. Biol Reprod, 102(3): 705–716.
Spanier A J, Kahn R S, Kunselman A R, et al. 2014. JAMA Pediatr, 168(12): 1131–1137.
Spector L G, Birch J. 2012. Pediatr Blood Cancer, 59(5): 776–779.
Stordal K, McArdle H J, Hayes H, et al. 2018. Sci Rep, 8(1): 9067.
Tan Y, Liu J, Deng Y, et al. 2012. Toxicol Lett, 214(3): 279–287.
Tan Y, Lu K, Li J, et al. 2018. Toxicol Lett, 295: 229–236.
Titus L, Hatch E E, Palmer J R, et al. 2019. Epidemiology, 30(5): 679–686.
Turedi S, Hanci H, Topal Z, et al. 2015. Electromagn Biol Med, 34(4): 390–397.
Veenendaal MV , Painter R C, de Rooij S R, et al. 2013. BJOG, 120(5): 548–553.
Vidaeff A C, Blackwell SC. 2011. Obstet Gynecol Clin North Am, 38(2): 205–214.
Vidyadharan V A, Betancourt A, Smith C, et al. 2022. Nutrients, 14(6): 1158.
Vinikoor-Imler L C, Davis J A, Meyer R E, et al. 2014. Environ Res, 132: 132–139.
Viteri O A, Soto E E, Bahado-Singh R O, et al. 2015. Am J Perinatol, 32(5): 405–416.
Walfisch A, Nikolovski S, Talevska B, et al. 2013. Arch Gynecol Obstet, 287(6): 1131–1136.
Wang G, He B, Hu W, et al. 2018. Toxicology, 408: 70–79.
Wu D M, He Z, Ma L P, et al. 2015. Toxicol Appl Pharmacol, 285(2): 89–97.
Xiao D, Kou H, Zhang L, et al. 2017. Arch Med Res, 48(1): 35–45.
Xiao D, Kou H, Gui S, et al. 2019. Front Endocrinol (Lausanne), 10: 34.
Xiao H, Wen Y, Pan Z, et al. 2018. Cell Death Dis, 9(6): 638.
Xie Z, Zhao Z, Yang X, et al. 2018. Toxicol Appl Pharmacol, 352: 107–118.
Xu D, Chen M, Pan X L, et al. 2011. Environ Toxicol Pharmacol, 32(3): 356–363.
Xu D, Wu Y, Liu F, et al. 2012a. Toxicol Appl Pharmacol, 264(3): 395–403.
Xu D, Zhang B, Liang G, et al. 2012b. PLoS One, 7(9): e44497.
Xu D, Xia L P, Shen L, et al. 2013. Acta Pharmacol Sin, 34(12): 1526–1534.
Xu D, Luo H W, Hu W, et al. 2018a. Faseb J: fj201701557R.
Xu D, Zhang C, He X, et al. 2018b. Toxicol Lett, 283: 39–51.
Yan Y E, Liu L, Wang J F, et al. 2014. Toxicol Appl Pharmacol, 277(3): 231–241.
York T P, Eaves L J, Neale M C, et al. 2014. Am J Obstet Gynecol, 210(5): 398–405.
Yu L, Zhou J, Zhang G, et al. 2018. Toxicol Appl Pharmacol, 352: 77–86.
Zeng H C, He Q Z, Li Y Y, et al. 2015. Environ Toxicol, 30(9): 1082–1090.
Zhang C, Xu D, Luo H, et al. 2014. Toxicology, 325: 74–84.
Zhang L, Li Y Y, Zeng H C, et al. 2011. Int J Toxicol, 30(2): 207–215.
Zhang L, Xu D, Zhang B, et al. 2013. Arch Med Res, 44(5): 335–345.
Zhang Q, Pei L G, Liu M, et al. 2020. Food Chem Toxicol, 135: 111057.
Zhang X, Shang-Guan Y, Ma J, et al. 2016. Br J Pharmacol, 173(14): 2250–2262.
Zhang Z, Chen H, Yu, P, et al. 2022. Toxicol Lett, 359: 65–72.
Zhu Y, Zuo N, Li B, et al. 2018. Toxicology, 400–401: 9–19.
Zhu Y, Chen H, Zhao X, et al. 2019. Toxicol Lett, 314: 63–74.

Chapter 2
Molecular Mechanism of Fetal Developmental Toxicity

Yu Guo and Qingxian Li

Abstract Adverse environments during pregnancy, such as exposure to xenobiotics, can disrupt normal fetal development and lead to developmental toxicity. Various physiological and pathological factors can interact to allow xenobiotics to enter the fetal blood circulation through the maternal-placental pathway and induce developmental toxicity through abnormal cell proliferation and apoptosis, oxidative stress, drug metabolism activation damage, and abnormal epigenetic modifications. Fetal developmental toxicity can result in fetal distress, neonatal asphyxia, perinatal death, and increased susceptibility to various chronic diseases in adulthood. This chapter presents an overview of xenobiotics with fetal developmental toxicity and analyzes their toxic effects and mechanisms at the cellular level, providing a theoretical and experimental basis for exploring the intrauterine origin and causative factors of fetal-originated disease.

1 Introduction

Developmental toxicity refers to harmful effects induced by prenatal and postnatal exposure to harmful factors, including malformation, functional defects, and death of offspring until they reach adulthood. Exogenous physical and chemical factors, such as stress, radiation, toxins, drugs, and environmental pollutants, can induce developmental toxicity. Among them, drugs, toxicants, and environmental pollutants ingested from the surrounding physical environment are collectively referred to as xenobiotics. The generation of toxic effects on fetal development can interfere

Y. Guo (✉)
Department of Pharmacology, School of Basic Medical Science, Wuhan University, Wuhan, Hubei, China
e-mail: guoy@whu.edu.cn

Q. Li
Division of Joint Surgery and Sports Medicine, Department of Orthopedic Surgery, Zhongnan Hospital of Wuhan University, Wuhan, Hubei, China

H. Wang et al. (eds.), *Fetal Origin of Diseases*,
https://doi.org/10.1007/978-981-97-5730-5_2

with the normal development of individuals at multiple levels, including molecular, cellular, and tissue levels, through different mechanisms of toxicity. Experimental studies have shown that fetal developmental toxic effects due to xenobiotics often have multipathway, multiorgan, and multitarget effects, and some of the developmental toxic effects may continue after birth, leading to increased susceptibility of adult offspring to various chronic diseases (Liu et al. 2018; Pei et al. 2017; Shangguan et al. 2017; Tan et al. 2018). This chapter introduces the common phenotypes of developmental toxicity, the classification of xenobiotics with developmental toxicity, and elucidates the mechanisms of occurrence and factors influencing their toxic effects at the molecular level. This information provides theoretical and experimental main references for fetal developmental toxicity studies and the establishment of early developmental toxicity evaluation systems.

2 Common Phenotypes of Fetal Developmental Toxicity

Developmental toxicity is characterized by structural malformations, growth retardation, functional defects, and even death of the organism during development, as depicted in Fig. 2.1. This includes ① structural malformation, where some organs exhibit morphological and structural abnormalities at birth; ② growth retardation, which is characterized by a slow development process and delayed physical and intellectual development after birth, leading to increased susceptibility to chronic diseases such as metabolic syndrome in adulthood; ③ functional defects, where abnormalities or defects occur in the physiology, biochemistry, immunity, metabolism, behavior, and neurological activities of the fetus; and ④ stillbirth, where organ cell apoptosis, necrosis, and even organism death occur. For example, exposure to dibutyl phthalate (DBP) causes developmental toxicity, pericardial edema, structural malformations, and functional alterations in zebrafish (Sun and Li 2019). *Fumonisin* B_1 (FB_1), one of the most common mycotoxins and contaminants in stale corn and other agricultural products, induces developmental retardation in rats and mice, increases the incidence of exencephalic and facial malformations, and increases embryonic mortality in mice following maternal exposure (Lumsangkul et al. 2019).

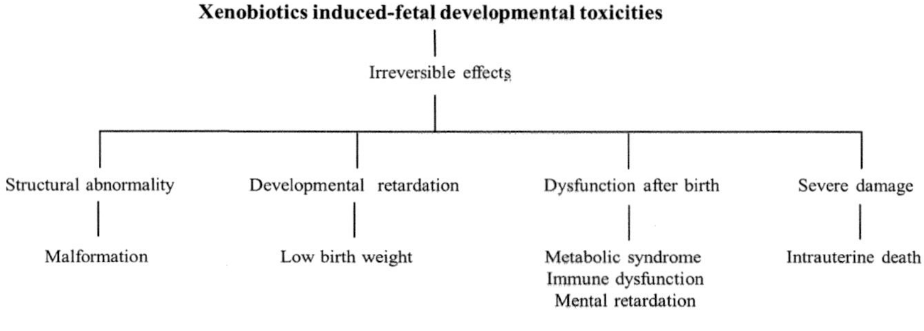

Fig. 2.1 Fetal developmental toxicity of xenobiotics and their manifestations

3 Xenobiotics with Fetal Developmental Toxicity

Xenobiotics with fetal developmental toxicity are commonly reported in the literature as drugs, environmental toxins, and poor diet. Xenobiotics are considered to be the primary non-genetic factors causing developmental abnormalities and congenital malformations.

Drugs: Drugs used during pregnancy have been a significant concern due to their potential safety risks. For instance, dexamethasone is widely used in clinical practice to treat various preterm birth-related conditions. However, several studies have shown that dexamethasone can cause low fetal birth weight and is a definite trigger for fetal developmental toxicity and increased susceptibility to multiple chronic diseases in adulthood (Lahti et al. 2015; Wang et al. 2020).

The United States Food and Drug Administration (USFDA) has developed a classification system for assessing the risks of drug use during pregnancy. The system comprises five categories, namely, A, B, C, D, and X. Category A drugs are considered the least harmful to the fetus, as there is no conclusive evidence of their harmful effects on the fetus during early pregnancy. Category B drugs have not shown any harm to the fetus in animal studies, clinical control experiments, or early pregnancy. Drugs in Category C have demonstrated adverse effects on the fetus in animal experiments, but no clinical control studies have been conducted; therefore, their use during pregnancy should be approached with caution, with the benefits to the pregnant woman and potential risks to the fetus being fully weighed. Category D drugs have sufficient evidence of harm to the fetus and should only be used in serious cases where the pregnant woman has a life-threatening condition or where all other drugs are ineffective. Finally, drugs in Category X have been confirmed to cause fetal abnormalities in all types of experiments and have almost no benefit except for the harm caused to the fetus. The drug instructions may include warnings about fetal risk, which are marked with an "m" suffix in the table.

In November 2007, the Society of Obstetricians and Gynecologists of Canada (SOGC) conducted a review of drugs/chemicals and infection factors that may pose teratogenic risks to humans. The findings of this review are presented in Table 2.1, as reported by Wilson et al. (2007). The objective of this review was to provide clinicians with practical guidelines for assessing the teratogenic risks of prescription and over-the-counter medications during pregnancy.

Environmental toxins: Environmental toxins pose a significant risk to fetal development, with air pollutants, heavy metals, and toxic organic compounds being the most common culprits. Epidemiological studies have shown that exposure to organophosphorus pesticides can cause embryonic and fetal teratogenicity or developmental defects in women before or after pregnancy (Todd et al. 2020). Additionally, maternal smoking during pregnancy has been strongly linked to low fetal birth weight, as demonstrated by clinical studies (Okah et al. 2005). A recent case-control study revealed that maternal exposure to organophosphate flame retardants (OPFRs), commonly used as flame retardants and plasticizers, was significantly associated with an increased risk of low fetal birth weight (Luo et al. 2020). In a rat model of

Table 2.1 Society of Obstetricians and Gynaecologists of Canada (SOGC) list of drugs/chemicals that are at risk of teratogenicity in humans (Wilson 2007)

Agents	Risk category	Fetal effects	Fetal risks	Maternal risks
Ethanol	D/X	Fetal alcohol syndrome (FAS), intrauterine growth retardation (IUGR), mental retardation, microcephaly, characteristic facies, congenital heart disease (CHD), joint/bone/skin defect	40% risk of FAES 6 drinks/day	–
Cocaine	C/X	IUGR, cerebral infarction, bowel Atresia, heart, limb, facial, genitourinary tract, vascular disruption	Fetal death	Abruption placenta
Toluene6	X	Toluene embryopathy similar to FAS	Maternal inhalation 10–100	–
Tetracycline	D	Tooth enamel hypoplasia, discoloration of deciduous teeth	Risk is mainly in second– and third trimester	–
Aminoglycosides			Risk in middle-late pregnancy	–
Streptomycin	Dm	Hearing loss-in rare cases with prolonged high-dose exposure		
Kanamycin	D			
Gentamycin	C			
Vancomycin	Bm			
Fluconazole	Cm	Brachycephaly, cleft palate, Arthrogryposis, CHD	Risk in the first trimester	Treatment of Coccidioidomycosis with a high dose
Efavirenz	Cm/D	Anencephaly, spina bifida	Risk in the first trimester	–
Metronidazole		No evidence of anomalies		–
Fluoroquinolones		Impaired cartilage formation in animal studies	Throughout pregnancy	–
Trimethoprim-sulam ethoxazole		Impaired bilirubin conjugation	3rd trimester	–

(continued)

Table 2.1 (continued)

Agents	Risk category	Fetal effects	Fetal risks	Maternal risks
Folic acid antagonist	Xm	Increased spontaneous abortion, craniofacial anomalies, skeletal anomalies, limb reduction defects, ectrodactyly, IUGR, stillbirth, neonatal death	30% risk with the first trimester – exposure (methotrexate) – a probable increased risk with first-trimester exposure	
Methotrexate	X			–
Aminoptera	Dm			–
Alkylating agents	Dm	IUGR, microphthalmia, cleft palate, genitourinary anomalies, limb reduction defects		
Busulfan Chlorambucil cyclophosphamide	Dm			
Phenytoin (hydantoin)	D	IUGR, mental retardation, microcephaly, facial, heart, hypoplastic nails/distal phalanges, neuroblastoma (increased risk)	10% syndrome 30% exposure effect	Genetic predisposition affects metabolism
Carbamazepine	Dm	Lumbo-sacral NTD (1%), facial, Fingernail hypoplasia, microcephaly, IUGR, developmental delay	1st-trimester exposure	–
Valproic acid	Dm	Lumbo-sacral NTD (1%) possible Fetal valproate syndrome	1st-trimester exposure	Maternal drug metabolism Modifies risk
Trimethadione	D	IUGR, CL ± CP, microcephaly, facial,	60–80% risk with – First-trimester exposure	–
Paramesasone	Dm	Mental retardation, ophthalmologic, limb, genitourinary		
ACE inhibitors (enalapril, captopril, lisinopril)	Cm/Dm	IUGR, renal tubular dysplasia, oligohydramnios, fetal morbidity, 30% joint contractures, pulmonary hypoplasia	Increased risk with second and third-trimester exposure	–
Danazol	Xm	Female virilization	Dose and gestational age dependent	–

(continued)

Table 2.1 (continued)

Agents	Risk category	Fetal effects	Fetal risks	Maternal risks
Diethylstilbestrol	Xm	Female vaginal, cervical, uterine effects, clear-cell adenocarcinoma		–
Letrozole	Xm	Bone, cardiac, and gastrointestinal malformation		–
Antithyroid (propylthiouracil, Methimazole, Carbimazole)	D	Hypothyroidism, cutis aplasia, methimazole embryopathy (choanal atresia, esophageal atresia, hypoplastic nipples, scalp defect, developmental delay)		–
Oral contraceptives	Xm	Female masculinization, neonatal hyperbilirubin	Risk 0.3%	–
Lithium	D	Congenital heart disease (Ebstein anomaly), neonatal CNS, and neuromuscular complications are increased no associated birth anomalies reported (Paxil: Cardiac malformations 2%) inconsistent fetal effects small studies show no risk	–	–
Selective 5-HT reabsorber (SSRI)	Cm		Risk/benefit with cautionary recommendation	–
Paroxetine	D		Risk/benefit with cautionary recommendation	–
Tricyclic antidepressants	D		–	–
Bupropion	Bm			
Lead	–	Reduced fetal growth	–	Increased spontaneous abortion and stillbirth
Organic mercury	–	Cerebral atrophy, microcephaly, mental retardation, seizures, blindness, spasticity	Exposure in any Trimester	
Polychlorinated biphenyls		IUGR, dermal pigmentation, developmental delay	–	Maternal neurotoxicity with fish and grain contamination

(continued)

Table 2.1 (continued)

Agents	Risk category	Fetal effects	Fetal risks	Maternal risks
Statin (HMG-CoA reductase inhibitors)	Xm	No associated birth anomalies established (hypothetical risk)	–	–
Methyl blue	Cm/D	Intra-amniotic exposure associated with possible bowel atresia	Dose-dependent	–
Misoprostol	Xm	Moebius syndrome, terminal transverse limb defects, arthrogryposis, CNS	–	–
Penicillamine	D	Cutis laxa/connective tissue anomaly	–	–
Thalidomide	X	Bilateral limb deficiency, anotia, microtia, cardiac, gastrointestinal	20% risk with exposure at 35–50 gestational days	–
Vitamin A (retinol, vitamin A1)	Cm Xm D/X	Microtia, craniofacial, microphthalmia, CL ± CP microtia, microphthalmia, craniofacial, cardiac, CL ± CP	Vitamin A risk may require >30,000 IU/day topical use appears to have no risk	–
Isotretinoin (Accutane)				–
Warfarin (Coumadin)			5–25% risk with first-trimester exposure	–

IUGR exposed to nicotine during pregnancy, multiple organs such as the adrenal gland, liver, and hippocampus, showed hypoplasia and dysfunction in fetal rats (Huang et al. 2015; Xu et al. 2015; Liu et al. 2012). These findings highlight the importance of avoiding exposure to environmental toxins during pregnancy to prevent adverse effects on fetal development.

Poor diet: Excessive consumption of alcohol, tea, coffee, and a low-protein or high-fat diet during pregnancy can also have adverse effects on fetal development. Alcohol is a well-known teratogen, and exposure to alcohol during pregnancy can result in various adverse fetal outcomes, including fetal alcohol syndrome (FAS). Moreover, alcohol exposure during pregnancy can lead to long-term effects on endocrine and metabolic functions in developing individuals. Furthermore, caffeine intake is common among pregnant women, but a 15-year prospective study found that caffeine exposure during pregnancy was significantly associated with an increased risk of obesity in offspring (Li et al. 2015). Excessive caffeine intake during pregnancy has also been shown to cause low birth weight and poor fetal development of multiple organs, including the skeleton, hippocampus, and adrenal glands, as confirmed by a series of animal experiments conducted by the authors' team (Tan et al. 2018; Xu et al. 2018; Chen et al. 2018). It is important to note that poor dietary habits during pregnancy can have far-reaching effects on fetal development, which can continue into adulthood and even be inherited for multiple generations (Zhao et al. 2020). Therefore, pregnant women should avoid these harmful dietary habits to ensure optimal fetal development and long-term health outcomes.

4 Factors Influencing Fetal Developmental Toxicity

Various factors can influence fetal developmental toxicity, including maternal and placental factors. During pregnancy, the mother provides nutrients to the fetus through the placenta. However, she can also transfer ingested xenobiotics or their metabolites to the fetus, which can result in fetal developmental toxicity. Changes in the maternal environment can also affect fetal growth and development, leading to developmental toxicity. Metabolic dysregulation of the maternal-placental-fetal unit is a critical factor in fetal developmental toxicity. Although developmental toxicity primarily occurs in the fetus and may be directly related to adverse intrauterine environmental factors such as xenobiotic exposure, changes in maternal and placental function can also have significant effects on fetal development. Therefore, it is essential to understand and address both maternal and placental factors to prevent fetal developmental toxicity and ensure optimal fetal development.

4.1 Maternal Factors

The biotransformation function of the mother, which includes phase I and phase II metabolic reactions, plays a crucial role in determining the yield of xenobiotics and their metabolites. The mother's biotransformation function can either eliminate most of the xenobiotics to reduce their amount in the fetus or produce active metabolites that can damage the placenta and indirectly cause fetal damage. For instance, some teratogens, such as phenytoin, can directly conjugate to some of the mother's phase II enzymes without undergoing phase I metabolism, reducing maternal elimination and increasing the number of xenobiotics transferred to the embryo or fetus, thereby increasing its teratogenic risk (Wells et al. 2004). Moreover, some xenobiotics, such as ethanol or sevoflurane, can directly inhibit fetal neural stem cell proliferation through the placenta, leading to impaired brain development (Vangipuram and Lyman 2012; Wang et al. 2018). Additionally, the mother's disease state can also affect normal fetal growth and development. For instance, intrahepatic cholestasis of pregnancy (ICP) is a common liver disease during pregnancy that can adversely affect normal fetal growth and development or even cause death. ICP is characterized by maternal pruritus, elevated serum bile acids, and/or transaminase levels. Human and animal studies have shown that ICP can cause neonatal hypoxic-ischemic encephalopathy, severe intraventricular hemorrhage, bronchopulmonary dysplasia, and necrotizing small bowel colitis (Herrera et al. 2018; Geenes et al. 2011).

4.2 Placental Factors

The placenta is a vital connection between the mother and the fetus, playing a crucial role in regulating the flow of nutrients from the mother to the fetus, exchanging hormones between the fetus and mother, and performing other functions. The placenta also serves as a barrier against harmful substances, although it cannot block all harmful substances from crossing into the fetus. Smoking, for example, can cause elevated maternal blood cadmium concentrations, which may accumulate in the placenta and cause fetal developmental toxicity (Menai et al. 2012). Adverse environmental conditions such as stress, hypoxia, and nutritional restriction during pregnancy can cause the placental glucocorticoid barrier to open, leading to fetal overexposure to maternal-derived glucocorticoids, which can affect fetal growth and development and mediate the development of distant diseases (Jobe 2020; Zhu et al. 2019). Exposure to xenobiotics such as caffeine, nicotine, and ethanol during pregnancy can also cause fetal overexposure to maternal glucocorticoids, exacerbating the stress state of fetal rats and leading to altered programming in their multiple organ development, as found in studies conducted by the authors' team (Xu et al. 2012a; b; Liang et al. 2011). Furthermore, caffeine intake during pregnancy has been shown to induce P53-dependent apoptosis in placental trophoblast cells, as previously reported by the authors' team (Huang et al. 2012). Excessive oxygen radicals in the organism can damage the placental villi, particularly the villi cell membrane and mitochondrial membrane, affecting the normal function of the placenta and limiting fetal growth and development.

5 Molecular Mechanisms of Xenobiotics Induced-Developmental Toxicity

Exogenous-induced fetal developmental toxicity can occur through various molecular mechanisms, including abnormal cell proliferation and apoptosis, oxidative stress, mitochondrial damage, exogenous metabolic activation damage, and abnormal epigenetic modifications.

5.1 Abnormal Cell Proliferation and Apoptosis

Cell proliferation is a fundamental process for the growth, development, reproduction, and inheritance of living organisms. However, the inhibition of cell proliferation caused by xenobiotic exposure during pregnancy can lead to fetal organ dysgenesis or even death. Studies have shown that exposure to tamoxifen during pregnancy can suppress fetal neural progenitor cell proliferation through the Wnt signaling pathway, resulting in pre- and postnatal neurodevelopmental defects in the fetus (Lee et al. 2020). Similarly, exposure to ethanol or sevoflurane during pregnancy can further inhibit fetal neural stem cell proliferation by suppressing the Wnt/β-catenin signaling pathway, thereby impairing brain development (Vangipuram and Lyman 2012; Wang et al. 2018). Moreover, Chen et al. (2018) found that caffeine exposure during pregnancy can lead to the suppression of adrenocortical cell proliferation before and after birth in male fetal rats by inhibiting the adrenal IGF1-related MAPK/ERK signaling pathway. In addition, maternal exposure to ketamine during pregnancy can impair the proliferation and differentiation of mouse embryonic stem cells by interfering with brain-derived neurotrophic factor synthesis and its downstream mTOR signaling (Liu et al. 2021). Furthermore, abnormal cell proliferation mediated by cellular autophagy is also involved in offspring organogenesis toxicity due to exposure to xenobiotics during pregnancy. Stasiak et al. (2016) suggested that dexamethasone impairs mouse spiny folliculogenesis and enhances follicular atresia by inducing autophagy or a combination of autophagy and apoptosis. Zhang et al. (2021) demonstrated that dexamethasone exposure during pregnancy activated autophagy in rat fetal liver cells, while the autophagy inhibitor bafilomycin was confirmed at the cellular level to reverse dexamethasone-induced inhibition of hepatocyte proliferation and cell cycle arrest in the G0 and G1 phases, thus confirming the involvement of autophagy in dexamethasone-induced inhibition of hepatocyte proliferation.

Caspases are crucial factors in the apoptosis pathway, and Caspase-3 can be activated through both the death receptor pathway and the mitochondrial pathway, resulting in irreversible apoptosis. In the context of IUGR due to feeding restriction in a sheep model, increased apoptosis in fetal liver cells may be responsible for impaired development (Liu et al. 2017). The oncogene p53, which normally slows down cell division, induces apoptosis when activated. In a rat model of IUGR due to caffeine exposure during pregnancy, the authors' team found that placental trophoblast damage may be due to chronic activation of the placental renin-angiotensin system, resulting in upregulation of AT1 and AT2 receptor expression and induction of p53-dependent trophoblast apoptosis due to structural and functional changes in the placenta (Huang et al. 2012). This may also be related to reduced expression of cytoleptin in the placental trophoblast due to caffeine exposure (Wu et al. 2015; Huang et al. 2012). Furthermore, the authors' team found that intrachondral osteogenic growth retardation in IUGR rats due to maternal caffeine exposure was associated with intrauterine glucocorticoid overexposure upregulating the mitogen-inducible gene 6, which inhibits the reduction of apoptosis mediated by the epidermal growth factor receptor pathway (Shangguan et al. 2017). Additionally, increased apoptosis in the brain may contribute to ultrastructural damage in the fetal rat brain in an IUGR rat model (Liu et al. 2011).

5.2 *Oxidative Stress and Mitochondrial Damage*

Oxidative stress is an imbalance between the oxidative and antioxidant states caused by the excessive production of highly reactive molecules, such as reactive oxygen species (ROS), in response to various harmful stimuli. The body's inability to remove these molecules in a timely manner results in oxidative damage to various macromolecules, such as sugars, lipids, proteins, and nucleic acids, that make up cell tissues, leading to structural or functional abnormalities in tissues and cells. Administration of the ROS scavenger acetylcysteine to experimental animals significantly reduced or alleviated both ROS levels and teratogenic manifestations due to 5-bromodeoxyuridine nucleoside. Exposure to 2,2′,4,4′,5-pentabromodiphenyl ether or silver nanoparticles during pregnancy increased levels of liver stress markers and caused hepatocyte damage in fetal rats (Blanco et al. 2012; Zhang et al. 2014). Hydroxyurea is a teratogen that can cause defects in the brain, jaws, or limbs. The administration of the free radical scavenger D-mannitol significantly inhibited the teratogenicity of hydroxyurea, and in vitro tests also found that the addition of superoxide dismutase (SOD) had a protective effect on the teratogenic effect of hydroxyurea, suggesting that ROS overproduction is one of the main mechanisms of hydroxyurea teratogenicity. The teratogenic effect of reaction arrest is related to oxidative stress, and its induced production of excessive ROS can cause abnormalities in the NF-κB signaling pathway, which further affects the expression of genes related to limb development, such as FGF-8, FGF-10, and Twist. Collectively, these findings suggest that oxidative stress, as a nonspecific form of damage, is a crucial factor in the mechanism of fetal developmental toxicity.

The endoplasmic reticulum (ER) is a crucial single-layer membrane organelle in cells that involves multiple signaling pathways, among which the most important is the unfolded protein response (UPR) that maintains ER homeostasis by reducing protein synthesis and increasing chaperone molecule synthesis. Endoplasmic reticulum stress (ERS) occurs when various stressors applied to cells induce the accumulation of misfolded and unfolded proteins in the ER lumen and disturb Ca^{2+} homeostasis, thereby activating the UPR and apoptotic signaling pathways. Enhanced placental ERS negatively regulates placental growth factor expression, which mediates IUGR onset (Mizuuchi et al. 2016) and alters placental morphology (Veerbeek et al. 2015; Kawakami et al. 2014). In rats, administration of the ERS inducer clindamycin on day 15.5 of gestation resulted in a preterm birth incidence of up to 75%, suggesting that pregnant rats are susceptible to ERS in late gestation. Clindamycin also dose-dependently reduced fetal rat and placental weights and clustered trophoblast cells in the placental labyrinth region, along with reduced fetal liver and placental glycogen content. Furthermore, clindamycin decreased hepatic glucose transporter 1 (GLUT1) expression and increased placental GLUT2 and GLUT3 expression while decreasing placental vascular endothelial growth factor receptor 1 (VEGFR-1) and VEGFR-2 expression. These findings suggest that ERS may cause structural and functional damage to the placenta and fetal liver by affecting GLUTs and VEGFR expression, ultimately leading to the development of IUGR (Kawakami et al. 2014).

Mitochondrial damage refers to mutations in mitochondrial DNA (mtDNA) or damage to mitochondrial membranes caused by various external stimuli, resulting in structural and functional disorders in mitochondria. ATP production is essential for meeting cellular energy demands, but mutations, reactions, or damage occurring in mtDNA can lead to severe consequences for mitochondrial respiratory function, resulting in cellular death and dysfunction. Excess ROS can lead to mitochondrial dysfunction by inducing mtDNA mutations and generating more ROS, creating a vicious cycle (Chen et al. 2020). Perinatal exposure to nicotine significantly increased protein expression of glutathione peroxidase 1 and manganese superoxide dismutase in the fetal pancreas, as well as islet ROS production and pancreatic protein carbonyl formation, particularly in the mitochondria of pancreatic cells (Bruin et al. 2008). Moreover, lower levels of mitochondrial protein complexes I, II, IV, and V proteins were found in cardiac cells of nicotine-exposed offspring during pregnancy (Barra et al. 2017). The potent carcinogen aflatoxin B1 can first attack mtDNA and promote apoptosis by covalently binding to mtDNA, interrupting mitochondrial biogenesis, and causing damage to the mitochondrial membrane (Xu et al. 2021).

5.3 *Metabolic Damage of Xenobiotics*

Xenobiotics can cause metabolic damage by being transformed into toxic metabolites within the body, leading to harm. Exogenous metabolic damage is mainly attributed to electrophile and oxygen radical metabolites, which can bind to nucleophilic macromolecules and cause damage to mitochondria, leading to cell apoptosis, necrosis, or gene mutation. CYPs play a crucial role in mediating the metabolic activation of xenobiotics, such as polycyclic aromatic hydrocarbons (PAHs) and polychlorinated biphenyls (PCBs), which can be converted into carcinogenic metabolites. The exposure of pyrrolizidine alkaloid (PA) during pregnancy has been found to result in developmental delay, preterm birth, and stillbirth in fetal rats. PA is metabolized by CYP3A isoforms to produce the electrophile metabolite dehydropyrrole, which is the main active intermediate exerting toxic effects. Acetaminophen is another example of a xenobiotic that is metabolized via CYP2E1 to produce quinone active intermediates, which can induce fetal developmental toxicity when they bind to developmentally relevant genes or proteins (Fig. 2.2) (Cederbaum et al. 2001; Guo et al. 2013, 2019; Lieber 1999; Luo et al. 2019; Reimers et al. 2004).

Living organisms contain various oxygen radicals such as epoxy radicals, hydroxyl radicals, and nitrogen dioxide radicals, which collectively with ozone,

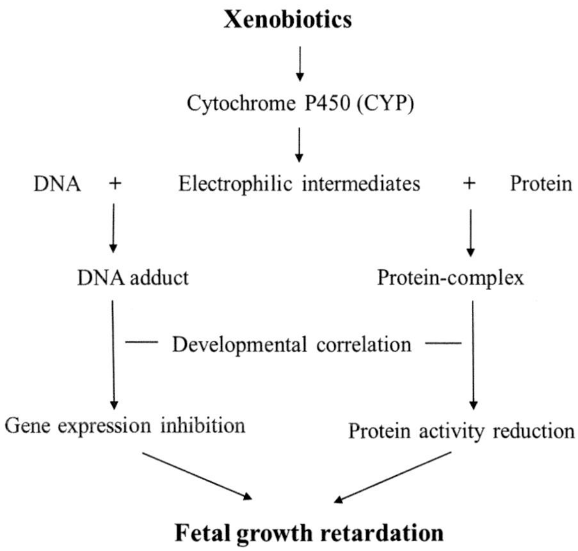

Fig. 2.2 Metabolic damage of xenobiotics mediated by electropilic metabolites

singlet oxygen, and hydrogen peroxide are referred to as "ROS." Free radical metabolites or ROS selectively target protons in unsaturated fatty acids, producing lipid alkoxides, leading to localized or widespread membrane damage, increased cell membrane permeability, and intracytoplasmic component leakage (Galdikova et al. 2015). These oxygen radical intermediates can also alter intracellular signal transduction or react with double bonds in lipids, proteins, and DNA to form covalent conjugates. Oxidative stress occurs when intracellular ROS surpasses the antioxidant capacity of the cell. Excess oxygen radicals can induce vascular smooth muscle contraction, platelet aggregation, and blood flow stasis, leading to chronic fetal ischemia, hypoxia, and growth retardation. The adverse pregnancy outcomes due to xenobiotic exposure during pregnancy are associated with the action of oxygen-free radicals in the body. A balance exists between oxidative and antioxidant mechanisms in a normal organism, and the imbalance between oxygen radical production and elimination can directly affect the stability of the internal environment of the pregnant woman and the fetus, leading to serious harm to normal fetal growth and development. Studies have identified prostaglandin H synthases (PHSs) and lipoxygenases (LPOs) as the enzymes mediating oxidative metabolism, which can oxidize xenobiotics to free radical intermediates through their own or related hydroperoxide activities. The expression of fetal PHSs and LPOs is high, and the oxygen radicals generated by catalyzing xenobiotics and their mediated oxidative stress can lead to fetal developmental toxicity (Wells et al. 2005). Studies have shown that maternal exposure to the fungicide azoxystrobin induces reactive oxygen-dependent apoptosis of neurons in the cerebral cortex of the offspring (Kang et al. 2021). Placental ischemia/hypoxia can result in increased rates of ROS, oxidative stress, and lipid peroxidation production, which may trigger pre-eclampsia and IUGR (Ozsurekci and Aykac 2016). On the other hand, antioxidant supplementation, such as curcumin and resveratrol during pregnancy, may significantly improve fetal growth restriction (Sebastiani et al. 2022).

5.4 *Abnormal Epigenetic Modifications*

Epigenetic modifications refer to reversible, heritable changes in gene expression that occur in the absence of alterations in DNA sequence, and mainly include DNA methylation, histone modifications, etc. Epigenetic modifications have been shown to be involved in fetal multiorgan developmental toxicity, growth retardation, or dysfunction resulting from multiple xenobiotic exposures during pregnancy, including adrenal (Liu et al. 2016; Yan et al. 2014), kidney (Li et al. 2019), testis (Liu et al. 2018), and long bones (Xiao et al. 2018). Intrauterine nicotine exposure has been found to result in adrenal impairment in fetal rats, possibly due to nicotine-induced methylation of the nt -377 site CpG of the Pax6 binding motif, and these phenomena persist for 5–10 generations at the cellular level (Wang et al. 2011). The mechanism of testicular developmental toxicity due to dexamethasone exposure during pregnancy may be through activation of GR that increases HDAC7 expression, which then synergistically decreases H3K9ac levels of StAR, leading to testicular hypoplasia (Liu et al. 2018). Dexamethasone can persistently activate RAS-mediated peak offspring by increasing H3K27ac levels of angiotensin I-converting enzyme, resulting in bone mass reduction and its intergenerational effects (Xiao et al. 2018). Dexamethasone exposure during pregnancy can also cause impaired hepatic cholesterol reversal transport function and the development of adult hypercholesterolemia in offspring rats by a mechanism related to hepatic activation of GR to enhance DGCR8/miR-148a expression, which leads to persistent low expression of LDLR (Li et al. 2020). High-dose BPA exposure during pregnancy has also been shown to increase DNA methylation at the CpG site of cyclic AMP-response element binding protein (CREB)-regulated transcriptional co-activator 1 produced in F2-generation mice, which in turn affects hippocampal neurogenesis and cognitive function in the offspring (Jang et al. 2012). All these studies suggest that abnormal epigenetic modifications are involved in the developmental toxicity of fetal multiple organs.

6 Summary and Prospect

This chapter presents a classification of xenobiotics with developmental toxicity and discusses the role of maternal and placental factors in fetal developmental toxicity. The possible cytotoxic mechanism of fetal developmental toxicity is also elucidated from the molecular level, providing a theoretical and experimental basis for the study of fetal developmental toxicity and the establishment of an early evaluation system (Fig. 2.3). However, the mechanisms of fetal developmental toxicity and fetal programming mechanisms of adult diseases remain unclear due to the lack

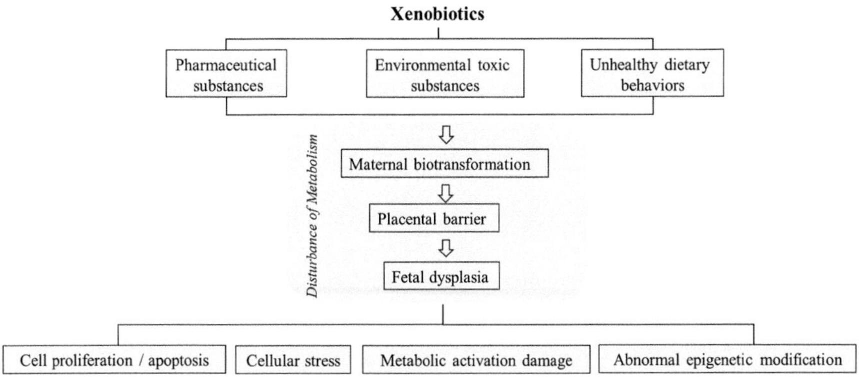

Fig. 2.3 The summary diagram of the molecular mechanism of developmental toxicity

of systematic clarification of organ function changes, programming mechanisms, gender differences, and key points of multigeneration inheritance in different periods before and after birth. Abnormal cell proliferation and apoptosis, oxidative stress and mitochondrial damage, exogenous metabolic activation damage, and abnormal epigenetic modifications are cellular pathophysiological processes that are closely related to fetal development and are potential targets for the prevention and treatment of fetal developmental abnormalities and their susceptibility to fetal-originated disease.

References

Barra N G, Lisyansky M, Vanduzer T A, et al. 2017. J Appl Toxicol, 37(12): 1517-1526.
Blanco J, Mulero M, Domingo J L, et al. 2012. Toxicol Sci, 127(1): 296-302.
Bruin J E, Petre M A, Lehman M A, et al. 2008. Free Radic Biol Med, 44(11): 1919-1925.
Cederbaum A I, Wu D, Mari M, et al. 2001. Free Radic Biol Med, 31(12): 1539-1543.
Chen G, Yuan C, Duan F, et al. 2018. Toxicol Appl Pharmacol, 341: 64-76.
Chen K, Lu P, Beeraka N M, et al. 2020. Semin. Cancer Biol, S1044-579X(20)30203-0.
Galdikova M, Sivikova K, Holeckova B, et al. 2015. J Environ Sci Health B, 50(10): 698–707.
Geenes V L, Lim Y H, Bowman N, et al. 2011. Placenta, 32(12): 1026–1032.
Guo Y, Ma Z, Kou H, et al. 2013. Toxicol Lett, 221(3): 212–218.
Guo Y, Xiao D, Yang X, et al. 2019. Reprod Toxicol, 85: 34–41.
Herrera C A, Manuck T A, Stoddard G J, et al. 2018. J Matern Fetal Neonatal Med, 31(14): 1913–1920.
Huang H, He Z, Zhu C, et al. 2015. Toxicol Appl Pharmacol, 288(1): 84–94.
Huang J, Zhou S, Ping J, et al. 2012. Clin Exp Pharmacol Physiol, 39(4): 357–363.
Jang Y J, Park H R, Kim T H, et al. 2012. Toxicology, 296(1–3): 73–82.
Jobe A H. 2020. J Pediatr, 217: 184–188.
Kang J, Bishayee K, Huh SO. 2021. Int J Mol Sci. 22(22): 12495.
Kawakami T, Yoshimi M, Kadota Y, et al. 2014. Toxicol Appl Pharmacol, 275(2): 134–144.
Lahti M, Eriksson J G, Heinonen K, et al. 2015. Psychological medicine, 45(5): 985–999.
Lee C M, Zhou L, Liu J, et al. 2020. Proc Natl Acad Sci U S A., 117(32): 19578–19589.

Li B, Zhu Y, Chen H, et al. 2019. Toxicology, 411: 32–42.
Li D K, Ferber J R, Odouli R. 2015. Int J Obes (Lond), 39(4): 658–664.
Li L, Hu W, Liu K X, et al. 2020. Toxicol Appl Pharmacol, 395: 114979.
Liang G, Chen M, Pan X L, et al. 2011. Exp Toxicol Pathol 63(7–8), 607–611.
Lieber C S. 1999. Alcohol Clin Exp Res, 23(6): 991–1007.
Liu J, Liu L, Chen H. 2011. Neuroscience, 181: 265–270.
Liu L, Liu F, Kou H, et al. 2012. Toxicol Lett, 214(3): 1307–1313.
Liu L, Wang J F, Fan J, et al. 2016. Int J Mol Sci, 17(9): 1477
Liu M, Chen B, Pei L, et al. 2018. Toxicology, 408: 1–10.
Liu S, Tao G, Zhou C, et al. 2021. Toxicol In Vitro, 72: 105091.
Liu Y, Ma C, Li H, et al. 2017. Theriogenology, 90: 204–209.
Lumsangkul C, Chiang H I, Lo N W, et al. 2019. Toxins (Basel). 11(2): 114.
Luo D, Liu W, Tao Y, et al. 2020. Environ Sci Technol, 54(6): 3375–3385.
Luo J, Yang X, Qiu S, et al. 2019. Toxicology, 418: 32–40.
Menai M, Heude B, Slama R, et al. 2012. Reprod Toxicol, 34(4): 622–627.
Mizuuchi M, Cindrova-Davies T, Olovsson M, et al. 2016. J Pathol, 238(4): 550–561.
Okah F A, Cai J, Hoff G L. 2005. Obstet Gynecol, 105(3): 543–550.
Pei L G, Chao Y, Guo Y T, et al. 2017. Reprod Toxicol, 71: 1150–1158.
Reimers M J, Flockton A R, Tanguay R L. 2004. Neurotoxicol Teratol, 26(6): 769–781.
Sebastiani G, Navarro-Tapia E, Almeida-Toledano L, et al. 2022. Antioxidants (Basel), 11(4): 648.
Shangguan Y, Jiang H, Pan Z, et al. 2017. Cell Death Dis, 8(10): e3157.
Stasiak M, Dobrowolski P, Tomaszewska E. 2016. Reprod Fertil Dev, 28(7): 1038–1048.
Sun G, Li Y. 2019. Chemosphere, 218: 76–82.
Tan Y, Lu K, Li J, et al. 2018. Toxicol Lett, 295: 229–236.
Todd S W, Lumsden E W, Aracava Y, et al. 2020. Neuropharmacology, 1;180: 108271.
Vangipuram S D, Lyman W D. 2012. Alcohol Clin Exp Res, 36(5): 788–797.
Veerbeek J H, Tissot Van Patot M C, Burton G J, et al. 2015. Placenta, 36(1): 88–92.
Wang T, Chen M, Liu L, et al. 2011. Toxicol Appl Pharmacol, 257(3): 328–337.
Wang Y, Yin S, Xue H, et al. 2018. Dev Biol, 435(2): 185–197.
Wang Y Z, Li Q X, Zhang D M, et al. 2020. BBA-Mol Cell Res, 1867(10): 118791.
Wells P G, Bhuller Y, Chen C S, et al. 2005. Toxicol Appl Pharmacol, 207(Suppl 2): 354–366.
Wells P G, Mackenzie P I, Chowdhury J R, et al. 2004. Drug Metab Dispos, 32(3): 281–290.
Wilson RD, SOGC GENETICS COMMITTEE, SOGC INFECTIOUS DISEASE COMMITTEE. 2007. J Obstet Gynaecol Can. 29(11): 911–917.
Wu Y M, Luo H W, Kou H, et al. 2015. Toxicol Appl Pharmacol, 289(1): 109–116.
Xiao H, Wen Y, Pan Z, et al. 2018. Cell Death Dis, 9(6): 638.
Xu D, Bai J, Zhang L, et al. 2015. Toxicol Res, 4(1): 112–120.
Xu D, Liang G, Yan Y E, et al. 2012a.Toxicology letters, 209(3): 282–290.
Xu D, Wu Y, Liu F, et al. 2012b. Toxicol Appl Pharmacol, 264(3): 1395–1403.
Xu D, Zhang C, He X, et al. 2018. Toxicol Lett, 283: 39–51.
Xu F, Li Y, Cao Z, Zhang J, et al. 2021. Ecotoxicol Environ Saf, 216: 112213.
Yan Y E, Liu L, Wang J F, et al. 2014. Toxicol Appl Pharmacol, 277(3): 231–241.
Yasemin Ozsurekci, Kubra Aykac.2016. Oxid Med Cell Longev. 2768365.
Zhang D M, Liu K X, Hu W, et al. 2021. Toxicology, 449: 152664.
Zhu P, Wang W, Zuo R, et al. 2019. Cell Mol Life Sci,76(1): 13–26.
Zhao Z, Qin J, Pei L G, et al. 2020. Toxicology, 442: 152533.

Chapter 3
Epigenetic Mechanism of Fetal-Originated Disease

Ying Ao and Liang Liu

Abstract Environmental factors that interfere with early life development have become the main cause of fetal developmental toxicity and susceptibility to multiple diseases in adulthood through changes in intrauterine programming, mainly related to abnormal epigenetic modifications. Epigenetics not only partially explains how the environment during pregnancy affects fetal physiological development but also elucidates the mechanism of fetal developmental toxicity related to the prenatal adverse environment, susceptibility to multiple diseases in adulthood, and even genetic effects. This chapter summarizes the recent research progress of worldwide scholars in epigenetics in intrauterine programming of fetal diseases, analyzes cases of multi-generation inheritance through epigenetic mechanisms, and clarifies the mechanism of susceptibility to fetal diseases and multi-generation genetic effects from the perspective of abnormal epigenetic modification. This information is crucial in clarifying the long-term harm caused by environmental factors and provides new ideas for exploring early prevention and treatment targets of fetal-originated disease.

1 Introduction

Epigenetic modification refers to heritable gene regulation that is independent of DNA sequence changes, including DNA methylation, DNA local configuration, nucleosome phase changes, histone modification, and special RNA molecules. Classical genetics holds that gene sequence determines the product of gene

Y. Ao (✉)
Department of Pharmacology, School of Basic Medical Science, Wuhan University, Wuhan, Hubei, China
e-mail: yingao@whu.edu.cn

L. Liu
Division of Joint Surgery and Sports Medicine, Department of Orthopedic Surgery, Zhongnan Hospital of Wuhan University, Wuhan, Hubei, China

H. Wang et al. (eds.), *Fetal Origin of Diseases*,
https://doi.org/10.1007/978-981-97-5730-5_3

expression, which is the basis of cell proliferation and differentiation, and plays a decisive role in the growth and development of the body. However, the same genome often shows different genetic phenotypes, and "gene determinism" cannot fully explain the phenotypic changes caused by environmental factors. The establishment of the theory of epigenetics provides a complementary viewpoint to classical genetics. As epigenetics has become the frontier and hotspot in life science research, more and more scholars are trying to associate epigenetics with developmental toxicity caused by environmental factors. Studies have shown that mammalian embryonic development is accompanied by the regulation of various epigenetic modifications (Mathers and McKay 2009). Epigenetic changes, such as DNA methylation and histone modification, caused by the prenatal adverse environment have significant effects on genome reprogramming during embryo or fetal development and the establishment of an early embryonic development model, resulting in long-term and even multigenerational genetic effects (Nafee et al. 2008). Hence, the investigation of epigenetics has significant implications for understanding the pathogenesis of disorders associated with abnormal development and differentiation. Additionally, it provides a theoretical foundation for preventing heritable damage caused by adverse environmental factors during pregnancy and exploring early prevention and treatment.

2 Epigenetic Modifications and Fetal Development under Physiological Conditions

Epigenetic modifications induced by environmental factors regulate not only the transcriptional process of coding genes by altering the composition of chromosomes, such as DNA methylation and histone modification, but also gene expression through non-coding RNA in the post-transcriptional modification stage and mRNA translation stage.

2.1 DNA Methylation and Fetal Development Under Physiological Conditions

DNA methylation, which involves the addition of a methyl group to the DNA sequence, is linked to gene transcriptional suppression. In eukaryotes, DNMTs, including DNMT1, DNMT3a, and DNMT3b, primarily mediate DNA methylation. While DNMT1 maintains the original methylation state, DNMT3a and DNMT3b can add a methyl group to specific unmethylated CpG islands, in addition to maintaining the methylation pattern. Conversely, the ten-eleven translocation (TET) family proteins, such as TET1, TET2, and TET3, oxidize 5-methylcytosine to generate 5-hydroxymethylcytosine, 5-formylcytosine, and 5-carboxylcytosine, thus facilitating demethylation (Hemberger et al. 2009).

The level of DNA methylation exhibits distinct patterns during different stages of embryonic development. There are two genome-wide DNA methylation reprogramming events that occur during gametogenesis and early embryonic development. The first stage of methylation reprogramming is necessary for the re-establishment and erasure of parental imprinted genes, while the second stage of methylation reprogramming is a prerequisite for fertilized eggs to acquire totipotency and generate new individuals. The level of DNA methylation undergoes dynamic changes throughout embryonic development, which include demethylation of the genome during the eight-cell stage of blastocyst formation and re-establishment of a new methylation pattern during the implantation period. Additionally, the level of DNA methylation also undergoes tissue-specific changes during the middle and late stages of embryonic development. DNA methylation during embryonic development plays several critical roles, such as inhibiting gene expression, inactivating the female X chromosome, regulating imprinted gene expression, silencing transposable elements, and regulating tissue-specific gene expression (Abdalla et al. 2009).

2.2 *Histone Modification and Fetal Development Under Physiological Conditions*

Eukaryotic histones are subjected to various covalent modifications, such as acetylation, methylation, phosphorylation, ubiquitination, and others. Among these modifications, histone acetylation and methylation are the most common ones. Histone methylation involves the transfer of a methyl group to the N-terminal lysine or arginine residues of histones H3 and H4 by histone methyltransferases. Currently, the most studied histone methylation sites include histone H3 lysine 9 (H3K9), H3K27, H3K36, H3K79, and H4K20. The impact of histone methylation on gene expression is complex and unpredictable, which may be attributed to the different positions of the methylated residues (Nielsen et al. 2001). Histone methylation undergoes a dynamic process during preimplantation development. After fertilization, H3K4me, H3K9me2/3, H3K27me1, and H4K20me3 can be detected in the maternal pronucleus. Studies on demethylases have also confirmed their contribution to the differentiation of pluripotent stem cells during development. Lysine-specific histone demethylase 1 (LSD1), a H3K4 demethylase, has been reported to play a role in this process. In mice, knockout of LSD1 leads to embryonic lethality (Wang et al. 2007), while in drosophila melanogaster, LSD1 mutation results in decreased sex-specific viability, abnormal ovarian development, and infertility (Di Stefano et al. 2007).

Histone acetylation refers to the acetylation or deacetylation of specific N-terminal residues of histones, which is mediated by histone acetyltransferases (HATs) and histone deacetylases (HDACs) and predominantly occurs on H3 and H4 proteins. Histone acetylation plays a crucial role in cell differentiation during embryonic development. It has been suggested that the activation of HDACs is essential for the development of the posterior lateral cord of the embryo, and HDAC inhibitors significantly impede early posterior lateral cord development (He et al. 2014). In embryonic stem cells, germline-specific genes that are suppressed or transcribed at low levels exhibit "bivalent" characteristics, i.e., they possess both active chromosome markers (H3K4me and H3K9ac) and inactive chromosome markers (H3K27me3). Based on this, it has been proposed that developmental regulatory genes with "bivalent" markers in embryonic stem cells can be rapidly inhibited or activated.

2.3 Noncoding RNAs and Fetal Development Under Physiological Conditions

Noncoding RNAs (ncRNAs) are RNA molecules that do not encode proteins directly. The traditional view holds that these RNAs are nonfunctional. However, recent research has revealed that these RNAs play a crucial role in regulating the expression of coding genes. Over 90% of the human genome is transcribed into ncRNAs. The current research on the regulation mechanism of epigenetics mainly focuses on small ncRNAs, medium ncRNAs, and long ncRNAs.

Small ncRNAs, such as microRNAs (miRNAs), can regulate the expression of developmental genes at various levels, including transcription, posttranscription, and translation. miRNAs can form an RNA-induced silencing complex with proteins, which can induce the degradation of target mRNA or inhibit its translation by binding to the specific sequence of target mRNA. Medium ncRNAs are involved in the responsive repair mechanism of DNA damage (Wang et al. 2011). Long ncRNAs can regulate histone modification, protein function activity, and RNA metabolism to modulate the expression of development-related genes (Tsai et al. 2010). Furthermore, a class of circular RNAs (circRNAs) discovered in recent years can act as "miRNA sponges" or bind to RNA-binding proteins to regulate target gene expression. CircRNAs are usually specifically expressed in particular tissues or at specific developmental stages, thereby playing a crucial regulatory role (Chen and Yang 2015). With the growing interest in ncRNAs in epigenetics, their functions are receiving increasing attention.

2.4 *Other Epigenetic Modifications and Fetal Development Under Physiological Conditions*

Apart from the various modifications and small molecules that directly impact gene expression, there exist several modifications or structures that indirectly participate in gene expression, such as the local configuration of DNA. As DNA unwinding is essential for DNA replication and transcription, the stability of local DNA double strands in the genome is considered to be related to gene regulation. Furthermore, the phase transition of nucleosomes and the HMG domain protein in nucleosomes can influence gene expression by regulating the localization of nucleosomes along DNA and the binding of other transcription factors to DNA. Studies have indicated that the location of nuclear corpuscles in fertilized eggs begins to shift one hour after fertilization, and this alteration is accompanied by changes in histone acetylation (Wang et al. 2022). Therefore, the change in nucleosome localization may also be involved in the programming of fetal-originated disease, but further investigations are necessary to confirm this hypothesis.

3 Abnormal Epigenetic Modifications and Fetal Developmental Toxicity

The generation and elimination of epigenetic markers occur frequently during development. Epigenetic reprogramming plays a crucial role in erasing parental DNA methylation markers and histone modification sites on a large scale to ensure that cells acquire totipotency, gradually establishing new epigenetic markers. Abnormal epigenetic modifications can have severe consequences on organ development. The insults that occur during early stages of development can persist after birth or even to the next generation, which may be due to acquired abnormal epigenetic marks escaping epigenetic reprogramming.

3.1 *DNA Methylations and Fetal Developmental Toxicity*

As the most extensively researched epigenetic modification, the relationship between DNA methylation and the development of individuals has been frequently reported. In particular, abnormal expression of X-type collagenase in human articular chondrocytes has been linked to the demethylation of the promoter region, causing developmental cartilage toxicity (Zimmermann et al. 2008). Additionally, Wang et al. (2011) found that nicotine decreased cortisol synthesis in primary and passing human fetal adrenal cells, which was linked to nicotine increasing the methylation rate at the −377 site in the steroidal synthesis acute regulatory protein (StAR) promoter to inhibit its expression. This effect persisted for 5 to 10 generations in vitro

(Wang et al. 2011). Similarly, Ping et al. (2012) observed a similar phenomenon in caffeine-treated adrenal cells, with effects that lasted for multiple generations in vitro and were accompanied by demethylation of the CpG islands in the StAR promoter. Moreover, caffeine directly inhibited the expression of 11β-hydroxysteroid dehydrogenase type II (11β-HSD2) in fetal hippocampal neurons, which was found to be related to the increase in total methylation rate and multi-site methylation rate in the 11β-HSD2 promoter −358 ~ 77 bp region (Xu et al. 2012).

Pancreatic duodenal homeobox gene-1 (Pdx-1) is a crucial regulator of early pancreatic development, differentiation, and function. However, the methylation of a highly conserved CpG site near the promoter region of Pdx-1 can be induced by exogenous substances, leading to the permanent silencing of Pdx-1 expression (Park et al. 2008). Additionally, protein restriction during pregnancy has been linked to the hypermethylation of liver X receptor α (LXR α) promoter, resulting in decreased LXR α expression in the liver on day 19.5 of pregnancy (van Straten et al. 2010). Moreover, prenatal nicotine exposure has been found to increase vascular angiotensin II receptor type 1a (AT1aR) mRNA levels and decrease angiotensin II receptor type 2 (AT2R) expression by reducing AT1aR promoter methylation and increasing DNA methylation at the AT2R promoter, leading to abnormal vascular development in adult offspring (Xiao et al. 2014). Additionally, maternal exposure to Di-(2-ethylhexyl)-phthalate (DEHP) during pregnancy has been linked to ovarian dysplasia in offspring, with methylation sequencing of offspring ovaries showing hypermethylation of 71 genes related to fertilization and reproductive development in the DEHP group (Xie et al. 2012). These studies indicate that adverse environmental factors during individual development can alter the DNA methylation of key genes, leading to developmental toxicity of tissues and organs.

3.2 Histone Modifications and Fetal Developmental Toxicity

Runt-related transcription factor 2 (Runx2) is a crucial transcription factor for cartilage development and endochondral ossification. HDACs have been found to regulate the H3K9ac levels and expression of Runx2 during osteogenic differentiation, leading to the inhibition of osteogenic differentiation (Fu et al. 2014). Previous studies have shown that caffeine exposure during pregnancy can lead to bone growth retardation in fetal rats, which is linked to the downregulation of histone methylation and expression of insulinlike growth factor-1 (IGF1) promoter in the liver (Tan et al. 2012). Additionally, prenatal nicotine or dexamethasone exposure has been found to alter the histone modification of key genes related to bone development, leading to abnormal bone development (Xiao et al. 2018; Xie et al. 2018). In vitro, nicotine has been shown to decrease H3K9ac and H3K14ac and the expression of key genes related to bone development in bone mesenchymal stem cells (BMSCs) via its receptors, thereby inhibiting the directional differentiation of chondrocytes (Tie et al. 2018).

Abnormal adrenal steroid synthesis has been linked to changes in HDACs' expression and abnormal histone acetylation modification. Previous studies have shown that both prenatal nicotine and caffeine exposure can increase the expression of HDACs in fetal adrenal tissue, leading to decreased levels of H3K9ac and H3K14ac in the promoter of steroidogenic factor 1 (SF1) gene and decreased expression of StAR and steroid hormone synthesis (Ping et al. 2014; Yan et al. 2014). Histone modification also plays a crucial role in the regulation of intrauterine gonadal development. Exposure to multiple exogenous substances during pregnancy can inhibit sex hormone synthesis by reducing the levels of H3K9ac, H3K14ac, and H3K27ac in the promoter region of steroid synthase in the testis and ovary, thus interfering with the intrauterine development of gonads (Liu et al. 2018, 2019; Fan et al. 2019; Zhang et al. 2020). Additionally, histone modifications have been associated with pancreatic β cell development and maintenance of glucose and lipid metabolism homeostasis (Hu et al. 2019; Kou et al. 2020), cholesterol metabolism and hypercholesterolemia (Zhou et al. 2019), fetal bone and articular chondroplasia (Xiao et al. 2019; Li et al. 2020; Wen et al. 2020; Wu et al. 2020; Zhao et al. 2020), renal podocyte dysplasia and susceptibility to adult glomerulosclerosis (Li et al. 2019; Zhu et al. 2019), temporal lobe epilepsy (Hu et al. 2020), and impaired placental cholesterol transport (Huang et al. 2019) (Table 3.1).

3.3 Noncoding RNAs and Fetal Developmental Toxicity

miRNA plays crucial roles in various biological processes. The overall reduction of miRNA in chondrocytes in the mouse skeletal system leads to lethal skeletal dysplasia, while miR-140 deletion in mice leads to growth defects of endochondral bone, resulting in dwarfism and craniofacial malformations (Nakamura et al. 2011). Prenatal caffeine exposure has been found to cause H-type vascular-associated long-bone dysplasia through the miR-375 signaling pathway (He et al. 2021). The miRNA let-7 family has been identified as molecular switches for adrenal cortisol synthesis, with functional binding sites with StAR mRNA and the ability to regulate its expression during the translation phase (Men et al. 2017). In protein-restricted offspring during pregnancy and lactation, the expression of miR-29 in hepatocytes is significantly increased, negatively correlating with the expression of IGFl in the liver. This negative correlation is more significant in cloned rat hepatocytes transfected with miR-29 (Sohi et al. 2015). Germ cell differentiation in mammals starts from primordial germ cells during embryonic development. Prenatal exposure to environmental toxins such as endocrine disruptors can lead to decreased levels of specific miRNAs (such as miR-23b and miR-21), a decrease in the number of primordial germ cells, and an increase in the rate of apoptotic cells in male mice, accompanied by a decrease in the fertility of adult males. Similarly, exposure to endocrine disruptors during pregnancy can lead to abnormal expression of miRNAs in the ovaries of female fetuses. For instance, exposure to BPA during pregnancy can cause early disturbance of the fetal ovarian transcriptome (miRNA and mRNA)

Table 3.1 Histone modifications regulate changes in tissue function and disease susceptibility

Adverse agents	Species	Organs/tissues	Genes	Enzymes	Sites	Outcomes	Main references
–	Rat	BMSCs	Runx2	HDAC8	H3K9ac↓	Inhibited osteogenic differentiation	Fu et al. (2014)
Caffeine	Rat	Liver	IGF1	–	H3K36me2, H3K4me2, H3K4me3↓	Skeletal growth retardation	Tan et al. (2012)
Dexamethasone/ ethanol/nicotine	Rat	Bone	ACE	P300	H3K27ac↑	Bone dysplasia	Xiao et al. (2018); Xiao et al. (2019); Wu et al. (2020)
Nicotine	Rat	Cartilage	TGFβR1 COL2A1	–	H3K9ac↓	Cartilage dysplasia	Xie et al. (2018)
Nicotine	Rat	Cartilage	SOX9	HDAC1	H3K9ac↓ H3K14ac↓	Cartilage dysplasia	Tie et al. (2018)
Nicotine	Rat	Adrenal NCI-H295A cell	SF1	Dnmt1 Dnmt3a HDAC1/2	H3K9ac/ H3K14ac↓ CpG↑	Inhibited steroidogenesis	Ping et al. (2014); Yan et al. (2014)
Nicotine	Rat	Ovary	P450	–	H3K9ac↓ H3K27ac↓	Ovarian dysplasia	Fan et al. (2019)
Dexamethasone	Rat	Testis	StAR	HDAC7	H3K9ac↓	Testis dysplasia	Liu et al. (2018)
Ethanol	Rat	Testis	3β-HSD	HDAC2	H3K14ac↓	Testis dysplasia	Zhang et al. (2020)
Caffeine	Rat	Liver, L02 cell	Srebf2, Hmgcr, Hmgcs1	–	H3K14ac↑ H3K27ac↑	Induced cholesterol synthesis	Hu et al. (2019)
Dexamethasone	Rat	Pancreas	AT2R	HDAC2	H3K14ac↓ H3K27ac↓	B cell dysfunction	Kou et al. (2020)
Nicotine	Rat	Liver, HepG2	Hmgcr, Scarb1, Ldlr	SP1, P300	H3K27ac↓	Hypercholesterolemia	Zhou et al. (2019)
Dexamethasone	Rat	Growth plate	UGDH	SP3/ HDAC1	H3K9ac↓	Growth plate dysplasia	Wen et al. (2020)

Caffeine	Rat	Cartilage	TGFβ, TGFβR1, SOX9, COL2A1,ACAN	HDAC2	H3K9ac↓	Cartilage dysplasia	Li et al. (2020); Zhao et al. (2020)
Dexamethasone	Rat	Kidney	AT2R	–	H3K9ac↓	Glomerulosclerosis	Li et al. (2019)
Caffeine	Rat	Kidney	KLF4	HDAC7	H3K9ac↓	Podocyte developmental toxicity	Zhu et al. (2019)
Dexamethasone	Rat	Hippocampus	ACE	p300	H3K27ac↓	Epilepsy	Hu et al. (2020)
Dexamethasone	Rat	Placental	LXRα	HDAC3	H3K9ac↓	Transport dysfunction	Huang et al. (2019)

BMSC bone marrow stromal cell, *HDAC* histone deacetylase, *H3K9ac* H3 lysine 9 acetylation, *TGFβR1* transforming growth factor-β receptor 1, *LXRα* liver X receptor α related to gonadal differentiation, folliculogenesis, and insulin homeostasis (Veiga-Lopez et al. 2013).

4 Epigenetic Modifications Mediate Susceptibility to Fetal-Originated Disease

The developing embryo (fetus) is highly susceptible to endocrine disruptors and other external environmental factors during the critical period of early development. During this period, the embryo (fetus) responds to the adverse environment by adapting to the changing needs of the environment. However, when the offspring are removed from these adverse environments, the adaptive changes may evolve into over-adaptation, leading to changes in developmental programming and functional homeostasis. The interaction between genes and adverse environment during pregnancy can cause epigenetic modifications that persist after birth, inducing developmental toxicity of various fetal systems and organs and resulting in susceptibility to a variety of fetal-originated adult diseases.

4.1 Abnormal Intrauterine Epigenetic Modification Program Susceptibility to Fetal-Originated Disease

Based on epidemiological investigations, it has been observed that low-birth-weight children are at an increased risk of developing coronary heart disease, hypertension, hyperlipidemia, obesity, and osteoarthritis in adulthood. Furthermore, a series of animal experiments conducted by the authors' team have demonstrated that prenatal exposure to adverse environments, such as caffeine, nicotine, and ethanol, can cause epigenetic modifications in multiple organs of offspring, thereby increasing their susceptibility to various diseases in adulthood, including osteoarthritis, osteoporosis, nonalcoholic fatty liver disease, hyperchloremia, and chronic kidney disease (Fig. 3.1). The TGFβ signal pathway has been identified as a crucial regulator for cartilage development. Specifically, prenatal exposure to dexamethasone has been shown to inhibit miR-92a-3p and increase HDAC2 expression, which in turn results in low acetylation of the TGFβ signal pathway, thus inhibiting cartilage development and increasing susceptibility to osteoarthritis in adults (Tie et al. 2022). Additionally, prenatal exposure to dexamethasone has been found to mediate the low activity programming of the PDGFRβ/FAK signaling pathway through miR-34c signal, leading to H-type vascular dysplasia and susceptibility to adult osteoporosis (Shangguan et al. 2021).

Alterations in the renin-angiotensin system (RAS) may result in congenital nephropathy and urinary tract abnormalities. The authors' team have conducted research indicating that fetal renal growth retardation and adult nephrotic syndrome,

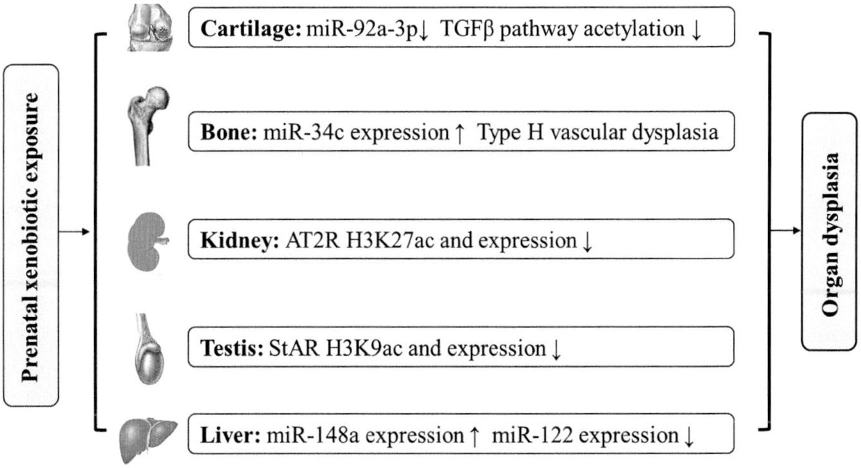

Fig. 3.1 The epigenetic changes of different organs in offspring by prenatal xenobiotic exposure

which may be associated with continuous reduction of H3K27ac of AT2R promoter region and expression, are induced by prenatal ethanol exposure (Zhu et al. 2018). Testicular dysplasia syndrome is an environment-related disorder that is associated with gene defects and polymorphism, adverse lifestyle, and exposure to environmental disruptors. The authors' team recent findings suggest that exposure to exogenous substances during pregnancy can cause abnormal testicular morphology, androgen synthesis dysfunction, and decreased spermatogenic function in adults, which is related not only to maternal glucocorticoid overexposure induced by caffeine and ethanol (Liu et al. 2019; Pei et al. 2019) but also to the direct action of substances, such as nicotine and dexamethasone (Liu et al. 2018; Zhang et al. 2020), and histone acetylation modification abnormalities in key genes. Furthermore, numerous studies have demonstrated that prenatal exposure to adverse environments may increase susceptibility to other diseases, such as non-alcoholic fatty liver disease (Liu et al. 2021) and hypercholesterolemia (Li et al. 2020), which are all associated with the direct epigenetic modifications of key genes or pathways induced by exogenous substances.

4.2 Epigenetic Mechanisms of Susceptibility to Fetal-Originated Disease Induced by Glucocorticoids

Glucocorticoids play a crucial role in the programming of intrauterine fetal development (Zhang et al. 2014; Chen et al. 2019). The activation and reduction of glucocorticoids are mediated by 11β-HSD1 and 11β-HSD2 oxidations in physiology, which promote fetal development. However, prenatal adverse environments have been shown to reduce the expression of 11β-HSD2 in the fetal liver and gastrocnemius muscle while increasing the concentration of blood glucocorticoid, 11β-HSD1, and glucocorticoid receptor (GR) expression (Fernandez-Twinn and Ozanne 2006). This can be attributed to the epigenetic modification, such as increased DNA methylation, in the promoter of placental 11β-HSD2, leading to decreased expression (Yu et al. 2018; Zhou et al. 2018). Consequently, adverse environments during pregnancy not only directly cause epigenetic changes but also weaken the placental barrier effect against maternal glucocorticoids by altering the expression of 11β-HSD2, resulting in fetal overexposure to maternal glucocorticoids.

Numerous studies have demonstrated that glucocorticoids can induce epigenetic modifications in target genes via various mechanisms, including DNA methylation, histone modification, and ncRNA (such as miRNAs). Maternal exposure to excessive glucocorticoids during pregnancy has been shown to regulate excitatory synaptic plasticity through miR-219 (Wibrand et al. 2010) and alter the level of IGF2 gene methylation (Greene et al. 2013), which is a significant factor in fetal depression and fetal diabetes, respectively. The author's team has conducted a series of studies that confirmed the epigenetic modification of vital genes in multiple tissues and organs due to maternal exposure to excessive glucocorticoids during pregnancy, such as exposure to coffee (Fig. 3.2). This can lead to organ dysfunction and increased susceptibility to various diseases. For instance, intrauterine high levels of corticosteroids caused by prenatal caffeine exposure led to reduced expression of IGF1 histones (H3K9ac and H3K27ac) in articular cartilage, thereby inhibiting the development of growth plate cartilage (Tan et al. 2018). Similarly, decreased KLF4 histone acetylation (H3K9ac) and expression in the kidney inhibited podocyte development (Zhu et al. 2019), while decreased IGF1 histone acetylation (H3K14ac) and expression in the testis inhibited testosterone synthesis (Pei et al. 2019). Furthermore, maternal exposure to excessive glucocorticoids during pregnancy downregulated Sirt1 expression and upregulated HMGCR histone acetylation

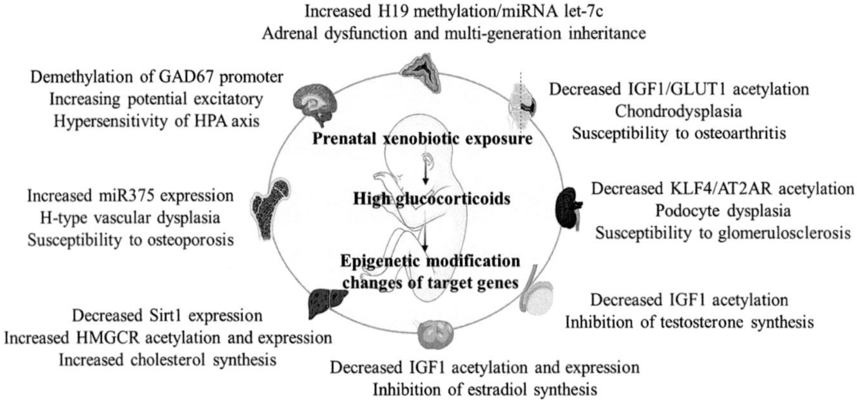

Fig. 3.2 The epigenetic programming changes in different organs by high levels of maternal glucocorticoids

(H3K9ac and H3K14ac) and expression in the liver, resulting in increased cholesterol synthesis (Xu et al. 2018).

5 Multigenerational Genetic Effects of Susceptibility to Fetal-Originated Disease Mediated by Epigenetic Modifications

Multigenerational heredity is a phenomenon that can be attributed to the changes in the epigenome of reproductive line cells caused by contemporary exposure to adverse environmental factors or the programmed effect of adverse environmental factors on each generation. In the past, it was commonly believed that epigenetic markers of parents would gradually disappear in the process of inheritance due to the reversibility of epigenetic modifications and the large-scale erasure of epigenetic markers to maintain genetic stability in physiology. However, recent studies have shown that the susceptibility to diseases caused by non-DNA sequence changes can also be inherited in multigenerational or intergenerational patterns. This is because there are certain regulatory mechanisms for the elimination and re-establishment of epigenetic markers that maintain the continuity of epigenetic markers, and some genes can "escape" epigenetic modification reprogramming, retaining their epigenetic markers. These epigenetic markers that "escape" reprogramming serve as the basis for the multigeneration inheritance of acquired phenotypes and transmit epigenetic information between generations through germ cells.

DNA methylation is one of the most significant epigenetic markers that can maintain the genomic "imprinting" of parents between generations in physiology. For instance, H19 and IGF1, as matrilineal and patrilineal imprinted genes, are differentially expressed in offspring due to DNA methylation, which may eventually lead to the occurrence of acquired phenotypes in subsequent generations. Moreover, there is increasing evidence that genomic DNA methylation markers can persist for several generations. For instance, researchers made mice afraid of acetophenone by electric shock and fed them normally, and the results showed that the methylation changes of the acetophenone receptor gene in mouse spermatozoa could continue to the F3 generation (Dias and Ressler 2014). Similarly, DNA methylation in the spermatozoa of F0 generation rats exposed to vinclozolin was altered in three consecutive generations of offspring (Guerrero-Bosagna et al. 2010). The reproductive system-dependent methylation establishment model proposes that germ cells can serve as the basis for intergenerational transmission of methylated memory.

In addition to DNA methylation, histone acetylation and ncRNAs are two other crucial forms of epigenetic modification. Research has demonstrated that cocaine exposure can cause changes in H3K27ac in spermatozoa and seminiferous tubules of male parents, and this alteration can also be transmitted to offspring through germ cells (Vassoler et al. 2013). Gaydos et al. used fluorescence labeling techniques to track chromosomes from germ cells (ovum and sperm) to fertilized eggs and then to mitotic embryonic cells, revealing the phenomenon and mechanism of histone acetylation passed on through germ cells (Gaydos et al. 2014). Similarly, TsRNAs were found to be highly enriched in mature spermatozoa of paternal mice fed with a high-fat diet, and the offspring of mating with normal female mice still displayed metabolic disorders such as obesity (Chen et al. 2016). Another study during the same period provided direct evidence for the transmission of tsRNAs through the reproductive system (Sharma et al. 2016). The author's team also demonstrated that miRNA320a-3p/RUNX2, miR-466b-3p/HDAC7, and miR-92a-3p/Jag1/Notch1 signal programming mediated the intergenerational inheritance of ovarian and testicular sex hormone synthesis inhibition and long bone dysplasia induced by prenatal dexamethasone exposure, respectively (Gong et al. 2021; Han et al. 2022; Liu et al. 2022). Overall, various adverse environments during fetal development not only affect the current phenotype of offspring but may also be transmitted to the next generation through epigenetics, leading to multigenerational changes or susceptibility to diseases.

6 Summary and Prospects

Epigenetic regulatory mechanisms are involved in the early development of organisms and the occurrence of diseases in adulthood and can even be transmitted in multiple generations (Fig. 3.3). DNA methylation, histone modification, and

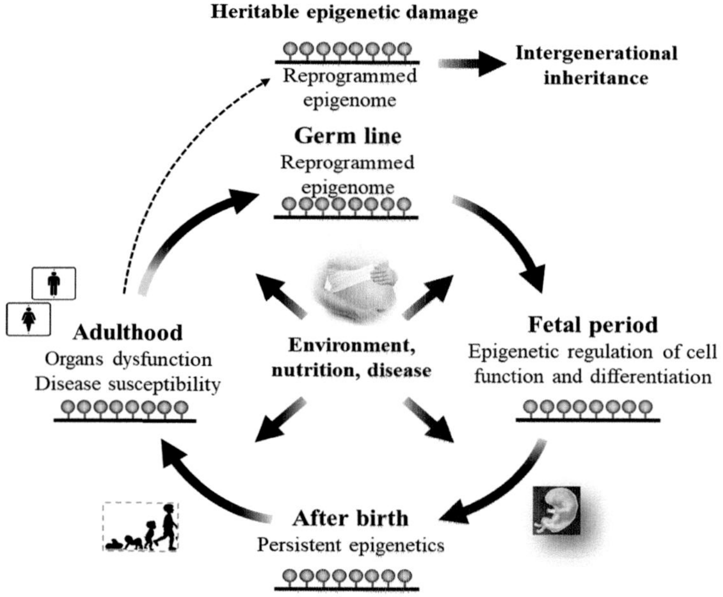

Fig. 3.3 The epigenetic modification changes and fetal developmental toxicities

ncRNAs may constitute the main carrier for the transmission of epigenetic information (acquired phenotype) across generations. However, the application of epigenetic modification in the diagnosis and treatment of diseases is still limited due to the lack of systematic clinical evidence. The clarification of epigenetic mechanisms for the susceptibility of offspring to adult diseases caused by adverse environments during pregnancy will help to explore biomarkers for early warning of these diseases, establish an early evaluation system for developmental toxicity, also determine early prevention targets and strategies, and guide the comprehensive prevention and treatment of fetal-originated diseases.

References

Abdalla H, Yoshizawa Y, Hochi S. 2009. J Reprod Dev, 55(4): 356–360.
Chen L L, Yang L. 2015. RNA Biol, 12(4): 381–388.
Chen Q, Yan M, Cao Z, et al. 2016. Science, 351(6271): 397–400.
Chen Y, He Z, Chen G, et al. 2019. Toxicology, 428: 152308.
Di Stefano L, Ji J Y, Moon N S, et al. 2007. Curr Biol, 17(9): 808–812.
Dias B G, Ressler K J. 2014. Nat Neurosci, 17(1): 89–96.
Fan G, Zhang Q, Wan Y, et al. 2019. Food Chem Toxicol, 128: 256–266.
Fernandez-Twinn D S, Ozanne S E. 2006. Physiol Behav, 88(3): 234–243.
Fu Y, Zhang P, Ge J, et al. 2014. Int J Biochem Cell Biol, 54: 68–77.
Gaydos L J, Wang W, Strome S. 2014. Science, 345(6203): 1515–1518.
Gong X, Zhang J, Ge C, et al. 2021. Pharmacol Res, 165: 105435.
Greene N H, Pedersen L H, Liu S, et al. 2013. Int J Epidemiol, 42(1): 186–193.
Guerrero-Bosagna C, Settles M, Lucker B, et al. 2010. PLoS One, 5(9):e13100.
Han H, Xiao H, Wu Z, et al. 2022. Exp Mol Med, 54(3): 298–308.
He H, Luo H, Liu L, et al. 2021. Faseb j, 35(2): e21370.
He Y, Wu J, Mei H, et al. 2014. Cell Prolif, 47(1): 91–104.
Hemberger M, Dean W, Reik W. 2009. Nat Rev Mol Cell Biol, 10(8): 526–537.
Hu S, Liu K, Luo H, et al. 2019. Toxicology, 418: 11–21.
Hu S, Yi Y, Jiang T, et al. 2020. Arch Toxicol, 94(9): 3201–3215.
Huang W, Zhou J, Zhang G, et al. 2019. Biochim Biophys Acta Mol Cell Biol Lipids, 1864(12): 158524.
Kou H, Gui S, Dai Y, et al. 2020. Toxicol Appl Pharmacol, 404: 115187.
Li B, Zhu Y, Chen H, et al. 2019. Toxicology, 411: 32–42.
Li J, Xiao H, Luo H, et al. 2020. Food Chem Toxicol, 140: 111279.
Li L, Hu W, Liu K, et al. 2020. Toxicol Appl Pharmacol, 395: 114979.
Liu H, He B, Hu W, et al. 2021. Biochem Pharmacol, 185: 114420.
Liu M, Chen B, Pei L, et al. 2018. Toxicology, 408: 1–10.
Liu M, Zhang Q, Pei L, et al. 2019. Epigenetics, 14(3): 245–259.
Liu Y, Liu M, Liu Y, et al. 2022. Biochem Pharmacol, 199: 115018.
Mathers J C, McKay J A. 2009. Adv Exp Med Biol, 646: 119–123.
Men Y, Fan Y, Shen Y, et al. 2017. Endocrinology, 158(2): 402–409.
Nafee T M, Farrell W E, Carroll W D, et al. 2008. Bjog, 115(2): 158–168.
Nakamura Y, Inloes J B, Katagiri T, et al. 2011. Mol Cell Biol, 31(14): 3019–3028.
Nielsen S J, Schneider R, Bauer U M, et al. 2001. Nature, 412(6846): 561–565.
Park J H, Stoffers D A, Nicholls R D, et al. 2008. J Clin Invest, 118(6): 2316–2324.
Pei L G, Zhang Q, Yuan C, et al. 2019. J Endocrinol, 242(1): M17–M32.
Ping J, Lei Y Y, Liu L, et al. 2012. Chem Biol Interact, 195(1): 68–75.

Ping J, Wang J F, Liu L, et al. 2014. Toxicology, 321: 53–61.
Shangguan Y, Wu Z, Xie X, et al. 2021. Biochemical Pharmacology: 114414.
Sharma U, Conine C C, Shea J M, et al. 2016. Science, 351(6271): 391–396.
Sohi G, Revesz A, Ramkumar J, et al. 2015. Endocrinology, 156(9): 3069–3076.
Tan Y, Liu J, Deng Y, et al. 2012. Toxicol Lett, 214(3): 279–287.
Tan Y, Lu K, Li J, et al. 2018. Toxicol Lett, 295: 229–236.
Tie K, Wu M, Deng Y, et al. 2018. Stem Cell Res Ther, 9(1): 98.
Tie K, Zhao Z, Wu Z, et al.2022. Biochem Pharmacol, 203: 115196.
Tsai M C, Manor O, Wan Y, et al. 2010. Science, 329(5992): 689–693.
van Straten E M, Bloks V W, Huijkman N C, et al. 2010. Am J Physiol Regul Integr Comp Physiol, 298(2): R275–282.
Vassoler F M, White S L, Schmidt H D, et al. 2013. Nat Neurosci, 16(1): 42–47.
Veiga-Lopez A, Luense L J, Christenson L K, et al. 2013. Endocrinology, 154(5): 1873–1884.
Wang J, Scully K, Zhu X, et al. 2007. Nature, 446(7138): 882–887.
Wang Y, Chen J, Wei G, et al. 2011. Nucleic Acids Res, 39(12): 5203–5214.
Wen Y, Shi H, Wu Z, et al. 2020. Faseb j, 34(9): 12834–12846.
Wibrand K, Panja D, Tiron A, et al. 2010. Eur J Neurosci, 31(4): 636–645.
Wu Z, Pan Z, Wen Y, et al. 2020. Food Chem Toxicol, 136: 111083.
Xiao D, Dasgupta C, Li Y, et al. 2014. PLoS One, 9(9): e108161.
Xiao H, Wen Y, Pan Z, et al. 2018. Cell Death Dis, 9(6): 638.
Xiao H, Wen Y, Pan Z, et al. 2019. Faseb j, 33(11): 12972–12982.
Xie X, Gao Y, Zhang Y, et al. 2012. Zhonghua Yu Fang Yi Xue Za Zhi, 46(9): 840–844.
Xie Z, Zhao Z, Yang X, et al. 2018. Toxicol Appl Pharmacol, 352: 107–118.
Xu D, Wu Y, Liu F, et al. 2012. Toxicol Appl Pharmacol, 264(3): 395–403.
Xu D, Luo H W, Hu W, et al. 2018. Faseb j, 32(10): 5563–5576.
Yan Y E, Liu L, Wang J F, et al. 2014. Toxicol Appl Pharmacol, 277(3): 231–241.
Yu L, Zhou J, Zhang G, et al. 2018. Toxicol Appl Pharmacol, 352: 77–86.
Zhang C, Xu D, Luo H, et al. 2014. Toxicology, 325: 74–84.
Zhang Q, Pei L G, Liu M, et al. 2020. Food Chem Toxicol, 135: 111057.
Zhao Z, Qin J, Pei L, et al. 2020. Toxicology, 442: 152533.
Zhou J, Liu F, Yu L, et al. 2018. Toxicol Appl Pharmacol, 344: 1–12.
Zhou J, Zhu C, Luo H, et al. 2019. FASEB J, 33(1): 1110–1123.
Zhu Y, Zuo N, Li B, et al. 2018. Toxicology, 400–401: 9–19.
Zhu Y, Chen H, Zhao X, et al. 2019. Toxicol Lett, 314: 63–74.
Zimmermann P, Boeuf S, Dickhut A, et al. 2008. Arthritis Rheum, 58(9): 2743–2753.

Chapter 4
Neuroendocrine Programming Mechanism of Fetal-Originated Disease

Hao Kou and Liang Liu

Abstract Physiological development is dependent on multiple neuroendocrine axes, such as the hypothalamic-pituitary-adrenal axis, growth hormone/insulinlike growth factor 1 axis, hypothalamic-pituitary-gonadal axis, hypothalamic-pituitary-thyroid axis, and renin-angiotensin system, to adapt to environmental changes and maintain homeostasis. However, increasing clinical and animal studies suggest that prenatal adverse environments can adversely affect multiple neuroendocrine axes in fetuses, which are closely associated with postnatal developmental programming, homeostasis disturbance, and susceptibility to diseases in adults. Among various hypotheses of fetal-originated disease, the programming change in intrauterine neuroendocrine development is currently well recognized. This chapter summarizes the developmental programming changes and the functional interaction of neuroendocrine axes and the relationships between neuroendocrine axes and the occurrence and development of fetal-originated disease.

1 Introduction

The neuroendocrine axes, which comprise the hypothalamic-pituitary-adrenal (HPA) axis, growth hormone-insulinlike growth factor 1 (GH-IGF1) axis, hypothalamic-pituitary-gonadal (HPG) axis, hypothalamic-pituitary-thyroid (HPT) axis, and renin-angiotensin system (RAS), play a crucial role in maintaining individual homeostasis and the growth and development of various systems and organs. Dr. Barker's theory of the intrauterine origin of adult diseases, based on a large-scale epidemiological investigation in the 1990s, proposed that prenatal adverse

H. Kou (✉)
Department of Pharmacy, Zhongnan Hospital of Wuhan University, Wuhan, Hubei, China
e-mail: kouhao007@whu.edu.cn

L. Liu
Division of Joint Surgery and Sports Medicine, Department of Orthopedic Surgery, Zhongnan Hospital of Wuhan University, Wuhan, Hubei, China

H. Wang et al. (eds.), *Fetal Origin of Diseases*,
https://doi.org/10.1007/978-981-97-5730-5_4

environments affect the development of fetal endocrine axes, leading to functional changes that are related to postnatal developmental programming, homeostasis changes, and susceptibility to diseases in adults. This theory has been supported by numerous clinical and animal studies (Martin-Gronert and Ozanne 2007; Shen et al. 2014; Wang et al. 2014; Zhang et al. 2016; Xia et al. 2020). In addition to the classical neuroendocrine axes, other neuroendocrine axes have been proposed during fetal development, such as the glucocorticoid-insulinlike growth factor-1 (GC-IGF1) axis, neuroendocrine-immune axis, and hypothalamic-pituitary-fat axis. The relationships and network regulations between these axes are complex and require further investigation.

2 Neuroendocrine Developmental Programming Mechanisms in Fetal-Originated Disease

There are several hypotheses regarding the mechanism underlying the link between low birth weight and adult susceptibility to diseases resulting from adverse prenatal environments. These hypotheses include the "programming of intrauterine neuroendocrine development," "mitochondrial dysfunction and oxidative stress," "nutrition and postnatal development," and "imprinted genes and their acquired modification." Among these, the "programming of intrauterine neuroendocrine development" hypothesis has been widely accepted by the academic community. This hypothesis suggests that prenatal adverse environments can program the development of multiple organs before and after birth, leading to homeostasis changes and increased susceptibility to various diseases in adulthood.

2.1 Intrauterine Programming Mechanism of Multiple Organ Development

Intrauterine growth retardation (IUGR) refers to the restriction of fetal growth and potential impairment of tissue and function development. The underlying mechanism of multiple organ dysplasia in IUGR offspring induced by prenatal adverse environments is still not fully understood, and various hypotheses have been proposed, including "thrifty phenotype," "developmental plasticity," and "predictive adaptive changes."

The "thrifty phenotype" hypothesis was first proposed by David Barker and Hales (Hales and Barker 1992). According to this theory, a fetus experiencing early malnutrition undergoes short-term compensatory adaptations to an adverse intrauterine environment, which then becomes permanent "programming." As a result, the fetal body size becomes smaller, and the structure and function of some organs are altered, making them more vulnerable to the nutrient-rich environment after birth, thereby increasing their risk of developing insulin resistance. Vaag et al. suggested that in the face of unfavorable environments (such as malnutrition), the fetus changes its metabolic processes, redistributes limited energy, and limits energy consumption in secondary organs to ensure the development of key organs such as the liver, resulting in a "thrifty phenotype." These changes can persist for a long time or even a lifetime, increasing individual susceptibility to a variety of chronic diseases in adulthood. Similarly, the "developmental plasticity" hypothesis suggests that fetuses have multidirectional developmental potential early on, enabling them to grow and reproduce in different environments by regulating various phenotypes. However, this hypothesis may not apply to all environments, as some adverse environments can lead to malformations, illnesses, and even death (Stamps and Frankenhuis 2016).

The "predictive adaptive change" hypothesis is another determinant of species' adaptability. Many organisms exhibit phenotypic plasticity to ensure survival in variable environments. In some cases, environmental changes are highly predictable, such as in mammals, where physiological changes during pregnancy ensure normal fetal growth and development. These adaptive changes include endocrine signals, such as the gradual increase in corticotropin release in the HPA axis during the third trimester of pregnancy to facilitate labor (Douglas 2011). Additionally, many mechanisms protect embryonic development from maternal high GC levels, including hydroxysteroid 11β-dehydrogenase type II (11β-HSD2). However, these protective mechanisms may fail in cases of maternal long-term stress or the use of synthetic GC, resulting in fetal overexposure to GC and neuroendocrine changes that persist later in life, leading to increased susceptibility to multiple diseases, such as type 2 diabetes and metabolic syndrome. These hypotheses suggest that the body maximizes homeostasis through various regulations and procedural changes in response to external environmental stimuli.

2.2 *Postnatal Programming Mechanism of Neuroendocrine Development*

Intrauterine programming alteration refers to the process of permanent changes in tissue morphology and function caused by injury during early development. These changes in tissue and organ function or gene expression patterns typically last from development to adulthood, or even throughout the entire lifespan, leading to various effects in adulthood (Meaney et al. 2007). Among the various hypotheses, the "programming of intrauterine neuroendocrine development" proposed by Fowden in 2005 is widely recognized. This hypothesis suggests that adverse intrauterine environments cause changes in the developmental programming of the fetal HPA axis, resulting in a slowed growth rate and increased sensitivity of peripheral tissues to metabolic hormones to ensure fetal survival in a nutrient-deficient environment after birth (Martin-Gronert and Ozanne 2007). However, after birth, the offspring may experience "catch-up growth" and fat deposition due to overnutrition, which ultimately increasing the risk of insulin resistance. Moreover, the increased sensitivity of the HPA axis to stress after birth can accelerate the onset of metabolic disorders and even lead to more severe metabolic disorders in adulthood.

The authors' team has confirmed that prenatal adverse environments, including extraneous exposure and food restriction, can stimulate maternal HPA axis activity and open the placental GC barrier, leading to fetal overexposure to maternal GC and low birth weight. These phenomena are commonly associated with multiple organ development programming, homeostasis changes, and susceptibility to diseases in postnatal offspring. The authors' team was the first to propose that "intrauterine maternal GC overexposure" caused by prenatal adverse environments is the key factor leading to low birth weight, changes in multiple organ development programming, and adult susceptibility to multiple diseases. In other words, maternal GC can program the susceptibility of offspring to multiple diseases. The authors' team comprehensively and systematically summarized the "two programming and two strikes" mechanism underlying the occurrence of multiple fetal-originated diseases and highlighted that the core of endocrine programming is the developmental programming change in the GC-IGF1 axis (Zhang et al. 2014; Chen et al. 2019).

An adverse prenatal environment can lead to programming changes in the fetus. When exposed to high levels of maternal GC, the fetus undergoes compensatory changes in the function of multiple organs. These adaptive changes ultimately protect the development of crucial organs, such as the liver and brain, while attenuating the development of noncritical organs, such as bone and cartilage, to ensure fetal survival. This results in a thrifty phenotype and weak metabolic capacity. After birth, offspring experience catch-up growth by redeploying and redistributing nutrients and energy under nutrient-rich conditions. However, this also leads to further programming changes in metabolic function, exacerbating abnormal tissue and organ morphology and glucose and lipid metabolic dysfunction, accelerating the depletion of functional potential, and ultimately increasing susceptibility to multiple metabolic diseases (Shen et al. 2014; Kamei et al. 2011). When these offspring experience environmental "strikes," such as a high-fat diet or chronic stress, it can induce or exacerbate multiple fetal diseases (Xia et al. 2014).

3 Developmental Programming of Neuroendocrine Axes and Fetal-Originated Disease

Multiple neuroendocrine axes play a critical role in maintaining homeostasis and normal development. When the fetus encounters adverse environments, the endocrine axes undergo adaptive changes to maintain fetal normal development and functional homeostasis (Fig. 4.1). These changes may persist and increase susceptibility to diseases.

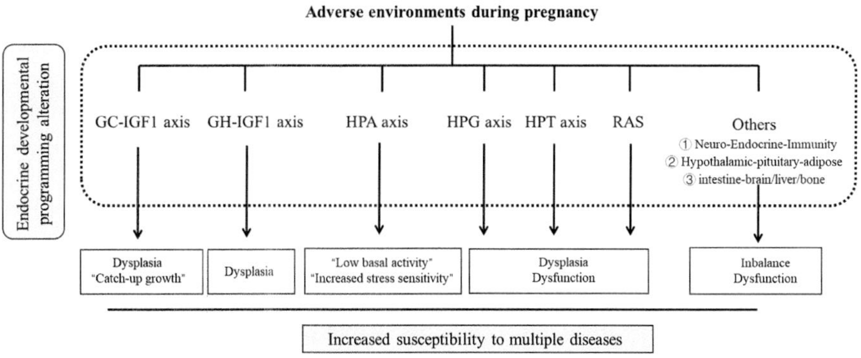

Fig. 4.1 Changes in the neuroendocrine axes and susceptibility to diseases caused by adverse environments during pregnancy

3.1 Glucocorticoid/Insulinlike Growth Factor 1 Axis

IGF1 plays a crucial role in the process of morphological and functional differentiation. During the fetal stage, IGF1 in the fetal circulation is primarily synthesized and secreted by the liver, whereas local autocrine/paracrine IGF1 in various tissues and organs is predominantly established during the middle and late stages of pregnancy. Subsequently, hepatic IGF1 expression continues to increase after birth, peaks during adolescence, and then maintains a certain level until it gradually decreases during aging (Agrogiannis et al. 2014). Previous studies conducted by the author's team have shown that IUGR fetal rats caused by prenatal adverse environments, such as xenobiotic exposure and food restriction, exhibit a common phenomenon of "maternal GC overexposure," which, in turn, leads to a simultaneous reduction in the levels of IGF1 in fetal blood and multiple tissues. Moreover, it has been observed that after birth, the levels of serum corticosterone decrease and the expression of IGF1 in the liver increases in the offspring of IUGR. Subsequently, under a high-fat diet, the catch-up growth and metabolic disorders become more severe (Shen et al. 2014; Wang et al. 2014; Zhang et al. 2016). In adult chronic stress, GC metabolism in the hippocampus and peripheral tissues is activated, and glucose and lipid metabolism exhibit GC-dependent changes (Liu et al. 2012; Xu et al. 2012a; Xia et al. 2014).

In peripheral tissues with high metabolic rates, such as the adrenal gland and long bone, intrauterine maternal high GC levels inhibit the expression of IGF1, which subsequently leads to the inhibition of fetal adrenal steroidal synthesis and a decrease in serum corticosterone levels (He et al. 2019). This, in turn, induces an increase in IGF1 expression in bone tissue, resulting in catch-up growth in bone length and a shortage of bone mass, thus increasing susceptibility to osteoporosis in adulthood (Shangguan et al. 2018). However, in tissues with low metabolic rates, such as the kidneys and testes, changes in GC-IGF1 axis programming may exacerbate the development of glomerulosclerosis in adult offspring and inhibit the synthesis of testosterone in the testes (He et al. 2019; Pei et al. 2019). The author's team also observed that changes in GC-IGF1 axis programming caused by prenatal alcohol and nicotine exposure were transmitted to F2 offspring through epigenetics, continuously affecting the liver lipid metabolism of multiple generations of offspring (Hu et al. 2020a, b). Based on these findings, the author's team has innovatively proposed the new concept of the "GC-IGF1 axis." The team has systematically elucidated the intrauterine programming mechanism of the "GC-IGF1 axis" in IUGR offspring that makes them susceptible to multiple diseases in adulthood. Moreover, the team has pointed out that programming changes in the "GC-IGF1 axis," as the physiological compensation axis, mediate IUGR offspring's IUGR, postnatal catch-up growth, and susceptibility to multiple diseases in adulthood.

3.2 Growth Hormone/Insulinlike Growth Factor 1 Axis

The GH-IGF1 axis plays a crucial role in various physiological processes, including regulating metabolism, cell proliferation, and differentiation. The effect of GH is mediated by IGF1. GH regulates the secretion of IGF1, and IGF1, in turn, inhibits the release of GH through negative feedback, thereby maintaining normal IGF1 levels to ensure normal growth and physiological responses after birth. Previous studies have demonstrated that IGF1 activates RNA polymerase, regulates RNA and DNA synthesis, and promotes the proliferation and differentiation of various cells. However, prenatal fetal tissue lacks GH receptors, and its growth mainly depends on blood IGF1 levels. As mentioned earlier, IGF1 in the fetal bloodstream during intrauterine development primarily originates from the liver, and the expression level of IGF1 in the liver directly influences fetal birth weight, organ structure, and functional development (Agrogiannis et al. 2014).

Studies have revealed that the homozygous fragment of the IGF1 encoding gene is partially lost in IUGR fetuses, leading to IGF1 deficiency. Knockout of IGF1 or insulinlike growth factor 1 receptor (IGF1R) in the liver can significantly reduce fetal birth weight and body length (Walenkamp et al. 2013). Numerous studies have confirmed that the occurrence of IUGR is closely associated with reduced IGF1 levels. Catch-up growth in IUGR offspring is frequently accompanied by an increase in IGF1 levels, which is the main stimulus for catch-up growth (Kamei et al. 2011). However, catch-up growth further exacerbates abnormal tissue and organ function and disrupts sugar and lipid metabolism (Shen et al. 2014), which is closely linked to the onset of diabetes, fatty liver, and other diseases. In small gestational age infants, stimulation with growth hormone-releasing hormone led to an increased GH secretion response. These offspring had abnormal GH in childhood, with a low amplitude and high secretion baseline, leading some scholars to propose the programming hypothesis of the GH-IGF1 axis (Laron 1995).

3.3 Hypothalamic-Pituitary-Adrenal Axis

The HPA axis comprises the hypothalamus, pituitary, and adrenal gland. The paraventricular nucleus of the hypothalamus produces antidiuretic hormone (ADH) and corticotropin-releasing hormone (CRH), which promote the release of adrenocorticotrophic hormone from the anterior pituitary. GC are synthesized by the adrenal cortex under the regulation of ACTH (cortisol in humans, corticosterone in rodents). The hippocampus is the high negative feedback regulation center of the HPA axis.

A growing body of evidence suggests that susceptibility to multiple diseases in postnatal life caused by adverse intrauterine environments is associated with abnormal HPA axis function. IUGR is a common adverse outcome of fetal HPA axis dysfunction. Offspring with IUGR caused by prenatal exposure to alcohol, food restriction, or maternal malnutrition typically exhibit elevated blood ACTH or GC concentrations in adulthood, as well as stress hypersensitivity in the hypothalamus, delayed negative feedback in the hippocampus (Vieau et al. 2007), glucose and lipid metabolism disorders, and the onset of multiple diseases in adulthood (Xia et al. 2020). Previous studies have revealed that exposure to exogenous substances (such as caffeine, nicotine, ethanol, and dexamethasone) and dietary restriction during pregnancy reduces the expression of placental 11β-HSD2, resulting in fetal exposure to excessive maternal-derived GC, inhibition of fetal HPA axis development, and IUGR (Xu et al. 2011, 2012b; Liu et al. 2012; Zhang et al. 2013; Xia et al. 2014). Moreover, alterations in IUGR can persist after birth, exhibiting low basal activity and stress hypersensitivity of the HPA axis (Zhang et al. 2013).

These changes in the HPA axis are closely associated with the glucocorticoid receptor (GR) and mineralocorticoid receptor (MR). Prenatal stress can lead to decreased expression of hippocampal GR in young and adult rats (Kapoor et al. 2006). It can also result in decreased GR binding sites in the hippocampus and reduced negative feedback of the paraventricular nucleus in porcine offspring, leading to increased HPA axis reactivity. Moreover, the expression of GR and MR in the hippocampus and pituitary gland decreases after GC administration during pregnancy. Prenatal nicotine and caffeine exposure also alter the expression of GR and MR in the hippocampus (Liu et al. 2012; Xu et al. 2012a). The GC-activating enzyme 11β-HSD1 and GR coexist in the same neuron in the hippocampus. Activation of 11β-HSD1 increases GC levels in the hippocampus, and GC can induce the activation of 11β-HSD1 through the activation of GR to form a positive feedback loop, further increasing GC levels (Yang et al. 2007). The author's team also found that exposure to caffeine and nicotine during pregnancy decreased the expression of the GC-metabolizing inactivated enzyme 11β-HSD2 in the fetal hippocampus, while the expression of the GC-activating enzyme 11β-HSD1 increased, resulting in the enhancement of the GC-activating system (11β-HSD1/11β-HSD2 expression ratio and GR expression) (Liang et al. 2011; Xu et al. 2012a). In rats exposed to dexamethasone during pregnancy, it was also found that GR activation could promote the transformation of glutamate to γ-aminobutyric acid (GABA), resulting in an imbalance of GABA afferents in the paraventricular nucleus of the hypothalamus, leading to low basal activity of the HPA axis in offspring before and after birth (Lu et al. 2020). The low basal activity and stress hypersensitivity of the HPA axis are core mechanisms underlying multiple disease susceptibility in offspring.

3.4 *Hypothalamic-Pituitary-Gonadal Axis*

The HPG axis plays a crucial role in the endocrine developmental programming of fetal-originated disease. In rodents, the development of the HPG axis occurs gradually during the late trimester and is fully established in the early postnatal period, whereas in humans, it can be established during the embryonic stage. Environmental factors have been demonstrated to affect the HPG axis at various stages, thereby exerting different effects on reproductive health in adulthood. Animal models have shown that adverse exposure during intrauterine and postpartum periods can lead to reproductive disorders. Stress during prenatal or postnatal periods can induce developmental programming changes in reproductive function. Exposure to multiple prenatal adverse environmental factors has been identified as a risk factor for polycystic ovary, ovarian cancer, and breast cancer (Basak et al. 2020). Although female infants with IUGR born during the famine in Denmark did not exhibit any reproductive dysfunction in adulthood, their offspring demonstrated low birth weight and high perinatal mortality (Lummaa and Tremblay 2003).

Low-calorie diets during pregnancy have been reported to cause ovarian DNA oxidative damage and delayed puberty in female fetal sheep (Murdoch et al. 2003), while in males, restricted testicular growth, delayed puberty, and abnormal testicular morphology, sperm count, and motility were observed (Toledo et al. 2011). Similar changes were observed in rats exposed to dexamethasone during pregnancy (Liu et al. 2022). Exposure to betamethasone during pregnancy in male sheep resulted in testicular morphological changes in adulthood, including a significant decrease in seminiferous tubule length, interstitial tissue, and testicular weight. In female offspring, a decrease in the number of follicles and germ cells and an increase in apoptosis rate were observed (Poulain et al. 2012). Delayed maturation or premature reproductive failure may result from changes in steroid production during the first stage of follicular growth and prepuberty. Maternal cadmium exposure during the third trimester of pregnancy reduced P450scc expression in the testes after birth; decreased testosterone levels, sperm number, and reproductive capacity in adulthood; and may also affect the F2 generation (Ji et al. 2011). Although there is a lack of direct evidence confirming the causal relationship between the HPG axis and the origin of disease development, the aforementioned studies suggest that changes in HPG axis programming during the intrauterine period may occur. Adverse prenatal environments may result in gonadal dysplasia and abnormal secretion of sex hormones in offspring, leading to related dysfunctions such as sex differentiation and decreased reproductive function.

3.5 *Hypothalamic-Pituitary-Thyroid Axis*

The HPT axis comprises thyrotropin-releasing hormone (TRH) secreted by the hypothalamus, thyrotropic-stimulating hormone (TSH) secreted by the pituitary, and thyroid hormone (TH) secreted by the thyroid. THs, including triiodothyronine (T3) and thyroxine (T4), is a crucial hormone for embryonic growth, particularly for brain differentiation and development. The negative feedback effect of TH is mainly mediated by thyroid hormone nuclear protein receptors in the pituitary and hypothalamus.

A population study has demonstrated that the density of nuclear TH protein receptors in the cerebral cortex of IUGR fetuses is significantly lower than that of normal fetal brains (Kilby et al. 2000). Clinical investigations have shown that IUGR fetuses exhibit significantly decreased levels of T4 and free T4 and significantly increased levels of TSH, which can persist after birth. Earlier reports have indicated that IUGR newborns have higher basal TSH levels, but the levels of T4 and T3 and the pituitary-thyroid responsiveness to TRH remain largely unchanged (Kilby et al. 1998). Setia et al. have reported that blood TSH and T4 levels are significantly and negatively correlated with increased insulin sensitivity (Setia et al. 2007), suggesting that TH may regulate the insulin sensitivity of IUGR offspring. Decreased insulin sensitivity is a major cause of several metabolic diseases and a common pathophysiological basis of metabolic syndrome. However, there is no evidence regarding how intrauterine changes in the HPT axis continue after birth and how they contribute to multiple disease susceptibility in adulthood.

3.6 *Renin-Angiotensin System*

RAS plays a crucial role in regulating blood pressure, water, and electrolyte balance. Local angiotensin (Ang) is necessary for the growth, differentiation, and development of the embryo (fetus). Adverse prenatal environments can lead to changes in RAS in circulation and multiple tissues. For instance, cord blood plasma renin levels are elevated in children who are small for gestational age, and cortisol injections in late pregnancy increase plasma angiotensin-converting enzyme (ACE) (Forhead et al. 1997). Clinical investigations have suggested that kidney volume and nephrons decrease in human IUGR fetuses and neonates, which may be due to changes in renal RAS (Schreuder 2006). Animal studies have revealed that protein restriction and placental disturbance during pregnancy result in changes in renal morphology and RAS, including decreased renin and angiotensin II (AngII) levels followed by activation of RAS, such as increased expression of renal ACE and AT1R (Barros et al. 2020). Previous studies by the authors' team have shown that prenatal caffeine exposure induces low expression of AT2R in the kidneys, leading to glomerulosclerosis in the offspring (Ao et al. 2015); alcohol exposure during pregnancy significantly inhibits the expression of AT2R in the renal classic RAS, while ACE2 and Mas are inhibited in the renal nonclassical RAS, which persists from the intrauterine period to adulthood (Zhu et al. 2018). RAS activation after birth can increase plasma renin activity and contribute to the development of hypertension.

RAS in the hippocampus plays a crucial role in regulating the development of the hippocampus, blood pressure, drinking behavior, salt intake, learning and memory, and hormone release. The upregulation of angiotensin and AT1R in the hippocampus and immature hippocampal development of fetal sheep exposed to cortisol during pregnancy suggest that RAS may be involved in hippocampal injury and intrauterine programming of hypertension (Reini et al. 2006). Previous studies by the authors' team have shown that prenatal dexamethasone exposure activates the classical RAS pathway in the fetal hippocampus, leading to the occurrence of partial temporal lobe epilepsy in adult offspring that persists through the F2 generation via epigenetics (S Hu et al. 2020). RAS is also expressed in bone tissue. Exposure to exogenous substances during pregnancy (caffeine, ethanol, dexamethasone, nicotine) leads to poor osteogenic differentiation before and after birth, osteogenic mineralization disorders, and low peak bone mass in adulthood by activating the RAS classical pathway (Wen et al. 2019; Xiao et al. 2019, 2020; Wu et al. 2020). Overall, adverse prenatal environments may affect the normal development of multiple organs through RAS, resulting in dysplasia that continues after birth and increases susceptibility to diseases, such as hypertension, chronic kidney disease, osteoporosis, and epilepsy.

4 Functional Interactions Among Neuroendocrine Axes

In both physiological and pathological conditions, the GC/GH-IGF1 axis, HPA axis, HPG axis, HPT axis, RAS, and other endocrine axes not only regulate fetal growth and development through their individual functions but also through complex interactions (Fig. 4.2). As a result, they systematically participate in fetal development programming changes.

GC/GH-IGF1, HPA, HPT, and HPG axes: The GC/GH-IGF1, HPA, HPT, and HPG axes are closely interconnected. GC is a product of the HPA axis, indicating a regulation between GC-IGF1 and the HPA axis. Studies conducted by the authors' team have confirmed a close relationship between the HPA axis and IGF1 (Zeng et al. 2010; Shen et al. 2014). IGF1 expression can be detected in almost all embryonic tissues from implantation, participating in the development of adrenal, thyroid, and gonadal organs (Agrogiannis et al. 2014). Therefore, there is a close relationship between the GC/GH-IGF1 axis and other endocrine axes such as the HPA, HPT, and HPG axes. Leal-Cerro et al. found that the level of GH in patients with Cushing syndrome was significantly lower than that in the control group (Leal-Cerro et al. 2002). Loke et al. also showed that recombinant GH reversed the inhibition of growth and development in children with nephrotic syndrome caused by GC (Loke et al. 1997). IGF1 plays a fundamental role in ovarian cell proliferation and

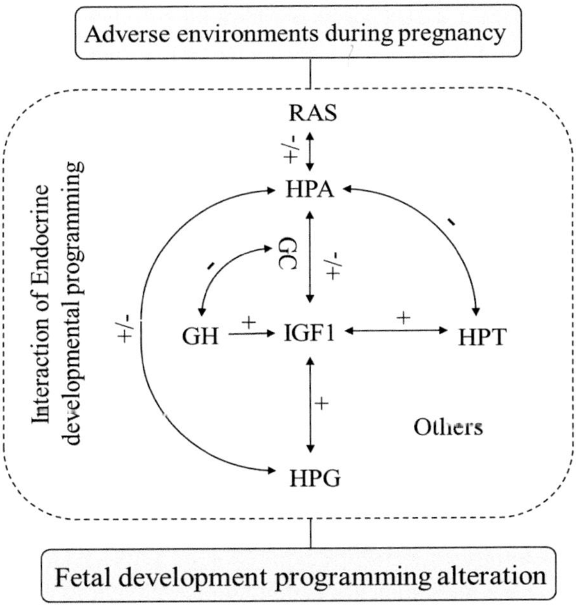

Fig. 4.2 The interaction of endocrine axes mediates alterations in fetal development programming induced by adverse environments during pregnancy

steroid hormone synthesis, with its production in the ovary being dependent on follicle-stimulating hormone and luteinizing hormone stimulation (Hyatt et al. 2007; Inder et al. 2010). TH is also essential for growth and development, particularly in the brain. Studies have indicated elevated serum IGF1 levels in patients with nodular goiter. Both TSH receptor and IGF1 receptor are G protein-coupled receptors, and they exhibit crosstalk and interference. Specifically, TSH exerts a partial effect on IGF1, and IGF1 promotes thyroid growth and stimulates TH synthesis and secretion (Krieger et al. 2020). Moreover, activation of the HPA axis has been correlated with decreased TSH levels and the inhibition of peripheral T4 to T3 transformation, implying a negative correlation between the HPA axis and the HPT axis.

HPG and HPA axes: Activation of the HPA axis has been proposed to inhibit the HPG axis in both men and women (Berga and Loucks 2005). Hyperandrogenemia is a crucial endocrine feature of polycystic ovary syndrome (PCOS). In normal women, androgens, including androstenedione, testosterone, and dehydroepiandrosterone (DHEA), are primarily produced by the adrenal gland, with DHEA secretion fluctuating day and night in response to adrenocortical hormone. In 20% to 30% of PCOS patients, excessive DHEA is associated with altered P450 cl7a activity, increased peripheral cortisol metabolism, and negative regulation of ACTH. The promoter of CRH contains an estrogen response element, and estrogen can stimulate CRH transcription in the hypothalamus. Conversely, the androgen response element in CRH inhibits CRH transcription. Studies have revealed that sex hormones can regulate GC production in rat adrenal glands. Testosterone can decrease corticosterone production, while estrogen has the opposite effect. Additionally, castration experiments have indicated that estrogen deficiency increases GR expression in the pituitary gland and enhances cellular response to circulating GC.

RAS and HPA axis: AngII in the brain can stimulate ACTH secretion, while circulating AngII can act on periventricular tissues and increase CRH secretion. The gradual maturation of the HPA axis during fetal development is related to the elevation of ACTH and cortisol. Studies have suggested that GC can stimulate Ca^{2+} influx into vascular smooth muscle cells, promote peripheral vascular resistance, activate RAS, and result in hypertension. Recent research by the author found that prenatal dexamethasone exposure decreased peak bone mass in offspring during adulthood, which could be passed down to the F2 generation. The mechanism behind this effect involves dexamethasone activating GR and recruiting CCAAT-enhancer-binding proteins alpha (C/EBPα) adenovirus E1A-associated cellular p300 into the nucleus to increase acetylation of the ACE promoter region and activate RAS, ultimately leading to continuous RAS activation and reduced osteoblast differentiation and bone formation ability in bone mesenchymal stem cells (BMSCs) (Xiao et al. 2018). Furthermore, exposure of muscle smooth muscle cells to glucocorticoids, especially dexamethasone, for 12 and 72 hours increased ACE activity due to increased mRNA expression of ACE. Overall, it is suggested that the RAS and HPA axes influence and promote each other's expression.

Furthermore, a high concentration of GC can impede macrophage phagocytosis and antigen processing, promote leptin secretion in human adipocytes, and cause insulin resistance. Intestinal microflora can also have an impact on multiple organs, including the brain, bone, and liver, through various pathways, such as the nervous system, immune system, tryptophan metabolism, and microbial metabolites (e.g., short-chain fatty acids, branched-chain amino acids, and peptidoglycan). In summary, during development, the endocrine systems interact in a complex network, enabling the fetus to respond optimally to adverse environments and maintain its development.

5 Summary and Prospect

In addition to the previously mentioned endocrine axes and systems, other scholars have proposed potential endocrine axes, including the neuro-endocrine-immune axis (Liu et al. 2015), hypothalamic-pituitary-fat axis (Cromi et al. 2009), intestinal-brain axis, intestinal-liver axis, and intestinal-bone axis (Gage et al. 2016; Wesolowski et al. 2017). However, current perspectives suggest that the intrauterine environment has lifelong effects on endocrine function, primarily through changes in neuroendocrine metabolism programming related to GC, low basal activity, and stress hypersensitivity of the HPA axis, and increased susceptibility to various chronic diseases in adulthood. Endocrine disorders caused by adverse environments during pregnancy may create a homeostasis that is unfavorable for fetal growth and development, resulting in hormone level imbalances and fetal growth restriction. In IUGR fetuses, changes in endocrine axis programming are linked to disease susceptibility in adulthood. In fetal-originated disease, changes in endocrine factors exhibit a reticular relationship, with GC programming changes serving as the core, and neuroendocrine metabolic programming changes mediated by GC serving as the primary mechanism of fetal-originated disease. However, the characteristics of endocrine axis changes in fetuses and after birth, as well as the interaction between endocrine axes, require further investigation.

References

Agrogiannis G D, Sifakis S, Patsouris E S, et al. 2014. Mol Med Rep, 10(2): 579–584.
Ao Y, Sun Z, Hu S, et al. 2015. Toxicol Appl Pharmacol, 287(2): 128–138.
Barros M A V, de Brito Alves J L, Barros R G N, et al. 2020. Life Sci, 263: 118574.
Basak S, Das M K, Duttaroy A K. 2020. Birth Defects Res, 112(17): 1308–1325.
Berga S L, Loucks T L. 2005. Minerva Ginecol, 57(1): 45–54.
Chen Y, He Z, Chen G, et al. 2019. Toxicology, 428: 152308.
Cromi A, Ghezzi F, Raffaelli R, et al. 2009. Ultrasound Obstet Gynecol, 33(4): 421–426.
Douglas A J. 2011. Prog Neuropsychopharmacol Biol Psychiatry, 35(5): 1167–1177.
Forhead A J, Broughton Pipkin F, Sutherland M F, et al. 1997. Exp Physiol, 82(4): 761–776.
Gage S H, Munafò M R, Davey Smith G. 2016. Annu Rev Psychol, 67: 567–585.
Hales C N, Barker D J. 1992. Diabetologia, 35(7): 595–601.
He Z, Zhang J, Huang H, et al. 2019. Toxicol Lett, 302: 7–17.
Hu S, Yi Y, Jiang T, et al. 2020. Arch Toxicol, 94(9): 3201–3215.
Hu W, Wang G, He B, et al. 2020a. Toxicology, 432: 152378.
Hu W, Yuan C, Luo H, et al. 2020b. Toxicol Lett, 331: 167–177.
Hyatt M A, Budge H, Walker D, et al. 2007. Endocrinology, 148(10): 4754–4760.
Inder W J, Jang C, Obeyesekere V R, et al. 2010. Clin Endocrinol (Oxf), 73(1): 126–132.
Ji Y L, Wang H, Liu P, et al. 2011. Toxicol Lett, 205(1): 69–78.
Kamei H, Ding Y, Kajimura S, et al. 2011. Development, 138(4): 777–786.
Kapoor A, Dunn E, Kostaki A, et al. 2006. J Physiol, 572(Pt 1): 31–44.
Kilby M D, Gittoes N, McCabe C, et al. 2000. Clin Endocrinol (Oxf), 53(4): 469–477.
Kilby M D, Verhaeg J, Gittoes N, et al. 1998. J Clin Endocrinol Metab, 83(8): 2964–2971.
Krieger C C, Neumann S, Gershengorn M C. 2020. Pharmacol Ther, 209: 107502.
Laron Z. 1995. Drugs, 50(4): 595–601.
Leal-Cerro A, Soto A, Martínez M A, et al. 2002. Clin Endocrinol (Oxf), 57(6): 745–749.
Liang G, Chen M, Pan X L, et al. 2011. Exp Toxicol Pathol, 63(7–8): 607–611.
Liu L, Liu F, Kou H, et al. 2012. Toxicol Lett, 214(3): 307–313.
Liu Y, He S, Zhang Y, et al. 2015. Am J Reprod Immunol, 74(1): 26–37.
Liu Y, Liu M, Liu Y, et al. 2022. Biochem Pharmacol, 199: 115018.
Loke K Y, Yap H K, Zhou X, et al. 1997. J Pediatr, 130(5): 793–799.
Lu J, Li Q, Xu D, et al. 2020. Toxicol Lett, 331: 33–41.
Lummaa V, Tremblay M. 2003. Proc Biol Sci, 270(1531): 2355–2361.
Martin-Gronert M S, Ozanne S E. 2007. J Intern Med, 261(5): 437–452.
Meaney M J, Szyf M, Seckl J R. 2007. Trends Mol Med, 13(7): 269–277.
Murdoch W J, Van Kirk E A, Vonnahme K A, et al. 2003. Reprod Biol Endocrinol, 1: 6.
Pei L G, Zhang Q, Yuan C, et al. 2019. J Endocrinol, 242(1): M17-M32.
Poulain M, Frydman N, Duquenne C, et al. 2012. J Clin Endocrinol Metab, 97(10): E1890- E1897.
Reini S A, Wood C E, Jensen E, et al. 2006. Am J Physiol Regul Integr Comp Physiol, 291(6): R1708- R1716.

Schreuder M, Delemarre-van de Waal H, van Wijk A. 2006. Kidney Blood Press Res, 29(2): 108–125.
Setia S, Sridhar M G, Koner B C, et al. 2007. Clin Chim Acta, 376(1–2): 37–40.
Shangguan Y, Wen Y, Tan Y, et al. 2018. Am J Pathol, 188(12): 2863–2876.
Shen L, Liu Z, Gong J, et al. 2014. Toxicol Appl Pharmacol, 274(2): 263–273.
Stamps J A, Frankenhuis W E. 2016. Trends Ecol Evol, 31(4): 260–268.
Toledo F C, Perobelli J E, Pedrosa F P, et al. 2011. Reprod Biol Endocrinol, 9: 94.
Vieau D, Sebaai N, Léonhardt M, et al. 2007. Psychoneuroendocrinology, 32 Suppl 1: S16–20.
Walenkamp M J, Losekoot M, Wit J M. 2013. Endocr Dev, 24: 128–137.
Wang L, Jin F, Qin A, et al. 2014. Mol Cancer, 13: 139.
Wen Y, Shangguan Y, Pan Z, et al. 2019. Toxicol Appl Pharmacol, 363: 1–10.
Wesolowski S R, Kasmi K C, Jonscher K R, et al. 2017. Nat Rev Gastroenterol Hepatol, 14(2): 81–96.
Wu Z, Pan Z, Wen Y, et al. 2020. Food Chem Toxicol, 136: 111083.
Xia L P, Shen L, Kou H, et al. 2014. Toxicol Lett, 226(1): 98–105.
Xia L, Jiao Z, Pei L, et al. 2020. Reprod Toxicol, 94: 48–54.
Xiao H, Wen Y, Pan Z, et al. 2018. Cell Death Dis, 9(6): 638.
Xiao H, Wen Y, Pan Z, et al. 2019. Faseb j, 33(11): 12972–12982.
Xiao H, Wu Z, Li B, et al. 2020. Br J Pharmacol, 177(20): 4683–4700.
Xu D, Chen M, Pan X L, et al. 2011. Environ Toxicol Pharmacol, 32(3): 356–363.
Xu D, Liang G, Yan Y E, et al. 2012a. Toxicol Lett, 209(3): 282–290.
Xu D, Wu Y, Liu F, et al. 2012b. Toxicol Appl Pharmacol, 264(3): 395–403.
Yang Z, Guo C, Zhu P, et al. 2007. J Endocrinol, 195(2): 241–253.
Zeng C, Wang R, Li D, et al. 2010. Hepatology, 52(5): 1702–1712.
Zhang C, Xu D, Luo H, et al. 2014. Toxicology, 325: 74–84.
Zhang L, Shen L, Xu D, et al. 2016. Reprod Toxicol, 65: 236–247.
Zhang L, Xu D, Zhang B, et al. 2013. Arch Med Res, 44(5): 335–345.
Zhu Y, Zuo N, Li B, et al. 2018. Toxicology, 400–401: 9–19.

Chapter 5
Glucocorticoid Programming Mechanism of Fetal-Originated Diseases

Hao Kou and Yawen Chen

Abstract Glucocorticoids play a critical role in regulating fetal development and maturation as well as determining the fate of the fetus after birth. The author's previous studies have demonstrated that adverse environments during pregnancy can compromise the placental glucocorticoid (GC) barrier, resulting in excessive maternal GC exposure in fetuses and inducing intrauterine growth retardation (IUGR). This can lead to developmental programming of multiple organs, changes in homeostasis, and susceptibility to related diseases through abnormal epigenetic modification. The team has further proposed that intrauterine maternal GC exposure is an "active compensatory" response of the mother to promote fetal survival in an adverse environment during pregnancy, inducing the development of a "thrifty phenotype." The team has innovatively proposed that "two-programming and two-strike" occurs in fetal-originated diseases caused by adverse environments during pregnancy. Among these, the "GC-insulin-like growth factor 1 axis" mediates IUGR, catch-up growth after birth, and susceptibility to multiple diseases in adulthood.

1 Introduction

Glucocorticoids (GCs) are a critical "switch" in body development, inducing changes in gene regulation and playing a vital role in normal fetal growth and developmental maturity. Early activation of this "switch" during intrauterine periods ensures that nutrient supply in utero is balanced with fetal growth, and fetal tissue is

H. Kou (✉)
Department: Department of Pharmacy, Zhongnan Hospital of Wuhan University, Wuhan, Hubei, China
e-mail: kouhao007@whu.edu.cn

Y. Chen
Department of Pharmacology, School of Basic Medical Science, Wuhan University, Wuhan, Hubei, China

H. Wang et al. (eds.), *Fetal Origin of Diseases*,
https://doi.org/10.1007/978-981-97-5730-5_5

mature enough to survive outside the womb at delivery. However, fetal development is susceptible to internal and external environmental stimuli. Glucocorticoid (GC) exposure during pregnancy can cause intrauterine growth retardation (IUGR), characterized by fetal growth and developmental limitations, multiple organ dysfunction, and low birth weight. The global incidence of IUGR ranges from 2.75% to 15.53%, with higher rates in developing countries. Epidemiological investigations have demonstrated that IUGR can cause fetal distress, neonatal asphyxia, perinatal death, and long-term harm to physical and intellectual development, leading to an increased risk of various chronic diseases in adulthood. This not only affects the quality of life but also leads to a series of family and social problems. Currently, there is no systematic theoretical framework for the mechanisms underlying chronic disease susceptibility in IUGR offspring exposed to adverse environments during pregnancy. This chapter reviews the GC programming mechanisms of fetal-originated diseases by reviewing the author's team's research and the latest international reports, providing strong experimental evidence and a theoretical basis for analyzing the occurrence and development of fetal-originated diseases.

2 Abnormal Exposure to Intrauterine GC Caused by Adverse Environments During Pregnancy

Basal GC levels during intrauterine periods play a critical role in regulating fetal tissue morphology and functional maturation. Maternal GCs are involved in early fetal growth and development under physiological conditions. However, abnormal intrauterine GC exposure can lead to abnormal fetal development.

2.1 Overexposure to GC in Utero Caused by Adverse Environments During Pregnancy

Numerous studies have reported that exposure to various adverse environments during pregnancy, including but not limited to xenobiotics, malnutrition, infection, hypoxia, and stress, can lead to an increase in intrauterine maternal GC levels (Pawluski et al. 2012; Cuffe et al. 2014; Correia-Branco et al. 2015). It has been observed that exposure to air pollution or bisphenol A during pregnancy can elevate blood cortisol levels in neonates, as evidenced by cohort studies (Giesbrecht et al. 2017; Khamirchi et al. 2020). Similarly, exposure to cocaine and cannabis during pregnancy has been shown to increase corticosterone levels in offspring rats (Franks et al. 2020). Furthermore, animal experiments conducted by the author's team have demonstrated that exposure to various xenobiotics, such as caffeine, nicotine, and ethanol, during pregnancy can result in excessive maternal GC exposure, leading to

IUGR and multiple organ development abnormalities (Chen et al. 2007; Liang et al. 2011; Xu et al. 2012b). These findings suggest that exposure to excessive maternal GC is a common phenomenon associated with adverse conditions during pregnancy.

2.2 *Low Exposure to GC in Utero Caused by Prenatal Administration of Synthetic GCs*

Data from the World Health Organization's maternal and child health survey, which was conducted in 359 institutions across 29 countries, revealed that the rate of prophylactic treatment with synthetic GCs for preterm infants between 22 and 36 weeks of gestation was 54% in all countries and as high as 91% in some countries (Vogel et al. 2014). Although maternal adrenal GC synthesis is enhanced during pregnancy, resulting in a 2–4 times higher serum-free glucocorticoid level in the third trimester of pregnancy compared to prepregnancy levels (Burke and Roulet 1970; Nolten and Rueckert 1981; Cousins et al. 1983), the fetal adrenal gland does not establish the de novo synthesis pathway of fetal adrenal GC until late pregnancy (Mesiano and Jaffe 1997). Consequently, the secretion of GC in the fetal adrenal gland is relatively less than that of the mother, and the GC in fetal blood during the intrauterine period mainly comes from the mother. However, the administration of synthetic GCs during pregnancy can inhibit maternal adrenal function, reducing the level of endogenous GC in maternal blood, and ultimately lead to fetal exposure to intrauterine low GC (Chen et al. 2021).

2.3 *Placental GC Barrier and Its Regulatory Mechanism*

The placenta is a vital organ in maintaining normal fetal development and plays a crucial role in regulating maternal GCs from entering the fetal circulation. The placental 11β-hydroxysteroid dehydrogenases (11β-HSDs) are responsible for mediating maternal-fetal GC transport. The glucocorticoid-metabolizing enzymes include 11β-HSD1 and 11β-HSD2. In a normal pregnancy, the function of the placental 11β-HSDs is mainly oxidation, with the activity of 11β-HSD2 being dominant. Studies have shown that 11β-HSD2 is expressed early in pregnancy and is highly expressed in the placenta and fetus during the second trimester, protecting fetal tissues from the adverse effects of high levels of maternal GCs. In contrast, 11β-HSD1 amplifies the effects of GC during circadian low periods, working together with 11β-HSD2 to ensure normal fetal development (Seckl and Chapman 1997). During the third trimester of pregnancy, 11β-HSD1 is expressed in the maze region of the placenta in rodents, potentially increasing the number of GCs reaching the fetus to induce terminal differentiation before delivery (Chapman et al. 2013).

Despite the presence of both 11β-HSD1 and 11β-HSD2 in placental tissue, 11β-HSD2 activity is the highest during pregnancy, forming a GC barrier between the mother and fetus. This barrier regulates the amount of maternal GCs entering the fetal circulation and maintains a stable and favorable intrauterine development environment, which protects normal fetal development (Salvante et al. 2017). However, studies in humans and rodents have shown that placental 11β-HSD2 activity is susceptible to a variety of adverse environments during pregnancy, such as xenobiotic exposure, poor diet, infection, hypoxia, and stress. These adverse environments can lead to exposure to maternal GC in developing fetuses (Lesage et al. 2001; Nolan et al. 2001; Morrison et al. 2012; Kirsten et al. 2013; Reynolds 2013). Prenatal exposure to adverse environments can increase maternal GC levels through maternal stress. Previous studies by our team have found that high levels of GCs during pregnancy can inhibit placental 11β-HSD2 expression, resulting in excessive exposure to maternal GC in fetuses (Chen et al. 2007; Liang et al. 2011; Xu et al. 2012b; Yu et al. 2018). These studies suggest that exposure to maternal GC during pregnancy is associated with down-regulated placental 11β-HSD2 expression in adverse environments.

Furthermore, the glycoprotein (P-gp) constitutes another GC barrier in the placenta. P-gp can reverse the concentration of GCs excreted to the mother, thus restricting the transplacental entry of maternal GC into the fetus and reducing fetal exposure to maternal GC. The expression of placental P-gp varies in different stages of pregnancy, with higher expression in early pregnancy than at term. P-gp protein expression in the placenta during the 13th to 14th week of gestation is about twice that at term. After high levels of P-gp expression in the second trimester, it decreases with an increase in gestational age, enabling the smooth introduction of maternal cortisol into the fetus, which is conducive to promoting fetal maturity and regulating labor during the labor stage. Studies have indicated that the use of cannabis during pregnancy can down-regulate P-gp expression, inhibiting P-gp-dependent efflux, weakening placental protection, and harming the fetus (Feinshtein et al. 2013). It has also been reported that exposure to lipopolysaccharides during pregnancy can reduce fetal protection against harmful xenobiotics in maternal circulation by inhibiting placental P-gp expression (Lye et al. 2015). Recently, our team found that placental P-gp expression in the IUGR rat model induced by caffeine exposure during pregnancy was associated with changes in the p-JNK/YB-1/P300 pathway. Combining clinical specimens and other animal models confirmed that this pathway is a common regulatory mechanism for decreased P-gp expression in IUGR fetuses. Finally, animal experiments confirmed that the induction of P-gp could reverse the occurrence of IUGR induced by caffeine exposure during pregnancy (Ge et al. 2021). These studies suggest that decreased placental P-gp expression in adverse environments during pregnancy may result in fetal exposure to maternal glucocorticoids GC by opening the GC barrier (Fig. 5.1).

3 Maternal GC Mediates Multi-Organ Programming and Multigeneration Inheritance

Epigenetics is a phenomenon that refers to alterations in gene expression that are heritable but do not involve any changes in the DNA sequence. Numerous studies have provided evidence that adverse prenatal environments can induce abnormal epigenetic modifications, leading to changes in the developmental programming of several organs in the offspring. The consequences of these modifications can also be inherited across multiple generations.

3.1 GC Mediates Epigenetic Regulation of Developmental-Related Genes

Excessive maternal exposure during pregnancy, resulting from adverse environmental conditions, is a major contributor to intrauterine developmental programming. Epigenetic regulation of functional genes is likely to be involved in the changes in intrauterine programming induced by high GC levels. For instance, in guinea pig models exposed to betamethasone during pregnancy, fetal and neonatal offspring exhibited changes in genomic DNA total methylation rates and altered gene expression in multiple organs such as the liver, adrenal gland, and kidney (Crudo et al. 2012). Furthermore, the authors' team demonstrated that maternal glucocorticoid GC exposure induced by caffeine intake during pregnancy could increase the histone acetylation level and expression of cholesterol synthesis key gene 3-hydroxy-3-methyl-glutaryl-coenzyme A reductase and fatty acid synthase in fetal liver. The epigenetic modifications and expression changes mediated by high GC levels have intrauterine programming effects that can persist after birth and into adulthood, leading to increased susceptibility to hypercholesterolemia and nonalcoholic fatty liver disease in adult offspring (Guo et al. 2018). Furthermore, high levels of GCs

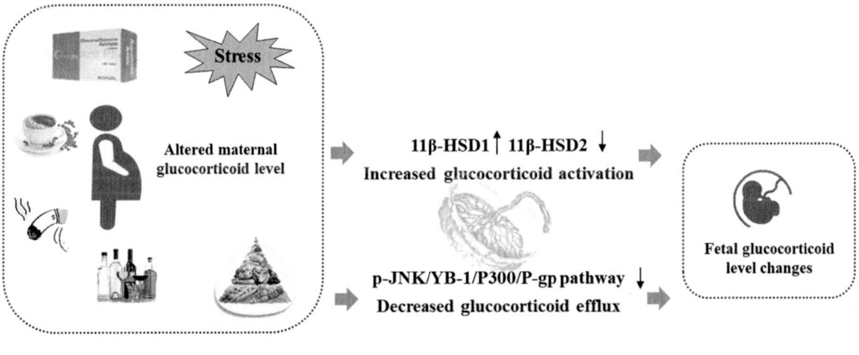

Fig. 5.1 Placental glucocorticoid barrier and its regulation mechanism

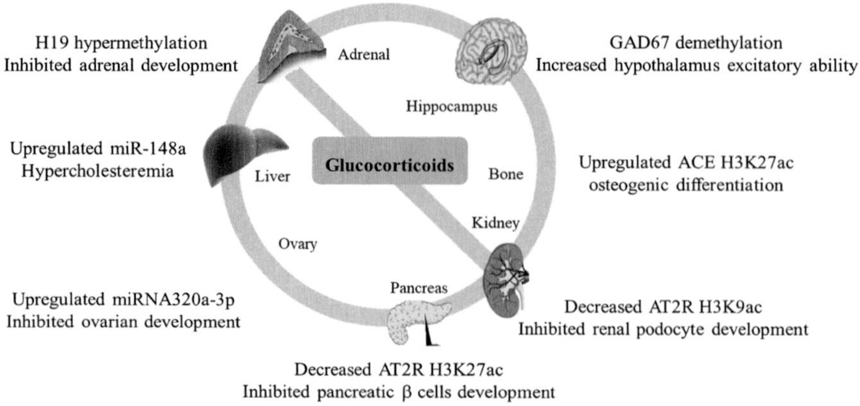

Fig. 5.2 Glucocorticoid-mediated epigenetic regulation of development-related genes

have been found to mediate epigenetic modifications of key genes involved in the development of multiple organs in offspring, such as the bone, kidney, and testis, leading to changes in developmental programming (Liu et al. 2019; Zhu et al. 2019; Xiao et al. 2020b). Additionally, miRNAs have been implicated in GC-mediated programming of downstream genes. For instance, hippocampal miR-134-5p has been shown to mediate the programming of the sex-determining region Y-box2 signaling pathway in offspring induced by prenatal dexamethasone exposure (Jiang et al. 2021). Similarly, miR-148a has been associated with low expression of low-density lipoprotein receptor in the liver of male offspring induced by prenatal dexamethasone exposure (Li et al. 2020). These findings suggest that abnormal epigenetic modifications are involved in the programming changes in multiple organs induced by maternal GC (Fig. 5.2).

3.2 *Epigenetics Participate in Programming Changes of Multi-Organ Development and Multigeneration Inheritance*

An increasing number of studies have demonstrated that intrauterine and early postnatal environmental changes can affect the DNA methylation and histone acetylation state of specific gene promoters. These changes can persist throughout the life of the offspring and can also be stably inherited by the next generation (Weaver et al. 2004; Lillycrop et al. 2005; Szyf et al. 2008; Nijland et al. 2010; Heard and Martienssen 2014). Programming intervention during pregnancy (F0 generation) can directly impact the development of the fetus in utero (F1 generation), and the germ cells forming the F2 generation can also be directly exposed to the prenatal adverse environments. Therefore, it can be argued that only effects observed in offspring beyond the F2 generation can truly be considered transgenerational and not a direct consequence of the original exposure (Skinner et al. 2011; Guerrero-Bosagna and Skinner 2012). For instance, studies have reported that betamethasone exposure during pregnancy can lead to changes in the total methylation rate of multi-organ genomic DNA in fetal and neonatal mice, which can also persist into the F2 generation (Crudo et al. 2012). Likewise, articular cartilage damage in female offspring induced by prenatal caffeine exposure was associated with decreased H3K9ac levels in the TGF-β signaling pathway and persisted into the F3 generation (Zhao et al. 2020). Furthermore, miRNAs have also been implicated in GC programming changes in multiple organ development. It has been reported that malnutrition during pregnancy can increase the blood corticosterone concentration in offspring, inhibiting miR-449a and corticotropin-releasing hormone receptor 1 (CRHR1) expression, leading to hypothalamus-pituitary-adrenal (HPA) axis dysfunction (Nemoto et al. 2015). Excessive maternal GC caused by maternal malnutrition have long-term effects on offspring carotid artery remodeling through miR-29c (Khorram et al. 2015). Recently, the authors' team found a "contemporary programming" effect of maternal GC on adrenal steroidogenesis induced by caffeine exposure during pregnancy. The mechanism involves the regulatory interaction between the methylation level of the imprinted gene H19 promoter region and miRNA let-7 expression mediated by maternal GC (He et al. 2021). Additionally, prenatal dexamethasone exposure has been shown to cause abnormal expression of the ovarian steroid synthase system in adult offspring of both F1 and F3 generations. This mechanism is related to decreased miR-320a-3p expression induced by activation of the ovarian glucocorticoid receptor (GR) in offspring (Gong et al. 2021) (Table 5.1).

4 Maternal GC Mediates Developmental Programming Changes in Multiple Endocrine Axes

The intrauterine environment has a profound and sustained impact on the developmental process of the organism. It is currently believed that "intrauterine programming" mainly represents the adaptive response of the body to adverse intrauterine environments, which is a compensatory protective mechanism for the developing fetus. However, this mechanism also increases the susceptibility to various diseases later in life.

4.1 Programming Changes in the GC-Insulin-like Growth Factor 1 (GC-IGF1) Axis

IGF1 is a major regulator that mediates the adaptive response to changes in the internal and external environment during the intrauterine period (Halmos and Suba 2019). Fetal IGF1 levels during intrauterine development are directly related to the

Table 5.1 Prenatal glucocorticoid exposure causes programming changes in multi-organ development and multigenerational inheritance through epigenetic modification

Epigenetic modification	Alteration	Outcomes	Main references
DNA methylation	DNA total methylation	Liver, adrenal, and kidney dysplasia	Crudo et al. (2012)
Histone modification	TGFβ H3K9ac	Articular damages	Zhao et al. (2020)
	IGF1 H3K27ac	Ovarian dysfunction	Lv et al. (2021)
	ACE H3K27ac	Low peak bone mass	Xiao et al. (2018)
	P450scc H3K9ac	Adrenal steroidogenic dysfunction	Cao et al. (2021)
Noncoding RNA	miR-148a	Liver dysplasia	Li et al. (2020)
	miRNA let-7	Low adrenal steroidogenic function	He et al. (2021)
	miR-320a-3p	Low ovary steroidogenic function	Gong et al. (2021)
	miR-29c	Carotid remodeling	Khorram et al. (2015)
	miR449a	HPA axis dysfunction	Nemoto et al. (2015)

TGFβ transforming growth factor beta, *IGF1* insulinlike growth factor 1, *ACE* angiotensin-converting enzyme, *P450scc* cytochrome P450 side-chain cleavage enzyme, *H3K9ac* histone 3 lysine 9 acetylation, *HPA* hypothalamus-pituitary-adrenal

degree of embryonic development, and changes in fetal GC concentration induced by the intrauterine environment can lead to changes in blood IGF1 levels (Fowden 2003). The authors' team investigated the serum corticosterone concentration and the expression of the IGF1 signaling pathway and steroidal synthase system in the blood and tissues (such as the liver, adrenal gland, and skeletal muscle) of a rat model exposed to various xenobiotics (such as caffeine and ethanol) during pregnancy. The results showed that an increase in fetal blood corticosterone levels led to the inhibition of the IGF1 signaling pathway and eventual organ dysplasia. However, with a decrease in serum corticosterone levels in the early postnatal stage, the IGF1 signaling pathway was gradually enhanced, and IGF1 signaling eventually caught up with the control rats in adult life (Shen et al. 2014; Huang et al. 2015; He et al. 2019). The negative regulatory effect observed between GC levels and blood/multi-tissue IGF1 levels suggests that high levels of intrauterine maternal GC program the expression of fetal multi-tissue IGF1 and the functional changes of downstream signaling pathways, namely the "GC-IGF1 axis" programming changes. However, the expression of the IGF1 signaling pathway varied across different tissues, with significant differences observed in tissues, sex, and time (before and after birth). The multi-tissue "GC-IGF1 axis" programming differences observed in the adverse intrauterine environment may be mainly related to individual regulation and adaptation caused by changes in tissue function and sex hormone levels. This is the result of IUGR fetal "thrifty phenotype" programming.

The action of GCs on corticosteroid receptors in fetal tissue is known to depend on circulating GC levels. High levels of GCs can bind to the GR receptor only after the saturation of the mineralocorticoid receptor (MR) (Chapman et al. 2013). Under physiological conditions, GCs are activated by 11β-HSD1 and inactivated by 11β-HSD2, and there is positive feedback regulation between GCs, 11β-HSD1, and 11β-HSD2. Several studies have shown that high levels of GCs can inhibit the expression of IGF1 in various tissues (Hyatt et al. 2007; Inder et al. 2010). The CCAAT enhancer-binding proteins (C/EBP), including α and β family members, are important transcription factors that cooperate with GCs and their receptors to rapidly regulate downstream gene expression (Chapman et al. 2013). The previous study reported on the inhibitory effect of C/EBP-α on the expression of insulin-like growth factor 1 receptor (IGF1R), as demonstrated by Zeng et al. (2010). Additionally, Staiger et al. (2009) observed a decrease in circulating IGF1 levels upon knockout of liver C/EBP-β in mice. Notably, both skeletal muscle C/EBP-β expression and blood IGF1 concentration were found to be increased in IUGR rats following birth, as reported by Zheng et al. (2011). Collectively, these findings suggest the involvement of GC-activated systems, including 11β-HSD1, 11β-HSD2, MR, GR, C/EBP-α, and C/EBP-β, in the programming of the GC-IGF1 axis and subsequent development of multiple organs.

4.2 Programming Changes in the HPA Axis

Offspring exposed to adverse environmental conditions during pregnancy exhibit a phased change in the basal activity of the HPA axis after birth. Population-based studies have demonstrated that severe psychosocial stress experienced by mothers during pregnancy, such as car accidents, divorce, and bankruptcy, results in lower basal cortisol levels in their offspring during adulthood, as reported by Entringer et al. (2009). Likewise, animal experiments involving guinea pigs exposed to RF light during early and middle stages of pregnancy revealed a reduction in basal blood glucocorticoid levels in their offspring at one month, followed by elevated levels at two months, with similar changes observed in fecal GC content (Schopper et al. 2012). A series of studies conducted by the author team further revealed that IUGR offspring, resulting from adverse environmental conditions during pregnancy, exhibit "time characteristic" changes in basal HPA axis activity after birth, characterized by an increase before weaning, a decrease in childhood, a surge in adolescence, a return to normal or even higher levels during adulthood, and eventually, a continuous state of elevated GC levels, as reported by Liu et al. (2012) and Xu et al. (2012a). The hippocampus serves as a high regulatory center of the HPA axis, with GR (but not MR) primarily expressed in the fetal hippocampus. Yang et al. (2007) reported that fetal hippocampal GR is a sensitive target for the action of high levels of circulating glucocorticoids, indicating that its expression serves as a marker for the establishment of HPA axis negative feedback regulation function, with its degree of expression representing the sensitivity of this function. Following birth, hippocampal MR and GR work in a complementary manner to adjust the body's adaptation to the internal and external environment. The maintenance of a relative balance between MR, GR, and downstream-related molecules is crucial for the normal activity of the HPA axis. Environmental changes that alter this balance inevitably lead to HPA axis dysfunction, which is characteristic of chronic stress, depression, and high levels of GC exposure. The author's team found that caffeine or nicotine exposure during pregnancy caused a significant decrease in MR during the early postnatal period (1–7 days after birth) and a significant increase during the later period (35–100 days after birth), with an increased MR/GR expression ratio observed (Liu et al. 2012; Xu et al. 2012a). These changes in hippocampal MR were positively correlated with basal HPA axis activity, indicating that hippocampal MR plays a role in establishing positive feedback regulation function of the HPA axis after birth in IUGR offspring and maintaining basal HPA axis activity within the normal range.

Adverse environments during pregnancy can result in hypersensitivity of the HPA axis stress response in offspring after birth, as reported by Davis et al. (2011). An epidemiological study by Entringer et al. (2009) found that prenatal severe psychosocial stress could increase cortisol levels induced by adrenocorticotropic hormone in offspring. Prenatal exposure to synthetic GCs resulted in persistent enhancement of the HPA axis stress response in offspring, as reported by Khalife et al. (2013). Similarly, alcohol exposure during pregnancy can lead to depression and anxiety disorders in offspring, accompanied by persistently increased HPA axis sensitivity, as reported by Hellemans et al. (2010). The author's team also found that prenatal exposure to xenobiotics resulted in high responsiveness to chronic stress in adulthood (Xu et al. 2012a; Xia et al. 2014). The hippocampus plays a key role in inhibiting the stress response of the HPA axis and promoting the recovery of hyperactive HPA axis function to the basal level under stress. Glutamic acid decarboxylase 67 (GAD67) plays a crucial role in balancing glutamic acid (Glu) and aminobutyric acid (GABA) levels and signal output. It has been reported that GR is co-expressed with GAD67 in GABA neurons and that both acute and chronic GC treatment can upregulate GAD67 expression, as reported by de Souza and Franci (2013) and Lu et al. (2020). These responses can attenuate glutamatergic signal input and Glu-GABA synaptic connections in the paraventricular nucleus (PVN) of the hypothalamus, thereby reducing the inhibitory effect of GABAergic nerves on the activity of CRH neurons in the hypothalamic PVN area and improving the potential excitability of this area. These changes can persist after birth and serve as the pathophysiological basis for high stress sensitivity of the HPA axis in offspring. The author's team has reported a series of findings (He et al. 2017; Xu et al. 2018; Lu et al. 2020) indicating that high stress sensitivity of the HPA axis in offspring rats, induced by exposure to xenobiotics (caffeine, nicotine, ethanol) during pregnancy, is associated with hypomethylation and high expression of the GAD67 promoter region in the hippocampus. Lu et al. (2020) further confirmed that intrauterine high levels of maternal GCs mediated the increase of GAD67 expression in the fetal hippocampus under ethanol exposure during pregnancy, promoting the conversion of Glu to GABA to balance the excitatory damage of these exogenous substances on the hippocampus. This programming of high GAD67 expression in the fetal hippocampus can persist into postnatal and even adulthood stages, weaken the negative feedback regulation of the hippocampus on HPA axis excitation, and ultimately mediate the occurrence of high HPA axis stress sensitivity.

4.3 Programming Changes in the Renin-Angiotensin System (RAS)

RAS components in the hippocampus play a crucial role in regulating hippocampal development, central control of blood pressure, regulation of drinking behavior, salt intake, learning and memory, and hormone release. Fetal sheep exposed to cortisol during pregnancy exhibit upregulation of hippocampal angiotensin and angiotensin type 1 receptor (AT1R), resulting in delayed hippocampal development, suggesting that RAS in the brain may contribute to hippocampal damage and intrauterine programming of hypertension. Previous studies by the author's team have demonstrated that dexamethasone exposure during pregnancy can activate the RAS classical pathway in the fetal hippocampus, leading to the occurrence of partial temporal lobe epilepsy in adult offspring, with this phenomenon potentially affecting the F2 generation through epigenetic inheritance (Hu et al. 2020). RAS also exists in bone tissue, and exposure to xenobiotics (such as caffeine, ethanol, dexamethasone, nicotine) during pregnancy can activate the RAS classical pathway, leading to poor osteogenic differentiation before and after birth, osteogenic mineralization disorders, and low peak bone mass in adulthood (Wen et al. 2019; Xiao et al. 2019, 2020b; Wu et al. 2020). In a rat model of excessive maternal GC exposure resulting from ethanol exposure during pregnancy, the fetal kidney exhibited inhibited glial cell line-derived neurotrophic factor signaling pathway. Renal AT2R expression was significantly inhibited in classical RAS, while renal angiotensin converting enzyme (ACE) 2 and Mas expression were inhibited in nonclassical RAS, with these changes persisting from intrauterine to adulthood (Zhu et al. 2018). Postnatal RAS activation leads to increased plasma renin activity and contributes to the development of hypertension. In conclusion, maternal GC exposure during pregnancy can program offspring susceptibility to multiple diseases through RAS.

5 Maternal GC Mediates Susceptibility to a Variety of Fetal-Originated Disease

Numerous epidemiological studies and animal experiments have provided evidence that the IUGR fetus is more vulnerable to various chronic diseases in adulthood, including metabolic and neuropsychiatric disorders.

5.1 *Maternal GC Induces Fetal Thrifty Phenotype Development*

Exposure to adverse environmental conditions during gestation can result in the programming of an intrauterine "thrifty phenotype" in offspring, which can increase their susceptibility to metabolic diseases in adulthood. The "thrifty phenotype" hypothesis was originally introduced by Hales and Barker (1992), positing that in response to an unfavorable intrauterine environment and consequent malnutrition, the fetus undergoes adaptive changes that may become permanent "programming." This programming renders individuals more sensitive to nutrient-rich environments after birth and increases the likelihood of insulin resistance. Vaag et al. (2012) further elucidated that in the face of adverse growth conditions such as malnutrition, the fetus redistributes its limited energy and limits secondary organ energy consumption to ensure the development of critical organs, such as the brain and liver. This "thrifty" adaptation is an initiative by the fetus to cope with adverse gestational environments and involves the development of multiple organs throughout the body, primarily to ensure survival during the dangerous period.

Utilizing a rat model of IUGR induced by prenatal exposure to xenobiotics, the authors' team conducted a series of studies and discovered that excessive maternal GC exposure in IUGR offspring led to lasting differential changes in the structural and functional development of various tissues and organs. For instance, exposure to excessive maternal GC enhanced liver triglyceride and cholesterol synthesis (Hu et al. 2020) as well as adrenal steroid synthesis (He et al. 2019). Conversely, it inhibited articular cartilage matrix synthesis (Shangguan et al. 2018), osteoblast differentiation of long bone (Pan et al. 2016), glomerular podocyte development (Ao et al. 2015), and ovary and testis development (Fan et al. 2019; Pei et al. 2019). The authors' team propose that the "thrifty phenotype" of fetal energy redistribution is a result of the "active opening" of the placental GC barrier and the high levels of maternal GCs that are necessary for fetal survival. The "first programming" resulting from high levels of maternal glucocorticoid induces significant functional differences in various tissues and organs to ensure the development of vital organs such as the brain and liver, while limiting energy consumption of secondary organs. This "GC-IGF1" axis programming, or the "second programming," also reduces the overall fetal metabolic rate by inhibiting the IGF1 signaling pathway of all tissues and organs, enabling survival through the critical period. However, the development of the "thrifty phenotype" involves "abnormal" epigenetic modifications that have intrauterine programming effects, which can persist after birth or throughout life. This genetic basis is responsible for the vulnerability of IUGR individuals to multiple tissue and organ damages and related diseases after birth (Fig. 5.3).

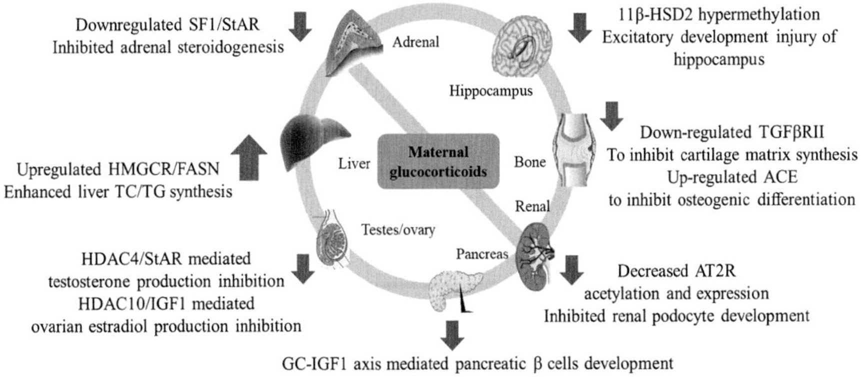

Fig. 5.3 Maternal glucocorticoid induces thrifty phenotypic development in fetuses

5.2 Maternal GC Induces the Susceptibility to a Variety of Fetal-Originated Disease

The intrauterine environment has a decisive and lasting impact on life development. Intrauterine programming is currently understood to be primarily an adaptive response of the body to adverse intrauterine conditions. These responses function as compensatory protective mechanisms for the developing fetus but also increase susceptibility to various diseases in adulthood. Multiple studies have suggested that the fetus can adapt to limited energy supply in adverse intrauterine environments to ensure cardiovascular and nervous system development. Although these adaptations are beneficial for the fetus to survive in the environment at that time, rapid catch-up growth in the early postnatal period increases the risk of metabolic syndrome and associated diseases in adulthood.

The authors' team highlighted that exposure to prenatal xenobiotics and food restriction can result in excessive maternal GC exposure in fetuses, leading to various organ abnormalities and increased susceptibility to multiple diseases in adulthood. Offspring rats exposed to maternal GC during pregnancy exhibited decreased glutamatergic neurons and increased GABAergic neurons in the hippocampus, resulting in heightened HPA axis sensitivity after birth and greater susceptibility to learning and memory impairment and temporal lobe epilepsy in adult offspring. In the liver, offspring exposed to maternal GC during pregnancy experienced poor liver development, low basal HPA axis activity, and high stress sensitivity after birth, along with GC-dependent glucose and lipid metabolism programming changes in peripheral tissues (Xu et al. 2013; Lu et al. 2018). These changes are related to alterations in intrauterine GC-IGF1 axis programming, which can increase the body's sensitivity to environmental factors such as high-fat diets and chronic stress (He et al. 2017), and can persist through to the F2 generation (Luo et al. 2014;

Kou et al. 2017). For instance, prenatal GC exposure was found to induce metabolic diseases like non-alcoholic fatty liver disease in adulthood (Hu et al. 2020). Consuming a high-fat diet after birth can accelerate the development of nonalcoholic fatty liver disease (Shen et al. 2014; Wang et al. 2014; Zhang et al. 2016). Exposure to excessive maternal GC during pregnancy has been found to increase total blood cholesterol levels in adult offspring rats (Xu et al. 2012a, 2013; Xia et al. 2014), as well as elevate the LDL-C/HDL-C ratio after birth with a high-fat diet (Qi et al. 2017), indicating a higher risk of cardiovascular and cerebrovascular diseases in adulthood. In bone, maternal glucocorticoid exposure during pregnancy can decrease the synthesis of articular cartilage matrix in offspring, leading to susceptibility to osteoarthritis in adult offspring. Stimulation of a high-fat diet or long-distance running after birth can further reduce articular cartilage quality and induce osteoarthritis (Luo et al. 2015; Ni et al. 2015a,b; Tan et al. 2016; Tie et al. 2016a,b; Zhao et al. 2020). Moreover, maternal GC during pregnancy can induce shorter long bone lengths and abnormal growth plate morphology in offspring rats, with long bone length exhibiting catch-up growth after birth, but with osteoblast differentiation ability and bone mass remaining low and osteoporosis occurring after adult castration (Shangguan et al. 2018; Xiao et al. 2020b). Lastly, maternal GC during pregnancy can lead to poor fetal kidney development and obvious glomerulosclerosis in adulthood (He et al. 2019).

5.3 The "Two-Programming and Two-Strike" Mechanism for the Occurrence of Fetal-Originated Disease

The author's team has recently proposed a "two-programming and two-strike" mechanism for fetal-originated diseases. The "first programming" is the differential programming among fetal tissues and organs caused by exposure to excessive maternal GC, based on the "thrifty phenotype" developmental model, which redistributes fetal energy to ensure survival. The "second programming" is the "GC-IGF1 axis" programming, which is related to fetal development before and after birth, caused by excessive maternal GC exposure. This "GC-IGF1 axis" programming can reduce fetal metabolic rate by inhibiting IGF1 expression and downstream signaling pathways in all tissues and organs, further promoting the development of the "thrifty phenotype." Both of these "programmings" are primarily related to excessive maternal GC exposure caused by adverse pregnancy environments (i.e., the first strike), which increases offspring sensitivity to postnatal environmental changes such as overnutrition, high mental stress, or excessive exercise. Postnatal environmental factors act as a second strike, accelerating or exacerbating the occurrence of fetal diseases, including nonalcoholic fatty liver disease, hypercholesterolemia, osteoarthritis, osteoporosis, and glomerulosclerosis (Xia et al. 2014; Luo et al. 2015; Ni et al. 2015a; Ni et al. 2015b; Tan et al. 2016; Tie et al. 2016a; Tie et al. 2016b; Zhang et al. 2016; Qi et al. 2017; Shangguan et al. 2018; He et al. 2019; Xiao et al. 2020b; Zhao et al. 2020) (Fig. 5.4).

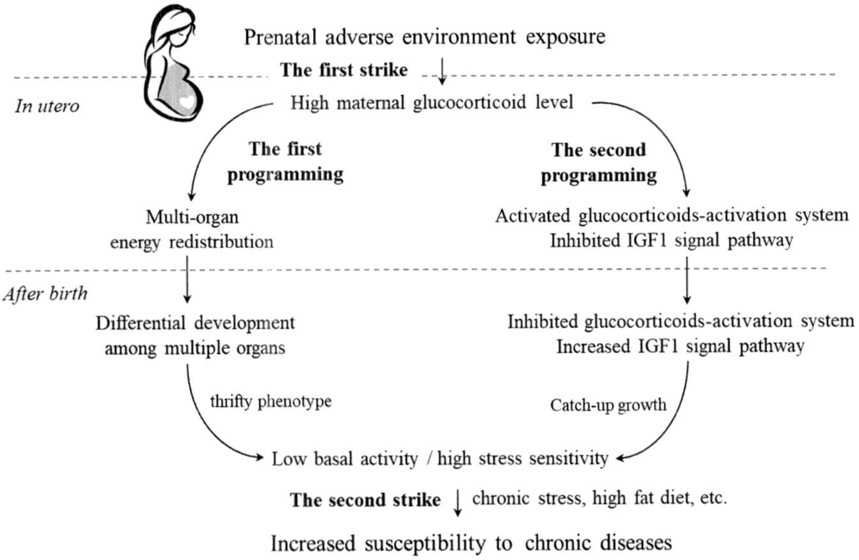

Fig. 5.4 The "two-programming and two-strike" mechanism for the occurrence of fetal-originated disease

6 Summary and Prospect

Excessive maternal GC exposure due to adverse pregnancy environments can have significant impacts on offspring growth, development, and health, including IUGR, changes in multi-organ developmental programming, and susceptibility to multiple chronic diseases in adulthood. The pathogenesis is primarily related to abnormal epigenetic modifications in offspring, which alter multi-organ development programming and endocrine axis development programming, leading to susceptibility to various diseases. As research on the intrauterine programming mechanism of fetal-originated disease advances, translational medicine is promoting the transformation of basic research findings into clinical practice or application. Translational research on fetal-originated disease aims to identify various biomarkers. The objective of this chapter is to identify the factors responsible for opening the placental GC barrier and to devise strategies to prevent its occurrence. Early diagnosis techniques based on neuroendocrine and metabolic programming changes can facilitate the prompt diagnosis and treatment of metabolic disorders of fetal origin. Additionally, early postnatal developmental plasticity can be utilized for drug intervention to partially reverse the adverse intrauterine programming changes, such as leptin intervention during the early postnatal period. Recent research by our team has demonstrated that early postnatal intervention with ACE can effectively reverse the susceptibility of adult offspring to osteoporosis caused by prenatal dexamethasone exposure (Xiao et al. 2020a). Furthermore, P-gp induction sodium ferulate can alleviate fetal weight loss caused by prenatal caffeine exposure (Ge et al. 2021). A balanced postnatal diet is essential not only to provide the necessary nutrients for rapid growth and development but also to avoid excessive catch-up growth, which may trigger the "second strike" of fetal-originated diseases after birth.

References

Ao Y, Sun Z, Hu S, et al. 2015. Toxicol Appl Pharmacol, 287(2): 128–138.
Burke C W, Roulet F. 1970. Br Med J, 1(5697): 657–659.
Cao J G, Chen Y W, Xia X, et al. 2021. Toxicol Lett, 351:78–88.
Chapman K, Holmes M, Seckl J. 2013. Physiol Rev, 93(3): 1139–1206.
Chen M, Wang T, Liao Z X, et al. 2007. Exp Toxicol Pathol, 59(3–4): 245–251.
Chen Y, Xia X, Fang M, et al. 2021. Sci Total Environ, 797: 149084.
Correia-Branco A, Keating E, Martel F. 2015. Reprod Sci, 22(2): 138–145.
Cousins L, Rigg L, Hollingsworth D, et al. 1983. Am J Obstet Gynecol, 145(4): 411–416.
Crudo A, Petropoulos S, Moisiadis V G, et al. 2012. Endocrinology, 153(7): 3269–3283.
Cuffe J S, Walton S L, Singh R R, et al. 2014. J Physiol, 592(14): 3127–3141.
Davis E P, Waffarn F, Sandman C A. 2011. Dev Psychobiol, 53(2): 175–183.
de Souza L M, Franci C R. 2013. Neurosci Lett, 534: 199–204.
Entringer S, Kumsta R, Hellhammer D H, et al. 2009. Horm Behav, 55(2): 292–298.
Fan G, Zhang Q, Wan Y, et al. 2019. Food Chem Toxicol, 128: 256–266.
Feinshtein V, Erez O, Ben-Zvi Z, et al. 2013. PeerJ, 1: e153.
Franks A L, Berry K J, DeFranco D B. 2020. J Neuroendocrinol, 32(1): e12786.
Fowden A L. 2003. Placenta, 24(8–9): 803–812.
Ge C, Xu D, Yu P, et al. 2021. BMC Med, 19(1): 311.
Giesbrecht G F, Ejaredar M, Liu J, et al. 2017. Environ Health, 16(1): 47.
Gong X, Zhang J, Ge C, et al. 2021. Pharmacol Res, 105435.
Guerrero-Bosagna C, Skinner M K. 2012. Mol Cell Endocrinol, 354(1–2): 3–8.
Guo Y, Luo H, Wu Y, et al. 2018. Reprod Toxicol, 79: 47–56.
Hales C N, Barker D J. 1992. Diabetologia, 35(7): 595–601.
Halmos T and Suba I, 2019. Orv Hetil, 160(45):1774–1783.
He H, Xiong Y, Li B, et al. 2019. Toxicol Lett, 311: 17–26.
He X, Lu J, Dong W, et al. 2017. Arch Toxicol, 91(12): 3927–3943.
He Z, Lv F, Ding Y, et al. 2017. Sci Rep, 7(1): 14825.
He Z, Zhang J, Huang H, et al. 2019. Toxicol Lett, 302: 7–17.
He Z, Zhang J, Chen G, et al. 2021. Sci Total Environ, 792: 148440.
Heard E, Martienssen R A. 2014. Cell, 157(1): 95–109.
Hellemans K G, Verma P, Yoon E, et al. 2010. Alcohol Clin Exp Res, 34(4): 633–645.
Hu S, Yi Y, Jiang T, et al. 2020. Arch Toxicol, 94(9): 3201–3215.
Hu W, Yuan C, Luo H, et al. 2020. Toxicol Lett, 331: 167–177.
Huang H, He Z, Zhu C, et al. 2015. Toxicol Appl Pharmacol, 288(1): 84–94.
Hyatt M A, Budge H, Walker D, et al. 2007. Endocrinology, 148(10): 4754–4760.
Inder W J, Jang C, Obeyesekere V R, et al. 2010. Clin Endocrinol (Oxf), 73(1): 126–132.
Jiang T, Hu S W, Dai S Y, et al. 2021. Cell Biol Toxicol, 5(21): 90–104.
Khalife N, Glover V, Taanila A, et al. 2013. PLoS One, 8(11): e81394.
Khamirchi R, Moslem A, Agah J, et al. 2020. Sci Total Environ, 713: 136622.
Khorram O, Chuang T D, Pearce W J. 2015. J Dev Orig Hlth Dis, 6(4): 342–349.
Kirsten T B, Lippi L L, Bevilacqua E, et al. 2013. PLoS One, 8(12): e82244.
Kou H, Shen L, Luo H W, et al. 2017. Reprod Toxicol, 74: 85–93.
Lesage J, Blondeau B, Grino M, et al. 2001. Endocrinology, 142(5): 1692–1702.
Li L, Hu W, Liu K, et al. 2020. Toxicol Appl Pharmacol, 395: 114979.

Liang G, Chen M, Pan X L, et al. 2011. Exp Toxicol Pathol, 63(7–8): 607–611.
Lillycrop K A, Phillips E S, Jackson A A, et al. 2005. J Nutr, 135(6): 1382–1386.
Liu L, Liu F, Kou H, et al. 2012. Toxicol Lett, 214(3): 307–313.
Liu M, Zhang Q, Pei L, et al. 2019. Epigenetics, 14(3): 245–259.
Lu J, Jiao Z, Yu Y, et al. 2018. Cell Death Dis, 9(6): 659.
Lu J, Li Q, Ma G, et al. 2020. Food Chem Toxicol, 141: 111419.
Luo H, Deng Z, Liu L, et al. 2014. Toxicol Appl Pharmacol, 274(3): 383–392.
Luo H, Li J, Cao H, et al. 2015. Sci Rep, 5: 17746.
Lv F, Fan GL, Wan Y, et al. 2021. Sci Total Environ, 789:147691.
Lye P, Bloise E, Javam M, et al. 2015. Am J Pathol, 185(6): 1666–1675.
Mesiano S, Jaffe R B. 1997. Endocr Rev, 18(3): 378–403.
Morrison J L, Botting K J, Soo P S, et al. 2012. J Pregnancy, 2012: 839656.
Nemoto T, Kakinuma Y, Shibasaki T. 2015. J Endocrinol, 224(2): 195–203.
Ni Q, Tan Y, Zhang X, et al. 2015a. Sci Rep, 5: 14711.
Ni Q, Wang L, Wu Y, et al. 2015b. Toxicol Lett, 238(2): 117–125.
Nijland M J, Mitsuya K, Li C, et al. 2010. J Physiol, 588(Pt 8): 1349–1359.
Nolan L A, Hart E J, Windle R J, et al. 2001. Endocrinology, 142(7): 2996–3005.
Nolten W E, Rueckert P A. 1981. Am J Obstet Gynecol, 139(4): 492–498.
Pan Z, Zhang X, Shangguan Y, et al. 2016. Toxicol Appl Pharmacol, 305: 234–241.
Pawluski J L, Brain U M, Underhill C M, et al. 2012. Psychoneuroendocrinology, 37(7): 1019–1028.
Pei L G, Zhang Q, Yuan C, et al. 2019. J Endocrinol, 242(1): M17-M32.
Qi Y, Luo H, Hu S, et al. 2017. Cell Physiol Biochem, 44(2): 657–670.
Reynolds R M. 2013. Psychoneuroendocrinology, 38(1): 1–11.
Salvante K G, Milano K, Kliman H J, et al. 2017. J Dev Orig Hlth Dis, 8(2): 149–154.
Schopper H, Palme R, Ruf T, et al. 2012. Gen Comp Endocrinol, 176(1): 18–27.
Seckl J R, Chapman K E. 1997. Eur J Biochem, 249(2): 361–364.
Shangguan Y, Wen Y, Tan Y, et al. 2018. Am J Pathol, 188(12): 2863–2876.
Shen L, Liu Z, Gong J, et al. 2014. Toxicol Appl Pharmacol, 274(2): 263–273.
Skinner M K, Manikkam M, Guerrero-Bosagna C. 2011. Reprod Toxicol, 31(3): 337–343.
Staiger J, Lueben M J, Berrigan D, et al. 2009. Carcinogenesis, 30(5): 832–840.
Szyf M, McGowan P, Meaney M J. 2008. Environ Mol Mutagen, 49(1): 46–60.
Tan Y, Wu Y, Ni Q, et al. 2016. Br J Nutr, 116(8): 1346–1355.
Tie K, Tan Y, Deng Y, et al. 2016a. Reprod Toxicol, 60: 11–20.
Tie K, Zhang X, Tan Y, et al. 2016b. FASEB J, 30(2): 785–797.
Vaag A A, Grunnet L G, Arora G P, et al. 2012. Diabetologia, 55(8): 2085–2088.
Vogel J P, Souza J P, Gulmezoglu A M, et al. 2014. Lancet, 384(9957): 1869–1877.
Wang L, Shen L, Ping J, et al. 2014. Toxicol Lett, 224(3): 311–318.
Weaver I C, Cervoni N, Champagne F A, et al. 2004. Nat Neurosci, 7(8): 847–854.
Wen Y, Shangguan Y, Pan Z, et al. 2019. Toxicol Appl Pharmacol, 363: 1–10.
Wu Z, Pan Z, Wen Y, et al. 2020. Food Chem Toxicol, 136: 111083.
Xia L P, Shen L, Kou H, et al. 2014. Toxicol Lett, 226(1): 98–105.
Xiao H, Wen Y, Pan Z, et al. 2018. Cell Death Dis, 9(6):638.
Xiao H, Wen Y, Pan Z, et al. 2019. FASEB J, 33(11): 12972–12982.
Xiao H, Wen Y, Wu Z, et al. 2020a. Bone, 141: 115578.
Xiao H, Wu Z, Li B, et al. 2020b. Br J Pharmacol, 177(20): 4683–4700.
Xu D, Wu Y, Liu F, et al. 2012a. Toxicol Appl Pharmacol, 264(3): 395–403.
Xu D, Zhang B, Liang G, et al. 2012b. PLoS One, 7(9): e44497.
Xu D, Xia L P, Shen L, et al. 2013. Acta Pharmacol Sin, 34(12): 1526–1534.
Xu D, Zhang C, He X, et al. 2018. Toxicol Lett, 283: 39–51.
Yang Z, Guo C, Zhu P, et al. 2007. J Endocrinol, 195(2): 241–253.
Yu L, Zhou J, Zhang G, et al. 2018. Toxicol Appl Pharmacol, 352: 77–86.
Zeng C, Wang R, Li D, et al. 2010. Hepatology, 52(5): 1702–1712.

Zhang L, Shen L, Xu D, et al. 2016. Reprod Toxicol, 65: 236–247.
Zhao Z, Qin J, Pei L, et al. 2020. Toxicology, 442: 152533.
Zheng S, Rollet M, Pan Y X. 2011. Epigenetics, 6(2): 161–170.
Zhu Y, Zuo N, Li B, et al. 2018. Toxicology, 400–401: 9–19.
Zhu Y, Chen H, Zhao X, et al. 2019. Toxicol Lett, 314: 63–74.

Chapter 6
Gut Microbiota Mechanism of Fetal-Originated Disease

Hui Wang, Xiaoqian Lu, and Chi Ma

Abstract The intestinal microbiota engages in an intricate interplay with the host, establishing reciprocal communication pathways between the microbial community and various organs, including the liver, brain, and bones. These communication corridors are known as physiological axes, encompassing the gut-liver axis, gut-brain axis, and gut-bone axis. Unfavorable conditions during gestation can induce alterations in the composition of the progeny's gut microbiota by modulating the physiological colonization process, thereby augmenting the risk of chronic disease manifestation later in life. This article delivers a comprehensive overview of the physiological regulation and colonization dynamics of the gut microbiota; elucidates the underlying mechanisms regulating the gut-liver, gut-brain, and gut-bone axes; and underscores their intimate correlations with susceptibility to a myriad of diseases originating in fetal life. The overarching goal is to investigate the potential of the gut microbiota as a prognostic marker for susceptibility to diseases of fetal origin and to furnish novel insights for the efficacious prevention and management of such maladies.

1 Introduction

The gut microbiota, a vast and intricate consortium of microbes dwelling in the gastrointestinal tract, serves a pivotal role in preserving the equilibrium of the gut microenvironment. This microbiota and its metabolites establish physiological conduits linking the intestinal tract with other organs, such as the liver, brain, and bone,

H. Wang (✉) · X. Lu
Department of Pharmacology, School of Basic Medical Science, Wuhan University, Wuhan, Hubei, China
e-mail: wanghui19@whu.edu.cn

C. Ma
Division of Joint Surgery and Sports Medicine, Department of Orthopedic Surgery, Zhongnan Hospital of Wuhan University, Wuhan, Hubei, China

H. Wang et al. (eds.), *Fetal Origin of Diseases*,
https://doi.org/10.1007/978-981-97-5730-5_6

thereby constituting a series of distinctive physiological axes—the gut-liver, gut-brain, and gut-bone axes. A salubrious prenatal milieu ensures the normal colonization of gut microbiota and the proper development of the offspring's multiple organs. However, exposure to adverse conditions during pregnancy can precipitate alterations in the composition of the offspring's gut microbiota. The Developmental Origins of Health and Disease (DOHaD) hypothesis posits that fetal developmental aberrations, instigated by adverse intrauterine environments, lead to congenital anomalies and chronic diseases in adulthood, a phenomenon referred to as fetal-originated disease. The early-life modifications in gut microbiota, along with their associated metabolites, and how they mediate the susceptibility and emergence of "fetal-origin diseases" in offspring, warrant meticulous exploration. This domain also represents a significant area of research. This article concentrates on the physiological colonization and pathological shifts in the gut microbiota composition, offering a comprehensive overview of the correlation between alterations in the gut-liver axis, gut-brain axis, and gut-bone axis, and susceptibility to fetal-originated diseases.

2 The Colonization Process of the Gut Microbiota and Its Determining Influences

The concept of gut microbiota colonization pertains to the process wherein microbial communities establish residence, proliferate, and flourish within the gastrointestinal tract. The colonization of gut microbiota in neonates predominantly transpires at the moment of birth, encompassing both vertical (maternal) and horizontal (environmental) modes of transmission (Milani et al. 2017; Moore and Townsend 2019). The gut microbiota during early life accomplishes its colonization and turnover amidst labor, breastfeeding, and weaning, achieving a relatively stable adult-like microbial architecture by the age of three. Consequently, the microbiota's immune and metabolic functionalities also gravitate toward stabilization.

2.1 The Enterotypes of the Gut Microbiota

Enterotypes have been characterized as "concentrated clusters within a multidimensional compositional landscape," exhibiting independence from age, gender, cultural heritage, or geographical location (Arumugam et al. 2011). Each enterotype is associated with a particular ensemble of dominant taxa. In an extensive analysis of 13,776 fecal microbiota samples from 1956 infants within their first three years of life across 17 countries, Zhao et al. delineated four enterotypes dominated by the genera *Firmicutes*, *Bifidobacterium*, *Bacteroides*, and *Prevotella* (Xiao et al. 2021). The *Firmicutes* and *Bifidobacterium* enterotypes correspond to nascent developmental stages marked by less stable community structures and diminished microbial maturity, whereas the *Bacteroides* and *Prevotella* enterotypes are associated with later stages, typified by heightened microbial diversity and robust community interactions. Ecological modeling and functional analyses suggest that the transition trajectory of gut microbiota from infancy to adulthood is discernible and forecastable. The *Firmicutes* and *Bifidobacterium* enterotypes progressively transition toward the Bacteroides and *Prevotella* enterotypes as individuals mature, accompanied by alterations in metabolic pathways. Despite the challenges in defining infant gut microbiota enterotypes and establishing their pivotal points, the concept of enterotypes enhances our comprehension of developmental shifts in gut microbiota composition.

2.2 The Colonization Process of the Gut Microbiota

The traditional view posits that the fetus resides in a sterile milieu, acquiring its inaugural microbiota through exposure to the birth canal during childbirth. Nevertheless, recent investigations have discovered bacterial presence within the placenta, amniotic fluid, and fetal meconium, alongside the detection of bacterial metabolites like acetates and propionates in fetal meconium specimens (Stinson et al. 2019). Consequently, the inception of neonatal gut microbiota colonization may occur in utero. The exploration of the human gut microbiota's genesis is an ongoing endeavor, leaving ample space for discourse and validation of findings derived from clinical studies and animal models. The encounter during childbirth is deemed a critical catalyst for the early colonization of neonatal gut microbiota. Infants born through natural childbirth interact with maternal vaginal, fecal, and skin microbiota, with the dominant colonizers being *Streptococcaceae*, *Lactobacillaceae*, and *Clostridiaceae*, accompanied by a significant presence of *Bifidobacterium* species (Jakobsson et al. 2014). Conversely, infants delivered via cesarean section are more prone to colonization by bacteria from the maternal skin, hospital staff, or the hospital environment, potentially including pathogenic bacteria such as *Enterococcus*, *Enterobacteriaceae*, and *Klebsiella* (Kim et al. 2019). During the initial months, infants predominantly derive their nutrition from either breast milk or formula. Upon ingesting breast milk, the resident bacteria can be transferred to the infant's intestinal tract, with *Bifidobacteria*, *Lactobacilli*, and other lactic acid metabolism-friendly bacteria emerging as the dominant strains (Milani et al. 2017). Infants who consume formula milk display elevated levels of *Clostridium* and *Bacteroides* species, and their gut microbiota maturity is comparatively diminished at ages 12 to 24 months (Bokulich et al. 2016). Postweaning, the infant's gut microbiota primarily originates from their diet and environmental factors. The weaning phase represents a transition in energy sources and dietary structure for infants. As solid foods and supplementary feeding are introduced, the proportion of breast milk in the diet diminishes. This results in a gradual decrease in the relative abundance of Bifidobacterium, countered by an increase in bacteria associated with carbohydrate utilization, amino acid metabolism, and vitamin biosynthesis, such as *Bacteroides*,

Streptococcus, and *Clostridium*. Following weaning, the dominant genus in the offspring's fecal microbiota becomes *Prevotella*, while the abundance of *Bifidobacterium* recedes (Khine and Rahayu 2020).

2.3 The Influencing Factors of Gut Microbiota Colonization

The method of delivery exerts a significant influence on the process of gut microbiota colonization in early life. The gut microbiota of infants delivered vaginally mirrors the maternal vaginal microbiota closely, whereas those birthed via cesarean section display a resemblance to the microbiota of the maternal skin surface (Milani et al. 2017). Gestational age also plays a pivotal role in shaping the colonization of early-life gut microbiota. The gut microbiota of full-term infants is predominantly colonized by *Enterococcus* and *Staphylococcus*, while preterm infants tend to harbor dominant groups such as *Bacillus* (Moles et al. 2013). The colonization of the offspring's gut microbiota is further swayed by factors including feeding patterns, living environments, and medical history. Breast milk is rich in nutrients, such as human milk oligosaccharides, which foster the colonization of beneficial bacteria like *Bifidobacterium* (Milani et al. 2017). In comparison to infants fed with formula, those breastfed exhibit elevated levels of *Bifidobacterium* and *Lactobacillus* in their gut, coupled with higher concentrations of short-chain fatty acids (SCFAs) (Bridgman et al. 2017). The infant's living environment also contributes significantly to the colonization of gut microbiota. Research indicates a marked distinction in the gut microbiota of children residing in rural areas versus those in urban settings, with a higher prevalence of beneficial bacteria capable of fermenting SCFAs observed in the gut of rural children (De Filippo et al. 2017). A study by Atara et al. suggests that the administration of antibiotics early in life can substantially alter the composition of gut microbiota, especially if the antibiotics are introduced within the first two years of life. This can result in a decline in beneficial bacteria like *Bifidobacterium* and impact postnatal development (Uzan-Yulzari et al. 2021). It is noteworthy that early-life antibiotic use can affect the composition of gut microbiota, thereby heightening the risk of metabolic syndrome, inflammatory bowel disease, asthma, and allergic diseases in adulthood (Bokulich et al. 2016). In sum, the colonization of early-life gut microbiota is subject to a multitude of influencing factors (Fig. 6.1).

2.4 The Relationship between the Gut Microbiota and Multi-Organ Physiological Regulation

The multifaceted roles of gut microbiota and its association with alterations in the liver, nervous system, and bone-related diseases have emerged as a significant area of research. The intricate anatomical and functional interplay between the gut and

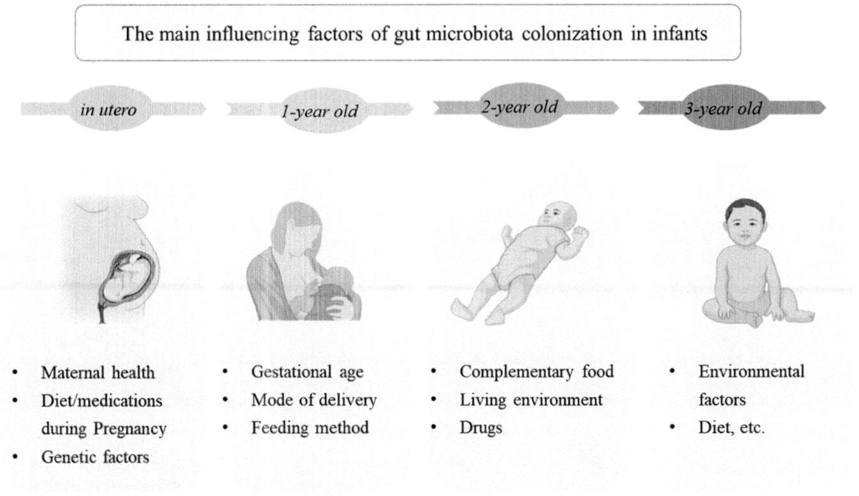

Fig. 6.1 The main influencing factors of the gut microbiota colonization in infants

liver is well-documented, connected via the portal vein circulation, with 70%–75% of the liver's total blood supply originating from the portal vein. Antigens and metabolic byproducts from the gut microbiota can be transported to the liver through the portal vein, establishing the "gut-liver axis" (Siebler et al. 2008). The "gut-brain axis" functions as a sophisticated communication network linking the gut and the brain. The gut microbiota can interact with the brain via numerous pathways, encompassing the nervous system, immune system, tryptophan metabolism, and microbial metabolites such as SCFAs, branched-chain amino acids, and peptidoglycans, thereby influencing brain function and behavior (Morais 2021). The "gut-bone axis" signifies the network connecting gut microbiota and bone health, involving microbial metabolites and immune regulation. The dialogue between the gut and bone is predicated on gut-derived metabolites such as serotonin and SCFAs and systemic inflammation induced by lipopolysaccharide (LPS) (Villa et al. 2017).

3 Abnormalities in the Gut Microbiota Composition and Susceptibility to Fetal-Originated Disease

Research has suggested that abnormal colonization of gut microbiota can result in an imbalance in the gut microbiota composition, subsequently impacting the progression of a variety of diseases originating from fetal development through the gut-liver, gut-brain, and gut-bone axes. These diseases encompass metabolic disorders, neurodevelopmental disorders, and osteochondral diseases (Fig. 6.2).

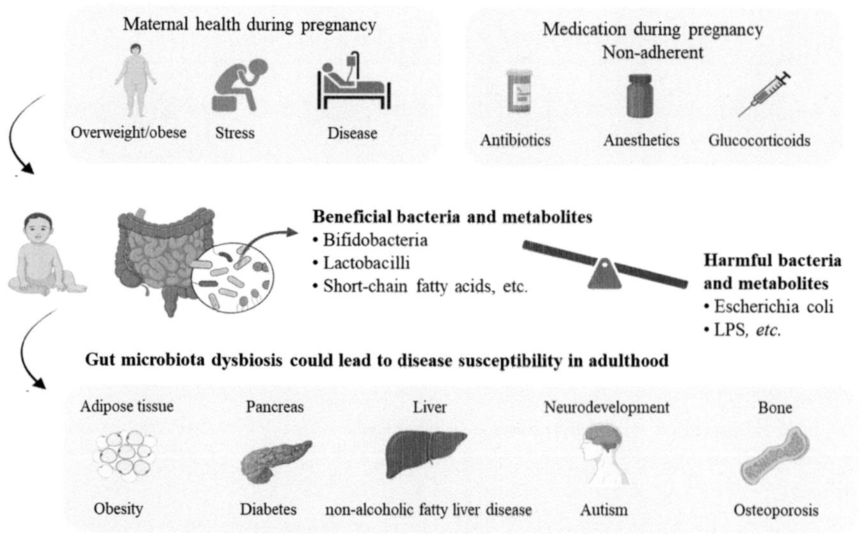

Fig. 6.2 Abnormal gut microbiota colonization and susceptibility to fetal-originated disease

3.1 Gut-Liver Axis Imbalance and Fetal-Originated Metabolic Diseases

During early life, the colonization of gut microbiota interplays with the body's endocrine, metabolic, and immune development. Disruptions to the gut microbiota in this critical period may have enduring effects, precipitating metabolic and immune abnormalities in the body, which ultimately contribute to the onset and progression of metabolic disorders in offspring, such as obesity, diabetes, and non-alcoholic fatty liver disease (NAFLD). These diseases, influenced by early-life metabolic programming, can be mitigated or even prevented by proactive avoidance of risk factors and early intervention (Dattilo et al. 2012; Giorgio et al. 2013). Gut microbiota may contribute to obesity in children via mechanisms such as increased energy intake, modulation of gut hormone secretion, chronic systemic inflammation, and insulin resistance. Animal studies have shown that cesarean sections elevate the risk of obesity in offspring rats, and it has been suggested that maternal obesity can enhance offspring's susceptibility to obesity through microbiota transmission (Cho et al. 2021). A plethora of population-based and animal studies have highlighted a connection between early-life antibiotic exposure and subsequent obesity, a consequence of alterations in gut microbiota diversity (Vallianou et al. 2021). For example, a study by Cox et al. (2014) demonstrated that early exposure to low-dose penicillin temporarily disrupts gut microbiota in mice, influencing host metabolism and fostering obesity development. Although the study found that gut microbiota can recover following the cessation of antibiotics, the aberrant metabolic phenotype persists. This suggests that modifying gut microbiota composition during pivotal developmental periods results in enduring metabolic consequences, underscoring the importance of early-life microbiota in both development and metabolic homeostasis.

Maternal obesity and diabetes are recognized as prevalent factors contributing to childhood obesity and other metabolic disorders such as NAFLD (Poston et al. 2011; Weng et al. 2013). Alterations in the gut microbiota composition of infants born to obese mothers can precipitate increased gut permeability, diminished macrophage phagocytic activity, and bacterial translocation to the liver, thereby sparking liver inflammation and initiating NAFLD (Soderborg et al. 2018). Transferring the gut microbiota from two-week-old offspring born to obese pregnant mice to germ-free mice results in escalated levels of blood cytokines, amplified endoplasmic reticulum stress in the liver, and heightened susceptibility to NAFLD (Soderborg et al. 2018). Another potential conduit linking maternal obesity and diabetes to offspring's susceptibility to NAFLD is the influence of maternal metabolic status on the early microbial colonization in offspring. Research has further unveiled that children with NAFLD display dysregulated gut microbiota and abnormal fecal bile acid levels. The dysregulation of the gut microbiota impacts the metabolism of primary bile acids into secondary bile acids through enzymes such as BSH, thereby reducing the concentration of secondary bile acids in feces (Yu et al. 2021). Taken together, these findings suggest the presence of gut microbiota regulatory mechanisms in fetal-originated NAFLD.

Diabetes is classified into type 1 diabetes (T1D) and type 2 diabetes (T2D). T1D is an autoimmune disorder targeting pancreatic beta cells, and its onset involves a complex interplay of genetic and environmental factors. The immune systems of genetically susceptible children can be incited by certain triggers, leading to the production of islet autoantibodies (IA) against pancreatic beta cells, significantly amplifying their predisposition to T1D. The gut microbiota is one such trigger that has been widely spotlighted (Zhang et al. 2022). Population studies have discovered that infants delivered via cesarean section bear a higher risk of T1D, and their gut microbiota is characterized by diminished α diversity, augmented *Bacteroides* abundance, and decreased SCFAs levels (Cardwell et al. 2008). Children with heightened susceptibility to T1D due to environmental exposure to antibiotics and novel nicotinic drugs exhibit reduced levels of butyric acid-producing bacteria and an increased abundance of potential pathogenic bacteria like *Escherichia* and *Shigella* in their gut microbiota (Xu et al. 2022). The functional association between the gut microbiota and the autoimmune progression of pancreatic islet cells offers significant insights into the role of the gut microbiota in the pathogenesis of T1D (Endesfelder et al. 2016). In the future, gut microbiota may potentially serve as biomarkers for early diagnosis of T1D (Zhou et al. 2021).

3.2 Gut-Brain Axis Imbalance and Fetal-Originated Neuropsychiatric Disorders

The succession of neonatal gut microbiota coincides with a critical period for neurodevelopment (Sharon et al. 2016). A growing body of evidence supports the concept that an adverse intrauterine environment can enhance the risk of low birth weight and susceptibility to neuropsychiatric disorders in offspring, such as Autism Spectrum Disorder (ASD) and depression (Gage et al. 2016). Notably, up to 90% of ASD patients exhibit gastrointestinal dysfunction (e.g., bloating, abdominal pain, alternating diarrhea, and constipation), and the severity of these symptoms exacerbates the core ASD symptoms (Bjørklund et al. 2020). ASD patients significantly diverge from healthy individuals in terms of their gut microbiota composition (Gondalia et al. 2012). Parallel findings have been observed in patients with depression, where shifts in the gut microbiota composition were recognized (Winter et al. 2018). These studies suggest a strong correlation between alterations in the gut microbiota composition and the onset of ASD and depression, indicating that regulation of the gut-brain axis may be a pivotal aspect in studying the pathogenesis and prevention of fetal-originated neuropsychiatric disorders.

Epidemiological studies imply that the onset of ASD is linked with adverse maternal exposures during gestation, encompassing exogenous substances, nutritional imbalances, stress, infections, and other detrimental environmental factors (Keil et al. 2010; Class et al. 2014). Our prior research also corroborated that adverse prenatal environments can precipitate intrauterine growth retardation (IUGR) in offspring (He et al. 2016). IUGR exhibits a significant correlation with the risk of ASD (Talmi et al. 2020), further underscoring its fetal origins. A proposed connection exists between ASD and gut microbiota (Geir et al. 2020), as up to 70% of ASD children endure gastrointestinal dysfunction (Valicenti-McDermott et al. 2006). The severity of gastrointestinal symptoms is proportional to the severity of ASD, suggesting that gut microbiota plays a role in ASD etiology (Liu et al. 2019). During critical phases of brain development, any deviations in normal physiological functioning can lead to improper neural network formation, subsequently resulting in neurodevelopmental disorders like ASD (Chagas et al. 2020). Microglia assume a central role in synaptic remodeling and the formation of neural circuits. The gut microbiota significantly modulates the maturation and function of microglia, with alterations in the gut microbiota leading to the expression of genes associated with immature states in microglia (Erny et al. 2015; Chu et al. 2019). The colonization of gut microbiota exhibits heritability and can be vertically transmitted from mothers, potentially spanning up to 10 generations (Moeller et al. 2018). The metabolite short-chain fatty acid (SCFA) originating from the maternal gut microbiota can permeate the placental barrier, influencing the development of embryonic sympathetic nerves and intestinal epithelium, thereby impacting the susceptibility of offspring to neurodevelopmental disorders in adulthood (Kimura et al. 2020). Animal studies have also suggested that dysregulation of maternal gut microbiota can precipitate abnormal neurodevelopment in offspring, culminating in ASD-like behaviors (Kim et al. 2017). These findings suggest that the maternal gut microbiota can influence offspring neurodevelopment or trigger the onset of fetal-originated neuropsychiatric disorders.

3.3 Gut-Bone Axis Imbalance and Fetal-Originated Osteochondral-Related Diseases

Our previous research revealed that osteochondral maladies such as osteoporosis (OP) and osteoarthritis are rooted in developmental origins (Deng et al. 2013; Xiao et al. 2020). A plethora of studies have corroborated the notion that the gut microbiota exerts a profound influence on bone health and functionality. OP is a systemic metabolic anomaly characterized by an attenuation in bone mass per unit volume, a disruption of the bone tissue microstructure, and an escalation in skeletal fragility. It poses a formidable public health challenge on a global scale. Sjögren et al. uncovered a significant correlation between gut microbiota and skeletal growth (Sjögren et al. 2012). Germ-free mice manifested an approximately 40% augmentation in bone density in the distal femur, ascribed to a diminution in osteoclasts, culminating in decreased bone resorption and loss (Sjögren et al. 2012). The gut microbiota can modulate OP and bone resorption by influencing the activity and functionality of osteoblasts and osteoclasts (Zaiss et al. 2019). Clinical investigations centered on maternal habits, and the bone mineral content of their newborns have unveiled that maternal smoking, alcohol consumption, or insufficient energy intake during gestation are risk factors that impinge on newborn bone mineral content (Godfrey et al. 2001). Moreover, maternal gut microbiota dysregulation, precipitated by antibiotic usage during pregnancy, can hinder embryonic bone growth (You et al. 2020). Consequently, gut microbiota dysbiosis in early life contributes to the genesis of fetal-originated OP (Li et al. Li 2020; Xu et al. 2020).

Metabolites originating from the gut microbiota primarily orchestrate the interaction between the gut and bone. For instance, bone cells express the serotonin receptor, and aberrant serotonin levels emanating from the gut microbiota can detrimentally affect bone formation in mice (Yadav et al. 2008). Studies have demonstrated that perinatal dietary supplementation with serotonin synthesis inhibitors can compromise infantile bone development (Weaver et al. 2018). In animal models subjected to ovariectomy, the inhibition of serotonin production could forestall bone loss (Cui et al. 2011). Augmented levels of LPS, a consequence of deleterious maternal lifestyle habits (such as smoking and alcohol consumption) during gestation, can impinge upon neonatal bone mass (Godfrey et al. 2001). Investigations involving antibiotic-treated and germ-free mice proposed that intermittent parathyroid hormone treatment kindles bone formation, a process implicating gut microbiota and butyrate (Li et al. 2020). The composition of the gut microbiota could potentially prognosticate the risk of OP, and modulating the gut microbiota or supplementing microbiota metabolites could emerge as innovative targets for the treatment of osteochondral disorders. Consequently, prioritizing a safe gestational period and ensuring a normal gut microbiota colonization in offspring post-birth are instrumental in averting osteochondral disorders in adulthood.

4 Preventive and Therapeutic Strategies for Fetal-Originated Disease Based on Abnormal Gut Microbiota Colonization

4.1 Supplementing with Probiotics

Research has ascertained that the colonization of gut microbiota in premature infants markedly lags behind that of full-term infants. For premature infants, the oral administration of a probiotic concoction comprising *Bifidobacterium* and *Lactobacillus* has proven to align their gut microbiota composition more closely with that of healthy full-term infants (Alcon-Giner et al. 2020). Consequently, for preterm and low-birth-weight infants, early administration of microbial formulations could foster a beneficial microbial milieu and curtail the colonization of pathogenic bacteria via niche occupation. There are currently probiotics available in the market for the treatment of metabolic disorders. *Bifidobacterium* TMC3115 can enhance the ratio of *Bifidobacterium* to *Clostridium* and diminish plasma total cholesterol and low-density lipoprotein cholesterol levels, thereby conferring a benefit to lipid metabolism. Studies also suggest that supplementation of probiotic capsules containing *Lactobacillus acidophilus*, *Lactobacillus casei*, and *Bifidobacterium* in diabetic patients ameliorates blood glucose control, inflammatory biomarkers, and oxidative stress following a 12-week treatment regimen. The gut microbiota status of mothers delineates the risk of neurodevelopmental disorders like ASD in their progeny (Lammert et al. 2018). Clinical investigations have underscored that probiotic intervention can ameliorate symptoms associated with depression (Liu et al. 2019). Animal studies have illustrated that prenatal probiotic supplementation could attenuate the incidence of ASD in progeny (Wang et al. 2019), and prenatal vitamin supplementation can enhance ASD-like behavior resulting from maternal exposure to environmental contaminants (Wang et al. 2019). Concurrently, probiotic supplementation can augment bone mass and stimulate bone development (Sadeghi 2014). Zebrafish larvae, when supplemented with *Lactobacillus rhamnosus*, displayed increased skeletal mineralization (Avella et al. 2012). Newborn zebrafish exposed to *Bacillus subtilis* and *Lactococcus lactis* could instigate an upregulation of osteogenic marker genes such as Runx2 and Sp7, concomitant with a significant increase in bone mass (Godfrey et al. 2001). These findings collectively suggest that probiotic supplementation can foster host growth and development, exert a positive influence on maintaining metabolic homeostasis, and ensure neural and skeletal development. There is an escalating acknowledgement of the role of probiotic therapy, and as an increasing number of novel strains are being discovered, their safety and efficacy warrant comprehensive investigation.

4.2 Supplementing with Microbiota-Related Metabolic Products

Specific dietary regimens can be metabolized by the gut microbiota to engender SCFAs and other metabolites that are instrumental for maintaining intestinal epithelial integrity. Research has established that the inclusion of cowpeas in infant complementary foods can serve as a remedy for growth retardation (Stephenson et al. 2017). Animal studies have revealed that supplementation of sodium butyrate in drinking water can mitigate liver inflammation and fat accumulation in mice with NAFLD (Zhou et al. 2017). In a pregnant mouse obesity model induced by a high-fat/high-sucrose diet, supplementing the maternal diet with prebiotic oligofructose could enhance glucose tolerance, insulin sensitivity, and reduce hepatic steatosis in offspring following prolonged high-fat feeding. These improvements were tethered to the enhanced early-life gut microbiota resulting from maternal prebiotic intake (Paul et al. 2019). Microbiota-derived SCFAs have been demonstrated to play a pivotal role in fostering the maturation and function of microglia, and supplementation of SCFAs in mice treated with antibiotics and germ-free mice can ameliorate abnormal microglial development (Boylan et al. 2008). Vitamin B12 functions as an ecological regulator of gut microbiota, and prenatal supplementation of vitamin B12 has been suggested as an efficacious method to diminish offspring susceptibility to ASD (Degnan et al. 2014). Moreover, studies have discovered that prenatal and postnatal vitamin D supplementation effectively manages blood LPS levels and enhances the trabecular microarchitecture of the lumbar spine and femur in male offspring fed a high-fat/high-sucrose diet (Villa et al. 2017).

4.3 Other Microbiota-Related Intervention Therapies

The microbiota in early life exerts a long-term influence on health, thereby necessitating preventive interventions to restore gut microbiota equilibrium. Research suggests that a procedure involving the insertion of gauze into the maternal vagina one hour prior to delivery, followed by using the same gauze to swab the skin, mouth, and anus of newborns delivered via cesarean section, can bolster the presence of *Lactobacilli* or *Bacteroides* in their skin, oral, and anal microbiota. This intervention can partially rectify the aberrant microbiota of infants delivered through cesarean section to a state akin to that of newborns born through natural delivery, indicating the feasibility of transferring vaginal microbiota to newborns delivered *via* cesarean section (Dominguez-Bello et al. 2016). The study also revealed that transplanting microbiota from lean individuals into obese males with metabolic syndrome ameliorated their glucose metabolism (Kootte et al. 2017). However, the safety and standardization of microbiota transplantation are currently subjects of rigorous investigation.

5 Summary and Prospect

In conclusion, there exists a multifaceted interaction between the host and gut microbiota, with the latter having the capacity to impact the functionality of various organs such as the liver, brain, and bones through diverse mechanisms. Alterations in the composition of gut microbiota may heighten susceptibility to fetal-originated disease via the gut-liver, gut-brain, and gut-bone axes. A salubrious pregnancy environment is paramount for facilitating the smooth colonization of gut microbiota in offspring, which is also of immense significance in mitigating the risk of fetal diseases. With the ongoing advancement of human microbiome research, our comprehension of gut microbiota has been enhanced, and novel methods for predicting microbiota-related diseases have been developed. Consequently, the identification and isolation of specific differential bacterial strains or metabolites in early life, and the development of personalized bacterial therapies predicated on various pathological features, could furnish new strategies for the early prevention and treatment of fetal-originated disease.

References

Alcon-Giner C, Dalby M J, Caim S, et al. 2020. Cell Rep Med, 1(5): 100077.
Arumugam M, Raes J, Pelletier E, et al. 2011. Nature, 473(7346): 174-180.
Avella M A, Place A, Du S J, et al. 2012. Plos One, 7(9): e45572.
Bjørklund G, Pivina L, Dadar M, et al. 2020. Neurosci Biobehav Rev, 118: 111-120.
Bokulich N A, Chung J, Battaglia T, et al. 2016. Sci Transl Med, 8(343): 343ra382.
Boylan S M, Cade J E, Kirk S F, et al. 2008. Br J Nutr, 100(4): 875-882.
Bridgman S L, Azad M B, Field C J, et al. 2017. Front Nutr, 4: 11.
Cardwell C R, Stene L C, Joner G, et al. 2008. Diabetologia, 51(5): 726-735.
Chagas L D S, Sandre P C, Ribeiro E R N C A, et al. 2020. Int J Mol Sci, 21(6): 2111.
Cho N A, Sales K M, Sampsell K, et al. 2021. Obesity (Silver Spring), 29(10):1664-1675.
Chu C, Murdock M H, Jing D, et al. 2019. Nature, 574(7779): 543-548.
Class Q A, Abel K M, Khashan A S, et al. 2014. Psychol Med, 44(1): 71-84.
Cox L M, Yamanishi S, Sohn J, et al. 2014. Cell, 158(4): 705-721.
Cui Y, Niziolek P J, MacDonald B T, et al. 2011. Nat Med, 17(6): 684-691.
Dattilo A M, Birch L, Krebs N F, et al. 2012. J Obes, 2012: 123023.
De Filippo C, Di Paola M, Ramazzotti M, et al. 2017. Front Microbiol, 8: 1979.
Degnan P H, Taga M E, Goodman A L. 2014. Cell Metab, 20(5): 769-778.
Deng Y, Cao H, Cu F, et al. 2013. Toxicol Appl Pharmacol, 269(1): 25-33.
Dominguez-Bello M G, De Jesus-Laboy K M, Shen N, et al. 2016. Nat Med, 22(3): 250-253.
Endesfelder D, Engel M, Davis-Richardson A G, et al. 2016. Microbiome, 4: 17.
Erny D, Hrabe de Angelis A L, Jaitin D, et al. 2015. Nat Neurosci, 18(7): 965-977.
Gage S H, Munafò M R, Davey Smith G. 2016. Annu Rev Psychol, 67: 567-585.
Geir B, Lyudmila P, Maryam D, et al. 2020. Neurosci Biobehav Rev, 118:111-120.
Giorgio V, Prono F, Graziano F, et al. 2013. BMC Pediatr, 13: 40.
Godfrey K, Walker-Bone K, Robinson S, et al. 2001. J Bone Mine Res, 16(9): 1694-1703.
Gondalia S V, Palombo E A, Knowles S R, et al. 2012. Autism Res, 5(6): 419-427.
He Z, Zhu C, Huang H, et al. 2016. Toxicol Res (Camb), 5(2): 388-398.
Jakobsson H E, Abrahamsson T R, Jenmalm M C, et al. 2014. Gut, 63(4): 559-566.

Keil A, Daniels J L, Forssen U, et al. 2010. Epidemiology, 21(6): 805-808.
Khine W W T, Rahayu E S. 2020. Gut Microbes, 12(1):1761240.
Kim H, Sitarik A R, Woodcroft K, et al. 2019. Curr Allergy Asthma Rep, 19(4): 22.
Kim S, Kim H, Yim Y S, et al. 2017. Nature, 549(7673): 528-532.
Kimura I, Miyamoto J, Ohue-Kitano R, et al. 2020. Science, 367(6481).
Kootte R S, Levin E, Salojärvi J, et al. 2017. Cell Metab, 26(4): 611-619.e616.
Lammert C R, Frost E L, Bolte A C, et al. 2018. J Immunol, 201(3): 845-850.
Li J Y, Yu M, Pal S, et al. 2020. J Clin Invest, 130(4): 1767-1781.
Li S, Mao Y, Zhou F, et al. 2020. Bone & joint research, 9(8): 524-530.
Liu R T, Walsh R F L, Sheehan A E. 2019. Neurosci Biobehav Rev, 102: 13-23.
Liu S, Li E, Sun Z, et al. 2019. Sci Rep, 9(1): 287.
Milani C, Duranti S, Bottacini F, et al. 2017. Microbiol Mol Biol Rev, 81(4):e00036-17.
Moeller A H, Suzuki T A, Phifer-Rixey M, et al. 2018. Science, 362(6413): 453-457.
Moles L, Gomez M, Heilig H, et al. 2013. PLoS One, 8(6): e66986.
Moore R E, Townsend S D. 2019. Open Biol, 9(9): 190128.
Morais L H, Schreiber H L t, Mazmanian S K. 2021. Nat Rev Microbiol, 19(4): 241-255.
Paul H A, Collins K H, Nicolucci A C, et al. 2019. Faseb j, 33(4): 5153-5167.
Poston L, Harthoorn L F, Van Der Beek E M. 2011. Pediatr Res, 69(2): 175-180.
Sadeghi A A. 2014. Probiotics Antimicrob Proteins, 6(3–4): 136–140.
Sharon G, Sampson T R, Geschwind D H, et al. 2016. Cell, 167(4): 915–932.
Siebler J, Galle P R, Weber M M. 2008. J Hepatol, 48(6): 1032–1034.
Sjögren K, Engdahl C, Henning P, et al. 2012. J Bone Miner Res, 27(6): 1357–1367.
Soderborg T K, Clark S E, Mulligan C E, et al. 2018. Nat Commun, 9(1): 4462.
Stephenson K B, Agapova S E, Divala O, et al. 2017. Am J Clin Nutr, 106(6): 1500–1507.
Stinson L F, Boyce M C, Payne M S, et al. 2019. Front Microbiol, 10: 1124.
Talmi Z, Mankuta D, Raz R. 2020. Autism Res, 13(4): 655–665.
Uzan-Yulzari A, Turta O, Belogolovski A, et al. 2021. Nat Commun, 12(1): 443.
Valicenti-McDermott M, McVicar K, Rapin I, et al. 2006. J Dev Behav Pediatr, 27(2 Suppl): S128–136.
Vallianou N, Dalamaga M, Stratigou T, et al. 2021. Curr Obes Rep, 10(3): 244–262.
Villa C R, Ward W E, Comelli E M. 2017. Crit Rev Food Sci Nutr, 57(8): 1664–1672.
Wang T, Zhang T, Sun L, et al. 2019. Ecotoxicol Environ Saf, 185: 109686.
Weaver S R, Fricke H P, Xie C, et al. 2018. Am J Physiol Endocrinol Metab, 315(6): E1133-e1142.
Weng S F, Redsell S A, Nathan D, et al. 2013. Pediatrics, 132(2): e414–421.
Winter G, Hart R A, Charlesworth R P G, et al. 2018. Rev Neurosci, 29(6): 629–643.
Xiao H, Wu Z, Li B, et al. 2020. Br J Pharmacol, 177(20): 4683–4700.
Xiao L, Wang J, Zheng J, et al. 2021. Genome Biol, 22(1): 243.
Xu Z, Xie Z, Sun J, et al. 2020. Frontiers in cellular and infection microbiology, 10: 160.
Xu Z R, Yuan X X, Chen R M, et al. 2022. World J Pediatr, 18(10): 671–679.
Yadav V K, Ryu J H, Suda N, et al. 2008. Cell, 135(5): 825–837.
You L, Zhu L, Li P Z, et al. 2020. Int J Mol Sci, 21(7): 2518.
Yu J, Zhang H, Chen L, et al. 2021. Front Cell Infect Microbiol, 11: 698852.
Zaiss M M, Jones R M, Schett G, et al. 2019. J Clin Invest, 129(8): 3018–3028.
Zhang L, Jonscher K R, Zhang Z, et al. 2022. Nat Commun, 13(1): 3551.
Zhou D, Pan Q, Xin F Z, et al. 2017. World J Gastroenterol, 23(1): 60–75.
Zhou H, Sun L, Zhang S, et al. 2021. Acta Diabetol, 58(3): 249–265.

Chapter 7
Sex Difference of Fetal-Originated Disease

Yu Guo and Pengxia Yu

Abstract Adverse conditions during pregnancy can lead to the development of fetal multi-organ function in a sex-specific manner, mediated by the maternal-placenta-fetal biological unit. This can result in sex differences in both birth outcomes and susceptibility to adult diseases. The susceptibility to fetal-originated disease caused by adverse prenatal environments is closely related not only to the sex-specific development of the fetal hypothalamic-pituitary-adrenal axis and its regulatory center, the hippocampus but also to the way in which the placenta regulates fetal development. This chapter focuses on the sex differences in fetal-originated disease, specifically fetal-originated metabolic and psychoneurotic diseases. It provides a comprehensive summary of the sex differences in the developmental programming of the hypothalamic-pituitary-adrenal axis and hippocampus as well as the placental adaptive changes in offspring under prenatal adverse environments. This chapter is important for understanding the sex differences in fetal birth outcomes and the long-term consequences thereof. It lays the groundwork for exploring personalized diagnosis and early prevention and treatment strategies for fetal-originated disease.

1 Introduction

Based on the theories of Developmental Origins of Health and Disease, adverse environmental exposures during pregnancy can have long-term effects on disease susceptibility in offspring, in addition to affecting fetal development and leading to poor outcomes during pregnancy. Studies have revealed that male fetuses are more vulnerable to prenatal adverse factors compared to female fetuses (Ingemarsson 2003; Orzack et al. 2015). Adverse environmental exposures during pregnancy can

Y. Guo (✉) · P. Yu
Department of Pharmacology, School of Basic Medical Science, Wuhan University, Wuhan, Hubei, China
e-mail: guoy@whu.edu.cn

H. Wang et al. (eds.), *Fetal Origin of Diseases*,
https://doi.org/10.1007/978-981-97-5730-5_7

also alter fetal developmental programming, leading to sex differences in the timing and severity of fetal-originated disease. However, the mechanisms that underlie sex differences in susceptibility to fetal-originated disease remain unclear. The hypothalamus-pituitary-adrenal (HPA) axis is a system that responds to stress by coordinating neuroendocrine and autonomic responses. Studies have demonstrated sex differences in many components of the HPA axis and their reactivity, which may be the basis for sex differences in the development of fetal-originated disease (Gifford and Reynolds 2017; Oyola and Handa 2017). The hippocampus, which regulates learning, memory, and neurobehavior, is the regulatory center of the HPA axis and may also play a crucial role in sex differences in neuropsychiatric diseases (Weinstock 2011). Furthermore, the placenta serves as a vital connection between the mother and the fetus, and differences in placental growth may contribute to sex differences in the programming of fetal development and its susceptibility to long-term diseases (Kalisch-Smith et al. 2017).

The molecular mechanisms that underlie the occurrence of sex differences in fetal-originated disease have not been extensively studied. The X and Y chromosomes have unique and specific genetic effects as well as downstream gonadal effects. Therefore, sex chromosomes and sex hormones associated with intrinsic sex may be the main determinants of sex-differentiated fetal-originated disease. Additionally, the development of fetal-originated disease is highly associated with overexposure to intrauterine glucocorticoid (GC) (O'Donnell and Meaney 2017; Facchi et al. 2020). There are sex differences in the sensitivity of developing fetal organs and systems to GC, which may lead to sex differences in fetal developmental programming. This chapter reviews studies on sex differences in the occurrence of fetal-originated disease, including fetal-originated metabolic and psychoneurological diseases, and emphasizes that these diseases have sex-differentiated origins. We provide a systematic summary of the sex differences in the developmental programming of the HPA axis and hippocampus, as well as the placental adaptive changes in offspring under adverse environments, and elucidate the potential molecular mechanisms underlying their occurrence. This chapter provides a theoretical basis for developing individualized diagnosis and early prevention strategies for fetal-originated disease.

2 Sex Differences of Fetal-Originated Disease

Fetal-originated disease refer to abnormalities in fetal development that are caused by gametogenic abnormalities and adverse factors during pregnancy. Studies have shown sex differences in the occurrence of fetal-originated disease, including metabolic diseases and neuropsychiatric diseases (Table 7.1) (Wang et al. 2014; Cartocci et al. 2019; Ji et al. 2020; Nemoto et al. 2021).

Table 7.1 Sex differences of fetal-originated disease

Research objects	Perinatal adverse environment	Testing time	Male offspring	Female offspring	Main references
Sprague Dawley rat	Restricted diet	PW3	Liver triglyceride and cholesterol levels↑, lipid synthesis genes expression↑	Liver cholesterol level↓	Choi et al. (2007)
Human	Famine	50 years old	NS	Gestational malnutrition is associated with higher BMI and waist circumference in female offspring	Ravelli et al. (1999)
Wistar rat	A high-fat diet	PM9	Mild glucose intolerance, hyperinsulinemia, and hyper-glucocorticoid. Perirenal adipocyte hypertrophy and hyperplasia. Adipogenic gene expression↑, adipogenesis inhibitory gene expression↓	NS	Lecoutre et al. (2016)
C57BL/6 N mice	A high-fat diet	PD21/70/120	Persistent obese phenotype with sex-specific adipocyte hypertrophy	Obesity in the early postnatal period is partially reversed later in life	Litzenburger et al. (2020)
Wistar rat	Prenatal caffeine exposure	PW24	Lipid synthesis↓	Increased lipid synthesis, oxidation, and output in the liver exacerbate NAFLD symptoms	Wang et al. (2014)

(continued)

Table 7.1 (continued)

Research objects	Perinatal adverse environment	Testing time	Male offspring	Female offspring	Main references
Human	–	25 years old	NS	Females with low birth weight have higher rates of abnormal blood glucose levels in adulthood	Al Salmi et al. (2008)
Wistar rat	Prenatal nicotine exposure	PW15	Hyperinsulinemia	NS	Nemoto et al. (2021)
C57BL/6 mice	Selenium deficiency diet	PD180	Insulin level↑, thyroxine level↓	Thyroxine level↑	Hofstee et al. (2020)
C57/BL6J mice	High-fat diet	PW20	Insulin secretion, islet area, and insulin concentration↓	NS	Yokomizo et al. (2014)
Human	Low-cholesterol diet	≤15 years old	High prevalence of ADHD	NS	Ji et al. (2017)
Wistar rat	Prenatal valproic acid exposure	PD 5/9/13/35/90	Social affective and cognitive deficits	Selective deficits in social play behavior and stereotypes	Melancia et al. (2018)
CrlFcen: CF1 mice	Prenatal valproic acid exposure	PD21/60	Decreased sociability and a lack of preference for social stimulation of new objects	NS	Kazlauskas et al. (2019)
Human	Prenatal acetaminophen exposure	From birth to age 21	High rates of autism	NS	Ji et al. (2020)
Human	Abnormal maternal blood lipids and circulating BCAAs	From birth to age 21	Increased risk of ASD	NS	Panjwani et al. (2020)
Human	Maternal obesity/ diabetes and abnormal BCAAs	–	Increased risk of ASD	NS	Panjwani et al. (2019)

(continued)

Table 7.1 (continued)

Research objects	Perinatal adverse environment	Testing time	Male offspring	Female offspring	Main references
Human	Low maternal weight or excessive weight gain during pregnancy	–	NS	Increased risk of eating disorders	Goodman et al. (2014)

PW postnatal week, *PD* postnatal day, *PM* postnatal months, *BMI* body mass index, *ADHD* attention deficit hyperactivity disorder, *GD* gestational day, *NAFLD* non-alcoholic fatty liver disease, *HDL-C* high-density lipoprotein-cholesterol, *BCAAs* branch chain amino acids, *ASD* autism spectrum disorder, – no related data, *NS* no significant difference

2.1 Fetal-Originated Metabolic Diseases

The incidence of metabolic diseases is increasing annually, and it has become a significant public health concern that affects global economic and social development. Several studies have demonstrated that maternal exposure to adverse environments during pregnancy can cause intrauterine growth retardation (IUGR) in offspring and lead to the occurrence of metabolic diseases in adulthood, with sex differences (Choi et al. 2007; Yang and Huffman 2013). Research has shown that deficient energy, protein, and micronutrient intake during pregnancy is associated with an increased risk of adult obesity in offspring, with female offspring being more susceptible than male offspring (Yang and Huffman 2013). Clinical studies have found that when mothers were malnourished during early gestation, their female offspring had significantly higher weight, body mass index, and waist circumference at the age of 50, whereas male offspring were not significantly affected (Ravelli et al. 1999). The occurrence of chronic diseases in the offspring of intrauterine malnutrition was not limited to the F1 female offspring but persisted in the F2 offspring (Painter et al. 2008). However, maternal obesity or overnutrition can also lead to metabolic diseases in offspring, showing adipose tissue dysfunction and obesity susceptibility in male offspring (Lecoutre et al. 2016; Litzenburger et al. 2020).

Nonalcoholic fatty liver disease (NAFLD) is among the most prevalent metabolic diseases. An Australian epidemiological study revealed that the NAFLD phenotype was more pronounced in young men than in women, owing to differences in fat distribution and adipocytokines between the sexes (Ayonrinde et al. 2011). Our team also demonstrated that caffeine exposure during pregnancy could result in IUGR in rats, whose offspring developed NAFLD on a high-fat diet, exhibiting a significant sex-specific phenotype. The female offspring displayed elevated lipid de novo synthesis, while the male offspring exhibited suppressed lipid oxidation (Wang et al. 2014).

Diabetes mellitus is a prevalent endocrine metabolic disease. Clinical research has found that women with low birth weight have a higher proportion of abnormal blood glucose in adulthood than men, indicating that women are at a higher risk of developing type 2 diabetes than men (Al Salmi et al. 2008). Animal studies have shown that a maternal high-fat diet during pregnancy can lead to insulin resistance and impairment of pancreatic β-cell function in offspring mice, which is more pronounced in males. This may be associated with increased islet oxidative stress and reduced blood estrogen levels (Yokomizo et al. 2014). Additionally, male rats exposed to nicotine during pregnancy exhibited hyperinsulinemia on a high-fat diet (Nemoto et al. 2021). Similarly, a perinatal maternal low-selenium diet resulted in elevated plasma insulin levels in male offspring but did not affect plasma insulin in female mice (Hofstee et al. 2020). The conflicting results obtained from animal experiments and population investigations may be due to differences in study subjects and maternal factors.

2.2 Fetal-Originated Neuropsychiatric Diseases

Neuropsychiatric diseases are a class of neurological disorders that are characterized by disruptions in behavior and mental activity. Research has revealed that male-prevalence diseases tend to be early-onset neuropsychiatric disorders, such as autism and attention deficit hyperactivity disorder (ADHD), while female-prevalence diseases tend to be mood disorders that manifest during adolescence, such as depression and eating disorders (Rutter et al. 2003).

The prevalence of depression is increasing significantly due to the accelerated pace of life. Studies have shown that the lifetime prevalence of major depression is twice as high in women than in men (Goldstein et al. 2019; Rubinow and Schmidt 2019). In addition to genetics and sociopsychological stress, depression is also associated with environmental factors, including maternal exposure during pregnancy. For example, maternal stress alters the sensitivity of the brain to stress hormones in a sex-specific manner in fetuses, which may account for the more frequent onset of depression in female offspring in adulthood (Weinstock 2007; Goldstein et al. 2019).

ADHD is a prevalent childhood psychological disease, with a higher prevalence observed in males than females (Polanczyk et al. 2007; Skounti et al. 2007). The etiology of ADHD is multifactorial, with adverse intrauterine environment being a significant contributor to offspring susceptibility. Maternal smoking during pregnancy, alcohol abuse, acetaminophen exposure, and negative childhood experiences are among the key environmental factors linked to ADHD (Knopik et al. 2006; Ystrom et al. 2017; Huang et al. 2018; Roberts et al. 2018). Additionally, low maternal cholesterol levels during pregnancy have been identified as a risk factor for ADHD in boys, but not girls (Ji et al. 2017). Elevated blood lead levels early in life have also been associated with an increased risk of ADHD in boys (Ji et al. 2018). These findings suggest an intrauterine origin of ADHD and a significant sex difference in its incidence.

Autism spectrum disorder (ASD) is a neurodevelopmental disorder characterized by speech and language impediments, interpersonal difficulties, limited interests, and stereotyped behaviors. The prevalence of ASD is higher in males than females, with a ratio of about 4:1 (Lai et al. 2015). Maternal metabolic status has been identified as a risk factor for ASD in children, with boys being at greater risk (Panjwani et al. 2019, 2020). Furthermore, drug use during pregnancy has been linked to an increased risk of ASD in offspring, with sex differences observed (Ji et al. 2020). Prenatal exposure to acetaminophen has also been associated with a higher incidence of ASD in male offspring than in females (Cartocci et al. 2019; Ji et al. 2020). Studies in rodents have shown that maternal exposure to valproic acid during pregnancy induces autism-like symptoms and affects brain cholesterol metabolism or basal cortisol levels in male offspring (Melancia et al. 2018; Cartocci et al. 2019; Kazlauskas et al. 2019).

Eating disorders are characterized by abnormal eating or weight control behaviors and are mainly classified as anorexia nervosa, bulimia nervosa, and binge eating disorder and are more prevalent in adolescents and young adults. Adverse prenatal or perinatal factors were significantly associated with the occurrence of eating disorders in offspring. A study found that factors strongly associated with anorexia nervosa included higher maternal age, pre-eclampsia and eclampsia, multiple births, hypoxic complications, prematurity, and small-for-gestational age or low-birth-weight babies, while bulimia nervosa was only associated with maternal stress (Marzola et al. 2021). It has been estimated that the prevalence of anorexia nervosa was about 10:1 in women and men, while the ratio of bulimia and binge eating disorder was about 3:1 (Treasure et al. 2020). A cohort study found that eating disorders were 15–20 times as prevalent in women as in men and were associated with low maternal weight or excessive weight gain during pregnancy (Goodman et al. 2014). It is suggested that the prevalence of eating disorders has sex differences and is strongly associated with prenatal and perinatal factors.

3 The Influence Factors of Sex Differences in Fetal-Originated Disease

The factors that contribute to sex differences in fetal-originated disease are multifaceted and intricate. One of the most probable mechanisms mediating the fetal origin of metabolic diseases is the altered developmental programming of the HPA axis. The development of the hippocampus is critical in regulating learning, memory, neurobehavior, and sex differences in neuropsychiatric diseases. Furthermore, the placenta serves as a vital maternal-fetal connection, and its sex-specific development has a broad impact on the fetus, potentially leading to altered intrauterine developmental programming of multiple fetal organs and systems.

3.1 Hypothalamic-Pituitary-Adrenal Axis

The HPA axis is a vital neuroendocrine axis that regulates the stress response. However, adverse prenatal environmental exposures can impair fetal HPA axis development, leading to long-term effects (García-Cáceres et al. 2010). Abnormal functional programming of the fetal HPA axis is associated with impaired brain development, altered neurobehavior, and increased susceptibility to metabolic disease (Moisiadis and Matthews 2014). Studies have suggested that the effects of prenatal stress on the maternal HPA axis and anxiety-like behaviors could be transmitted to offspring in a sex-dependent manner (Grundwald and Brunton 2015). Moreover, HPA axis reactivity was found to be higher in female offspring of prenatally stressed mothers compared to males, potentially due to differences in circulating estradiol levels (Gifford and Reynolds 2017; Oyola and Handa 2017). Major depression and cardiovascular disease comorbidity are significant contributors to morbidity and mortality worldwide and are more common in women than men. The overexposure of GC due to adverse prenatal environmental exposures may induce sex-dependent developmental programming changes in the fetal HPA axis, resulting in sex differences in mood, stress regulation, autonomic nervous system, and vasculature in the offspring as adults (Goldstein et al. 2014; Goldstein et al. 2016). Therefore, adverse prenatal environmental exposures can lead to sex differences in the development of HPA axis function in offspring, which, in turn, can lead to sex differences in HPA axis-related diseases.

3.2 Hippocampus

The hippocampus serves as a regulatory center of the HPA axis and maintains the body's internal environment homeostasis by negatively feedback regulating the basal and stress states of the HPA axis. Sex differences in hippocampal synaptic plasticity and individual cognitive function are observed in many diseases that are based on hippocampal function. Prenatal stress has been found to impair learned cognitive abilities in male rat offspring, while anxiety behaviors were more prevalent in females, possibly due to fetal overexposure to maternal corticosterone, resulting in reduced sex-dependent neurogenesis and dendritic morphology in the prefrontal cortex and hippocampal formation (Weinstock 2011). Additionally, maternal exposure to ethanol inhibited hippocampal brain-derived nerve growth factor-related pathways in adult female rat fetuses, suggesting a potential propensity for depression (Lu et al. 2018). Similarly, maternal exposure to bisphenol A (BPA) resulted in hippocampal neurotransmitter and neuroactive steroid dysfunction and depression-like behavior in male but not female offspring mice (Xin et al. 2018). Furthermore, selective serotonin reuptake inhibitors (SSRIs) are commonly used to treat mood disorders in pregnant women. However, a rat study found that hippocampal neurodevelopment and social-behavioral function were impaired in male offspring exposed to SSRIs, with no significant effect on females (Gemmel et al. 2019). These findings suggest that the occurrence of fetal-originated neuropsychiatric diseases is associated with sexually differential developmental abnormalities in the hippocampus of the offspring due to adverse environmental exposure.

3.3 Placenta

The placenta serves as the sole connection between the mother and fetus. In the Finnish Helsinki birth cohort, low-birth-weight fetuses were found to be susceptible to hypertension at 62 years of age, with sex differences associated with different placental areas and diameters (Eriksson et al. 2010). Studies in rodents have shown that the female placenta slows the development of the vascular system during adverse environmental exposures and becomes more efficient by increasing its adaptations, while the male placenta, despite increasing its surface area through labyrinthine zone expansion, does not have sufficient adaptive capacity to ensure normal male fetal development (Cuffe et al. 2014; Guo et al. 2020; van Abeelen et al. 2011). Furthermore, the placenta plays a critical role in regulating fetal brain development. Placental inflammation can lead to reactive astrocytic gliosis in the fetal hippocampus, resulting in fetal brain neuronal damage (Hawkins et al. 2018). A clinical study found increased expression of pro-inflammatory cytokines and immune pathway-related genes in the placenta of females with maternal disease (Osei-Kumah et al. 2011). Similarly, a rodent study found that a maternal high-fat diet led to increased placental macrophage activation and expression of cytokine genes in male mice (Kim et al. 2014). Additionally, the placenta is the largest endocrine organ during pregnancy and produces hormones that regulate fetal growth and development. Maternal combined with pre-eclampsia has been found to elevate serum androgen levels, and there are sex differences in placental aromatase expression, potentially helping prevent female fetuses from masculinization (Sathishkumar et al. 2012; Lan et al. 2020). These findings suggest that the placenta undergoes structural and functional adaptations in a sex-specific manner in response to changes in the altered intrauterine environment and the needs of fetal development during adverse pregnancy conditions.

4 Molecular Mechanisms Contributing to the Sexual Difference in Fetal-Originated Disease

The etiology of fetal-originated disease in relation to sex differences remains elusive. Intrinsic sex-related factors, namely sex chromosomes and sex hormones, have been posited as crucial triggers (Gabory et al. 2013). Additionally, excessive exposure to GC during fetal development has been implicated as a significant trigger for fetal-originated disease. Variations in fetal and placental sensitivity to GC may influence disparities in birth outcomes and predisposition to chronic diseases.

4.1 Sex Chromosomes

Although the X chromosome represents only 5% of the human genome, it is associated with approximately 15% of the genes linked to intellectual disability (Neri et al. 2018). Studies have demonstrated that individuals with Turner syndrome have smaller hippocampal, caudate nucleus, lenticular nucleus, thalamic nucleus, and parieto-occipital brain matter, indicating the crucial role of the X chromosome in thalamic and hippocampal development in humans (Murphy et al. 1993). Research has also shown that sex chromosomes regulate the sex-specific synthesis of growth hormones in the hypothalamus, and the effect of estradiol on neurogenin3 expression is mediated by the activation of estrogen receptor α (ERα) in hypothalamic neurons (Cisternas et al. 2020; Quinnies et al. 2015). The X chromosome has numerous genes associated with cognitive ability, and mutations in these genes can lead to mental disorders (Zechner et al. 2001). Specifically, the protocadherin-19 gene, located on the X chromosome and highly expressed in the hippocampus, has been linked to early-onset seizures and cognitive impairment in female heterozygotes (Hoshina et al. 2021). Additionally, some genes on the Y chromosome are expressed in the male brain and may impact the sex differentiation in the brain in a specific manner (Sekido 2014). Notably, prenatal exposure to lipopolysaccharides with autism-associated mutations and Y chromosomes has been shown to exacerbate certain autistic-like signs in mouse behavior (Magarinos et al. 2018). These findings suggest that sex chromosomes are involved in regulating hypothalamic and hippocampal development, and may contribute to sex differences in the occurrence of fetal-originated diseases.

The placenta shares genetic information with the fetus, which determines the sex differences in placental biochemistry, function, and signaling transduction. X chromosome-linked genes have been reported to regulate placental development (Hemberger 2002). In contrast to other fetal tissues and organs, inactivated X chromosomes in the placenta can be reactivated, which may contribute to sex differences in the placenta. Studies have shown that O-linked-N acetylglucosaminyltransferase, an X-linked gene, is expressed in the male placenta at half the level of the female placenta and is further reduced during prenatal stress, leading to sex-specific fetal growth restriction and vulnerability to neuropsychiatric diseases (Nugent et al. 2018). The mammalian Y chromosome is also involved in regulating placental development, as placentas with one X chromosome are larger than those with two X chromosomes, possibly due to the Y chromosome's role in accelerating the development of the preimplantation embryo, independently of gonadal or androgenic status (Burgoyne et al. 1995; Ishikawa et al. 2003). These findings suggest that sex chromosomes regulate placental development and may contribute to sex differences in long-term disease susceptibility through fetal developmental programming.

4.2 Sex Hormone

Alterations in sex hormone levels regulate the stress response and underlie sex differences in HPA axis function (Handa and Weiser 2014). Studies have shown that prenatal ethanol exposure has sex-specific effects on HPA axis activity before and after the birth of offspring, and that gonadotropins or hypothalamic-pituitary-gonadal (HPG) axis interactions with the HPA axis mediate the effects of prenatal ethanol exposure on HPA activity (Weinberg et al. 2008). Gonadectomy can reveal the effect of HPG axis alterations on HPA axis activity. In rodents, chronic stress-induced adrenal GC secretion is higher in females than in males, and ovariectomy reduces ACTH and corticosterone levels in females, which can be reversed by estradiol treatment (Handa and Weiser 2014). Androgens inhibit the HPA axis, and in adult rats, gonadectomy increases corticosterone levels, which can be restored by testosterone or dihydrotestosterone replacement treatment (Seale et al. 2004). In male rats, gonadectomy results in increased anxiety and depression-like behavior, which can be reversed by systemic testosterone treatment (Adler et al. 1999; Frye and Seliga 2001). These findings suggest that sex hormones may contribute to sex differences in fetal-originated diseases by affecting HPA axis activity.

Sex hormones are potent regulators of hippocampal structure and function, which can drive sex differences in hippocampal development and function. Prenatal ethanol exposure has been shown to affect the hippocampal 5-hydroxytryptamine system in female rat offspring, potentially increasing the risk of depression and anxiety disorders in adulthood through a mechanism related to the estrous cycle (Weinberg et al. 2008). In mice, the rate of hippocampal neuron production is faster in males than in females during the first postnatal week, and androgen receptor expression is higher in the cerebral cortex and hippocampus of males on the seventh postnatal day (Tsai et al. 2015; Zhang et al. 2008). Estrogens are conducive to glial cell differentiation and/or survival, whereas androgens are conducive to neuronal differentiation and/or survival (Zhang et al. 2008). Studies have found that ERα plays a role in neurotrophin receptor-mediated hippocampal neuroprotection in female mice, which can affect long-term memory and learning in offspring. Androgens and their receptors can cause androgenization of brain regions, affecting cognitive processing and responses to stress in the brain (Zafer et al. 2019). Taken together, these findings suggest that sex hormones may regulate hippocampal development in a sex-specific manner, influencing the cognitive function of neuronal circuits and leading to sex differences in adult disease susceptibility.

The placenta is an endocrine organ that produces high levels of estrogen and progesterone during late gestation and is capable of negative feedback regulation of the fetal HPG axis, which may have persistent adverse effects on the fetus and child (Lucaccioni et al. 2020). Disruption of placental sex hormone signaling pathways has been observed in complex pregnancies and is associated with differential expression of aromatase (Maliqueo et al. 2017). In normal pregnancy, aromatase levels are higher in the male placenta than in the female placenta, providing an effective barrier that prevents the transfer of testosterone from the fetal side to the maternal circulation (Sathishkumar et al. 2012). However, aromatase levels were elevated in the placentas of female fetuses with maternal obesity and combined pre-eclampsia and decreased in the placentas of male fetuses, which may be an adaptive protective mechanism that prevents high levels of testosterone in the maternal circulation from inducing masculinization in the female fetus (Sathishkumar et al. 2012; Maliqueo et al. 2017). Furthermore, androgen receptor expression in the placenta in maternal asthma and pre-eclampsia may play a critical role in regulating fetal placental growth and development as well as the sex-differentiated progression of long-term diseases (Sathishkumar et al. 2012; Meakin and Clifton 2019). These findings suggest that sex differences in placental sex hormone secretion can influence fetal development, leading to the sex-specific occurrence of long-term diseases.

4.3 Glucocorticoid Sensitivity

Prenatal GC exposure has sex-specific effects on the HPA axis in fetuses and adult guinea pigs (Owen and Matthews 2003). The glucocorticoid receptor (GR) and the mineralocorticoid receptor (MR) are the primary regulators of GC signaling in the brain. Studies in rodents have shown sex differences in the temporal and spatial expression of GR and MR in the fetal and early postnatal brain, resulting in differential sensitivity to GC exposure (Dean and Matthews 1999; Owen and Matthews 2003; Lan et al. 2017). Perinatal exposure to BPA has been found to cause female rat offspring to exhibit "anxiolytic-like" behavior, while male offspring exhibit increased depressive-like behavior, potentially by altering GR and MR expression and interfering with GR-mediated HPA axis feedback and MR-induced nNOS-CREB signaling pathways (Chen et al. 2014). Taken together, the available evidence suggests that the differential sensitivity of GC in the hypothalamus of male and female offspring under unfavorable conditions may result in sex-specific changes in HPA axis function and contribute to the development of fetal-originated disease.

Dean and Matthews (1999) reported that prenatal exposure to dexamethasone resulted in a significant increase in fetal blood corticosterone levels and hippocampal MR and GR expression in female guinea pigs, while males exhibited reduced blood cortisol levels and no significant changes in hippocampal MR and GR expression. McCabe et al. (2001) found that multiple courses of dexamethasone exposure upregulated MR expression in the hippocampus of female guinea pig fetuses, while males exhibited significantly increased GR expression. These findings suggest that prenatal dexamethasone exposure modifies hippocampal MR and GR expression during early offspring development, leading to sex-specific differences in GC sensitivity in the hippocampus. Furthermore, Ordyan et al. (2008) observed sex differences in hippocampal GR isoform expression in male and female rats, with changes in expression levels consistent with alterations in blood corticosterone levels during early postnatal development. These results imply that adverse prenatal conditions may regulate distinct mechanisms of MR and GR expression in the hippocampus of male and female offspring, possibly mediating varying levels of GC sensitivity and contributing to the regulation of sexual differentiation in hippocampal development and the occurrence of fetal-originated disease.

Lan et al. (2017) demonstrated that prenatal ethanol exposure in rats resulted in significantly elevated maternal blood GC levels and impaired placental glucocorticoid signaling, with a sex-specific effect. In addition, MR expression in male placentas was negatively correlated with maternal alcohol consumption. The GR gene undergoes selective splicing and translation to produce multiple GR isoforms, which mediate differences in GC sensitivity. Hodyl et al. (2010) found that the sex-specific correlation between cord blood cortisol and fetal weight in combined asthma of pregnancy was not due to altered placental GR expression levels, but rather related to GR isoform expression in the placenta. Overexposure to cortisol in male placentas may induce GC resistance by upregulating GRβ, GRα, and GRP in the nucleus, while female placentas may enhance GRα activity by reducing GRβ expression and promoting GRαD3 and GRα-C interaction, leading to increased GC sensitivity in female placentas (Saif et al. 2014). Moreover, Cuffe et al. (2017) reported that female placentas of prenatal dexamethasone-exposed mice were more sensitive to GC than male placentas, indicating that male and female placentas have different mechanisms of adaptation to high GC levels that are dependent on MR and GR isoforms. These adaptations may have implications for fetal birth outcomes and the susceptibility to long-term diseases.

5 Summary and Prospect

Chronic non-communicable diseases often have their origins in the intrauterine period and exhibit sex-specific differences in their susceptibility. This chapter provides a comprehensive review of the sex-specific differences in disease susceptibility in adult offspring under adverse prenatal conditions. It systematically elucidates the sex differences in altered HPA axis and hippocampal developmental programming, as well as the placental responses of different sexes. Furthermore, the chapter explores the critical roles and potential mechanisms of sex chromosomes, sex hormones, and GC sensitivity differences in sex-differentiated fetal and placental development (Fig. 7.1). Understanding fetal developmental programming, birth outcomes, and long-term consequences is crucial, highlighting the need to fully consider the impact of sex in future basic and translational clinical research and to recognize the contribution of biological sex in the Developmental Origins of Health and Disease.

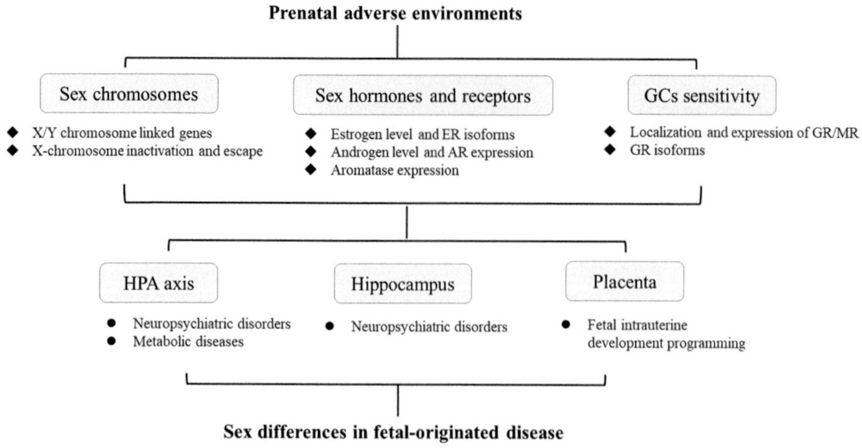

Fig. 7.1 Mechanism of sex differences in fetal-originated disease. *ER* estrogen receptor, *AR* androgen receptor, *GCs* glucocorticoids, *GR* glucocorticoid receptor, *MR* mineralocorticoid receptor

References

Adler A, Vescovo P, Robinson J K, et al. 1999. Neuroscience, 89(3): 939–954.

Al Salmi I, Hoy W E, Kondalsamy-Chennakesavan S, et al. 2008. Diabetes Care, 31(1): 159–164.

Ayonrinde O T, Olynyk J K, Beilin L J, et al. 2011. Hepatology, 53(3): 800–809.

Burgoyne P S, Thornhill A R, Boudrean S K, et al. 1995. Philos Trans R Soc Lond B Biol Sci, 350(1333): 253–260 discussion 260–251.

Cartocci V, Tonini C, Di Pippo T, et al. 2019. J Cell Physiol, 234(4): 4362–4374.

Chen F, Zhou L, Bai Y, et al. 2014. Brain Res, 1571: 12–24.

Choi G Y, Tosh D N, Garg A, et al. 2007. Am J Obstet Gynecol, 196(5): e471–e477.

Cisternas C D, Cabrera Zapata L E, Mir F R, et al. 2020. Sci Rep, 10(1): 8223.

Cuffe J S, Walton S L, Singh R R, et al. 2014. J Physiol, 592(14): 3127–3141.

Cuffe J S M, Saif Z, Perkins A V, et al. 2017. J Endocrinol, 234(2): 89–100.

Dean F, Matthews S G. 1999. Brain Res, 846(2): 253–259.

Eriksson J G, Kajantie E, Osmond C, et al. 2010. Am J Hum Biol, 22(3): 330–335.

Facchi J C, Lima T A L, Oliveira L R, et al. 2020. Metabolism, 104: 154047.

Frye C A, Seliga A M. 2001. Cogn Affect Behav Neurosci, 1(4): 371–381.

Gabory A, Roseboom T J, Moore T, et al. 2013. Biol Sex Differ, 4(1): 5.

García-Cáceres C, Lagunas N, Calmarza-Font I, et al. 2010. Psychoneuroendocrinology, 35(10): 1525–1535.

Gemmel M, De Lacalle S, Mort S C, et al. 2019. Neuropharmacology, 144: 70–81.

Gifford R M, Reynolds R M. 2017. Early Hum Dev, 114: 7–10.

Goldstein J M, Handa R J, Tobet S A. 2014. Front Neuroendocrinol, 35(1): 140–158.

Goldstein J M, Holsen L, Huang G, et al. 2016. Dialogues Clin Neurosci, 18(4): 425–436.

Goldstein J M, Hale T, Foster S L, et al. 2019. Neuropsychopharmacology, 44(1): 59–70.

Goodman A, Heshmati A, Malki N, et al. 2014. Am J Epidemiol, 179(7): 852–863.

Grundwald N J, Brunton P J. 2015. Psychoneuroendocrinology, 62: 204–216.

Guo J, Fang M, Zhuang S, et al. 2020. Ann Transl Med, 8(5): 233.

Handa R J, Weiser M J. 2014. Front Neuroendocrinol, 35(2): 197–220.

Hawkins S J, Crompton L A, Sood A, et al. 2018. Nat Nanotechnol, 13(5): 427–433.

Hemberger M. 2002. Cytogenet Genome Res, 99(1–4): 210–217.
Hodyl N A, Wyper H, Osei-Kumah A, et al. 2010. Thorax, 65(8): 677–683.
Hofstee P, McKeating D R, Bartho L A, et al. 2020. Nutrients, 12(1): 267.
Hoshina N, Johnson-Venkatesh E M, Hoshina M, et al. 2021. Science, 372(6539).
Huang L, Wang Y, Zhang L, et al. 2018. Pediatrics, 141(1): e20172465.
Ingemarsson I. 2003. BJOG, 110 Suppl 20: 34–38.
Ishikawa H, Rattigan A, Fundele R, et al. 2003. Biol Reprod, 69(2): 483–488.
Ji Y, Riley A W, Lee L C, et al. 2017. Brain Sci, 8(1): 3.
Ji Y, Hong X, Wang G, et al. 2018. J Pediatr, 199: 124–131.
Ji Y, Azuine R E, Zhang Y, et al. 2020. JAMA Psychiatry, 77(2): 180–189.
Kalisch-Smith J I, Simmons D G, Pantaleon M, et al. 2017. Biol Sex Differ, 8: 17.
Kazlauskas N, Seiffe A, Campolongo M, et al. 2019. Psychoneuroendocrinology, 110: 104441.
Kim D W, Young S L, Grattan D R, et al. 2014. Biol Reprod, 90(6): 130.
Knopik V S, Heath A C, Jacob T, et al. 2006. Psychol Med, 36(10): 1461–1471.
Lai M C, Lombardo M V, Auyeung B, et al. 2015. J Am Acad Child Adolesc Psychiatry, 54(1): 11–24.
Lan K C, Lai Y J, Cheng H H, et al. 2020. Reprod Biol Endocrinol, 18(1): 12.
Lan N, Chiu M P, Ellis L, et al. 2017. Neuroscience, 342: 167–179.
Lecoutre S, Deracinois B, Laborie C, et al. 2016. J Endocrinol, 230(1): 39–53.
Litzenburger T, Huber E K, Dinger K, et al. 2020. Clin Sci (Lond), 134(7): 921–939.
Lu J, Jiao Z, Yu Y, et al. 2018. Cell Death Dis, 9(6): 659.
Lucaccioni L, Trevisani V, Boncompagni A, et al. 2020. Front Pediatr, 8: 612235.
Magarinos A M, Schaafsma S M, Pfaff D W. 2018. Front Neuroendocrinol, 49: 86–90.
Maliqueo M, Cruz G, Espina C, et al. 2017. Int J Obes (Lond), 41(11): 1636–1645.
Marzola E, Cavallo F, Panero M, et al. 2021. Arch Womens Ment Health, 24(2): 185–204.
McCabe L, Marash D, Li A, et al. 2001. J Neuroendocrinol, 13(5): 425–431.
Meakin A S, Clifton V L. 2019. Placenta, 84: 63–68.
Melancia F, Schiavi S, Servadio M, et al. 2018. Br J Pharmacol, 175(18): 3699–3712.
Moisiadis V G, Matthews S G. 2014. Nat Rev Endocrinol, 10(7): 391–402.
Murphy D G, DeCarli C, Daly E, et al. 1993. Lancet, 342(8881): 1197–1200.
Nemoto T, Ando H, Nagao M, et al. 2021. Front Endocrinol (Lausanne), 12: 694336.
Neri G, Schwartz C E, Lubs H A, et al. 2018. Am J Med Genet A, 176(6): 1375–1388.
Nugent B M, O'Donnell C M, Epperson C N, et al. 2018. Nat Commun, 9(1): 2555.
O'Donnell K J, Meaney M J. 2017. Am J Psychiatry, 174(4): 319–328.
Ordyan N E, Galeeva A Y, Pivina S G. 2008. Bull Exp Biol Med, 146(2): 176–179.
Orzack S H, Stubblefield J W, Akmaev V R, et al. 2015. Proc Natl Acad Sci U S A, 112(16): E2102-E2111.
Osei-Kumah A, Smith R, Jurisica I, et al. 2011. Placenta, 32(8): 570–578.
Owen D, Matthews S G. 2003. Endocrinology, 144(7): 2775–2784.
Oyola M G, Handa R J. 2017. Stress, 20(5): 476–494.
Painter R C, Osmond C, Gluckman P, et al. 2008. BJOG, 115(10): 1243–1249.
Panjwani A A, Ji Y, Fahey J W, et al. 2019. Autism Res, 12(10): 1562–1573.
Panjwani A A, Ji Y, Fahey J W, et al. 2020. J Autism Dev Disord, 50(2): 540–550.
Polanczyk G, de Lima M S, Horta B L, et al. 2007. Am J Psychiatry, 164(6): 942–948.
Quinnies K M, Bonthuis P J, Harris E P, et al. 2015. Biol Sex Differ, 6: 8.
Ravelli A C, van Der Meulen J H, Osmond C, et al. 1999. Am J Clin Nutr, 70(5): 811–816.
Roberts A L, Liew Z, Lyall K, et al. 2018. Am J Epidemiol, 187(9): 1896–1906.
Rubinow D R, Schmidt P J. 2019. Neuropsychopharmacology, 44(1): 111–128.
Rutter M, Caspi A, Moffitt T E. 2003. J Child Psychol Psychiatry, 44(8): 1092–1115.
Saif Z, Hodyl N A, Hobbs E, et al. 2014. Placenta, 35(4): 260–268.
Sathishkumar K, Balakrishnan M, Chinnathambi V, et al. 2012. J Perinatol, 32(5): 328–335.
Seale J V, Wood S A, Atkinson H C, et al. 2004. J Neuroendocrinol, 16(12): 989–998.
Sekido R. 2014. Adv Genet, 86: 135–165.

Skounti M, Philalithis A, Galanakis E. 2007. Eur J Pediatr, 166(2): 117–123.
Treasure J, Duarte T A, Schmidt U. 2020. Lancet, 395(10227): 899–911.
Tsai H W, Taniguchi S, Samoza J, et al. 2015. Neurosci J, 2015: 525369.
van Abeelen A F M, de Rooij S R, Osmond C, et al. 2011. Placenta, 32(9): 694–698.
Wang L, Shen L, Ping J, et al. 2014. Toxicol Lett, 224(3): 311–318.
Weinberg J, Sliwowska J H, Lan N, et al. 2008. J Neuroendocrinol, 20(4): 470–488.
Weinstock M. 2007. Neurochem Res, 32(10): 1730–1740.
Weinstock M. 2011. Stress, 14(6): 604–613.
Xin F, Fischer E, Krapp C, et al. 2018. Horm Behav, 102: 93–104.
Yang Z, Huffman S L. 2013. Matern Child Nutr, 9 Suppl 1: 105–119.
Yokomizo H, Inoguchi T, Sonoda N, et al. 2014. Am J Physiol Endocrinol Metab, 306(10): E1163-E1175.
Ystrom E, Gustavson K, Brandlistuen R E, et al. 2017. Pediatrics, 140(5): e20163840.
Zafer D, Aycan N, Ozaydin B, et al. 2019. Neuroendocrinology, 109(3): 249–256.
Zechner U, Wilda M, Kehrer-Sawatzki H, et al. 2001. Trends Genet, 17(12): 697–701.
Zhang J M, Konkle A T, Zup S L, et al. 2008. Eur J Neurosci, 27(4): 791–800.

Chapter 8
Multigenerational Genetic Effect of Fetal-Originated Disease

Hui Wang and Yi Liu

Abstract Fetal diseases instigated by adverse environmental exposures during gestation are extensively documented. However, a comprehensive understanding of their multigenerational genetic impacts and underlying mechanisms is yet to be fully achieved. In this chapter, we critically examine the phenomenon of tissue and organ dysplasia—encompassing the liver, gonads, hippocampus, and bones—attributable to detrimental prenatal environments. We delve into the intricacies of epigenetic programming as a pivotal mechanism behind these effects. Additionally, we elucidate the role of germ cells, both oocytes and sperm, in the transference of these multigenerational genetic effects. This review aims to enrich our understanding of the long-term consequences of fetal-origin diseases and to inspire innovative approaches for early prevention and treatment strategies.

1 Introduction

Recently, the multigenerational genetic consequences of offspring phenotypic anomalies or diseases, triggered by suboptimal parental environments, have garnered increasing scholarly attention. The term "multigenerational genetic effect" describes the phenomenon wherein offspring exhibit similar traits or diseases as those of the F1 generation, despite being several generations removed from the parents exposed to adverse factors. This effect is intimately linked with epigenetic modifications in germline cells and their subsequent transmission (Burton and Greer 2022). Earlier research has established that exposure to detrimental environmental conditions during early life stages, including the intrauterine period or childhood, elevates the risk of various diseases in later life, with these effects potentially enduring across multiple generations (Yohn et al. 2015). For instance, offspring

H. Wang (✉) · Y. Liu
Department of Pharmacology, School of Basic Medical Science, Wuhan University, Wuhan, Hubei, China
e-mail: wanghui19@whu.edu.cn

H. Wang et al. (eds.), *Fetal Origin of Diseases*,
https://doi.org/10.1007/978-981-97-5730-5_8

exposed to lipopolysaccharide during gestation exhibited hypertension and displayed transgenerational genetic effects (Guan et al. 2022). Similarly, hyperglycemia during pregnancy has been linked to alterations in the methylation patterns of genes related to obesity and diabetes in F1 primordial germ cells, with these epigenetic changes persisting in F2 germ cells and F3 somatic cells (Ma et al. 2022). These findings collectively underscore the potential multigenerational genetic impact of offspring dysplasia induced by adverse intrauterine environmental conditions. This chapter systematically explores the multigenerational genetic effects and mechanisms of fetal-origin diseases, enhancing our comprehension of their long-term detriment and providing a foundational theory for improved eugenics and childbearing guidance.

2 Transgenerational Genetic Consequences of Fetal-Originated Disease

Multigenerational inheritance encompasses both intergenerational and transgenerational inheritance (see Fig. 8.1). Intergenerational inheritance implies that if the F1 generation is subjected to an adverse environment, this will manifest as a similar phenotype in the F2 generation. However, such phenotypic changes do not typically extend into the F3 generation (Nicolella and de Assis 2022). On the other hand, transgenerational inheritance is characterized by the occurrence of abnormal phenotypes or related diseases in F1 offspring, who were directly exposed to adverse factors during intrauterine or developmentally sensitive periods. Remarkably, these phenotypic traits persist in the F3 generation, even without direct exposure to the

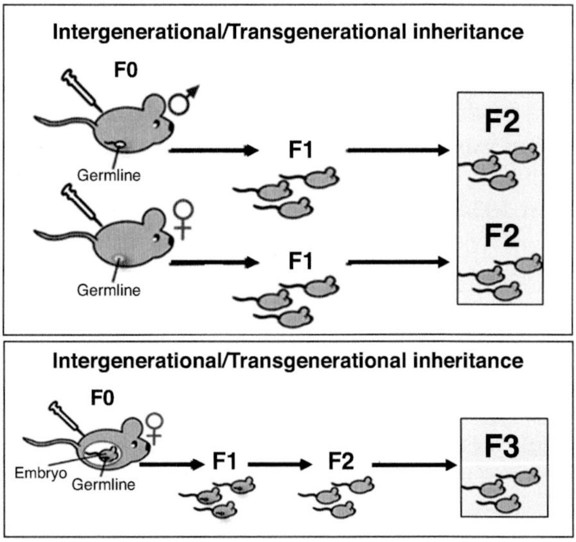

Fig. 8.1 Classification and dynamics of multigenerational genetic effects

same adverse factors (Boskovic and Rando 2018). The intrauterine period is a critical developmental stage where most tissues and organs attain their definitive physiological structures and functions, rendering them highly susceptible to external environmental influences (Wasterlain and Shirasaka 1994; Trefts et al. 2017; Estermann et al. 2021). Comprehensive studies, both domestically and internationally, have verified that exposure to adverse environmental conditions during pregnancy can predispose individuals to multiple diseases in adulthood. This is attributed to the intrauterine programming of organ development, potentially resulting in multigenerational genetic effects (He et al. 2019; Hojsager et al. 2022; Huang et al. 2022). For instance, our research team has demonstrated that prenatal exposure to exogenous substances like caffeine, nicotine, and ethanol can induce glucocorticoid-dependent alterations in serum glucose and lipid metabolism in offspring (F1 generation) postnatally. This effect can extend to the F2 generation and is correlated with changes in neuroendocrine metabolism programming induced by excessive maternal glucocorticoid exposure in utero (Luo et al. 2014; Hu et al. 2020a; Hu et al. 2020b). Consequently, it is evident that diseases originating in the fetal stage have significant multigenerational genetic impacts.

3 Adverse Prenatal Environments and their Multigenerational Genetic Impacts on Organ Dysplasia

We conducted a comprehensive review of the multigenerational genetic effects and underlying mechanisms of multi-tissue organ dysplasia in offspring, resulting from adverse prenatal environments. This review encompasses a detailed analysis of dysplasia in various organs, including the long bones and cartilage, gonads (testes and ovaries), adrenal glands, and the hippocampus.

3.1 Transgenerational Genetic Impacts on Bone Dysplasia

Bone-related conditions, notably osteoporosis and osteoarthritis, have increasingly been recognized as having intrauterine origins. Recent research has identified that adverse environmental exposures during pregnancy can precipitate multigenerational genetic susceptibilities to bone-related diseases. For instance, prenatal dexamethasone exposure (PDE) is linked to reduced cartilage thickness and a lower chondrocyte count in both sexes of offspring rats, alongside a suppression of TGFβ signaling pathways. Notably, these cartilage pathological alterations have been observed to persist from the intrauterine stage through to postnatal development, extending into the F2 and F3 generations. This underscores the multigenerational genetic influence of fetal osteoarthritis. Furthermore, a study by Liu et al. (2023) found that administering glucosamine to the F3 generation reactivated the TGFβ signaling in chondrocytes and ameliorated histopathological symptoms of osteoarthritis. This therapeutic response is associated with the in utero dexamethasone-induced downregulation of miR-92a-3p, a condition transmitted through F1 and F2 oocytes (Tie et al. 2022). Another significant finding is that intrauterine PDE can inhibit fetal osteoblast differentiation and cause long bone dysplasia and postnatal bone mass reduction through the miR-98-3p/JAG1/Notch1 signaling pathway. This inhibitory effect on the signaling pathways and the resultant pathological changes in long bones have also been observed to persist through the F2 and F3 generations, a phenomenon linked to sustained alterations in miR-98-3p in F1 and F2 oocytes (Han et al. 2022). In summary, these findings reveal the profound multigenerational genetic impacts of adverse environmental conditions during pregnancy on bone dysplasia, offering potential avenues for therapeutic interventions in offspring.

3.2 Transgenerational Genetic Impacts on Gonadal Dysplasia

The testicles in males and ovaries in females play a pivotal role in the synthesis and secretion of sex hormones as well as in the generation of germ cells. The endocrine functions of the gonads and the reserve capacities of primordial germ cells are both established during the intrauterine period. Extensive research has indicated that adverse intrauterine environments contribute to reduced reproductive function in offspring. Our team's previous studies (Liu et al. 2023) have demonstrated that in males, PDE directly affects the F1 testes, activating the glucocorticoid receptor (GR) in stromal cells. This leads to the suppression of miR-466b-3p expression, consequently upregulating histone deacetylase 7 (HDAC7) and subsequently reducing the acetylation levels and expression of histone 3 lysine 9 (H3K9) at the StAR promoter region. This results in impaired testosterone synthesis in the F1 generation. Moreover, the suppression of miR-466b-3p expression appears to bypass the reprogramming effects in F1 and F2 generation oocytes, transmitting to the F2 and F3 generation testes. This transmission manifests as a continued decline in miR-466b-3p expression, increased HDAC7 expression, and inhibited H3K9 acetylation and expression in the StAR promoter region in F2 and F3 testes, leading to transgenerational genetic effects on testosterone synthesis. Additionally, other studies (Lv et al. 2018; Gong et al. 2021; Gong et al. 2023) have revealed that PDE can disrupt ovarian functional development across multiple female offspring generations. This disruption is mediated by a decrease in the miR-320-3p/RUNX2/P450arom signaling pathway, inhibiting ovarian estrogen synthesis and reducing the number of primordial follicles. Concurrently, the MiR-17-5p/FOXL2/CDKN1B signaling pathway mediates the impairment of ovarian reserve function. These alterations can persist across generations, with sustained changes in the expression of miR-320-3p and miR-17-5p detectable in F1 and F2 oocytes (Gong et al. 2021; Gong et al. 2023). In summary, adverse environmental conditions during pregnancy can induce gonadal dysplasia in offspring and manifest multigenerational genetic effects.

3.3 Transgenerational Genetic Implications of Adrenal Dysplasia

The adrenal gland, a key component of the hypothalamic-pituitary-adrenal (HPA) axis, serves as an effector organ in response to external stimuli. The glucocorticoids it secretes are vital for maintaining endocrine equilibrium and the development of various organs. Exposure to neuroexcitatory substances such as caffeine, ethanol, and nicotine during pregnancy has been linked to a reduction in the adrenal cortex thickness in offspring, both pre- and post-birth, alongside an inhibition of steroid synthesis function. For instance, alcohol exposure during gestation triggers the activation of adrenal GR, upregulation of HDAC11, and suppression of SF1. This leads to a decrease in the acetylation level and expression of histone in the P450 cholesterol side chain cleavage (P450scc) promoter region, with these changes extending into the F2 generation (Cao et al. 2021). Additionally, high intrauterine glucocorticoids activate GR, suppressing StAR expression by enhancing let-7 expression and inducing H19 hypermethylation and low expression through the downregulation of CTCF and upregulation of DNMT3a/3b. This weakens the "molecular sponge" effect and augments the inhibitory impact of let-7 on adrenal StAR in offspring. Intriguingly, F1 and F3 female progeny exhibit low adrenal steroid synthesis, with increased function in the F2 generation. This variation may be attributable to "contemporary" fluctuations in intrauterine maternal glucocorticoid exposure, leading to adrenal reprogramming in the progeny (He et al. 2021). Conversely, PDE results in diminished adrenal steroid synthesis in F1, F2, and F3 generations, regardless of sex. This effect is linked to continuous programming changes in the imprinting gene H19/let-7 axis induced by intrauterine dexamethasone exposure, which are subsequently transmitted to progeny through F1 and F2 oocytes (He et al. 2023). In summary, adverse environmental conditions during pregnancy can lead to adrenal dysplasia in offspring, demonstrating multigenerational genetic effects.

3.4 Transgenerational Genetic Consequences of Hippocampal Dysplasia

The hippocampus, a crucial regulatory center of the HPA axis, is highly sensitive to the intrauterine environment. Research has shown that prenatal glucocorticoid exposure can induce neuroendocrine and behavioral alterations in the third generation of female guinea pigs. These changes are accompanied by significant alterations in the CpG methylation level of the RNA pol II binding region in several small non-coding RNA (snRNA) genes, potentially linked to abnormal hippocampal development across three generations (Constantinof et al. 2019). Additionally, transgenerational genetic effects have been observed on the expression of estrogen response-related genes in the hippocampus of offspring rats, induced by prenatal betaine supplementation. This is associated with changes in DNA methylation in the imprinting control region (ICR) of IGF2 (Yang et al. 2020). Moreover, the administration of methylhydroxyl carbinol acetate during pregnancy has been linked to a schizophrenic phenotype observed in the third trimester. This phenotype correlates with an increased function of the hippocampal dopamine system and a decrease in GABA transmission across the third trimester (Guidotti et al. 2012). A recent study by our team revealed that after administering a clinically equivalent dose of dexamethasone, offspring rats exhibited a decline in novel cognitive indices, as well as spatial learning and memory abilities. Remarkably, this learning and memory impairment persisted into the F2 and F3 generations, which was associated with alterations in the hippocampal miR-133a-3p/SIRT1/CDK5/NR2B signaling axis. Subsequent intervention experiments on zebrafish fertilized eggs indicated that the high specific expression of miR-133a-3p in oocytes might be a key factor for its transgenerational inheritance (Luo et al. 2023).

3.5 Transgenerational Genetic Influences on Various Organ Dysplasias

Beyond the aforementioned tissues and organs, adverse environmental exposures during pregnancy can also detrimentally impact the development of numerous other tissues and organs, manifesting multigenerational genetic effects. For instance, prenatal nicotine exposure in rats has been linked to alterations in liver glucose and lipid metabolism in offspring, with these changes being inheritable to the F2 generation (Hu et al. 2020). Exposure to prenatal tributyltin has been observed to increase the weight of white adipose tissue and the size and count of fat cells and to reprogram pluripotent mesenchymal stem cells toward adipocyte lineages at the expense of bone development in third-generation offspring. This is accompanied by the upregulation of genes associated with lipid storage/transport, lipogenesis, and lipolysis in the livers of three successive offspring generations (Chamorro-Garcia et al. 2013). Furthermore, prenatal lipopolysaccharide exposure has been shown to impair renal salt excretion capabilities in offspring, leading to hypertension in the F1, F2, and F3 generations and salt-sensitive hypertension in the F4 and F5 generations. This effect is linked to alterations in histone methylation levels in oocytes (Cao et al. 2022). Additionally, research has indicated that maternal obesity during pregnancy heightens the risk of obesity and cardiometabolic diseases in offspring. Notably, while this pathological change is less pronounced in the F1 generation, it becomes significantly more evident in the F2 and F3 generations. In conclusion, diseases originating in fetal development are characterized by multigenerational genetic impacts, as depicted in Fig. 8.2.

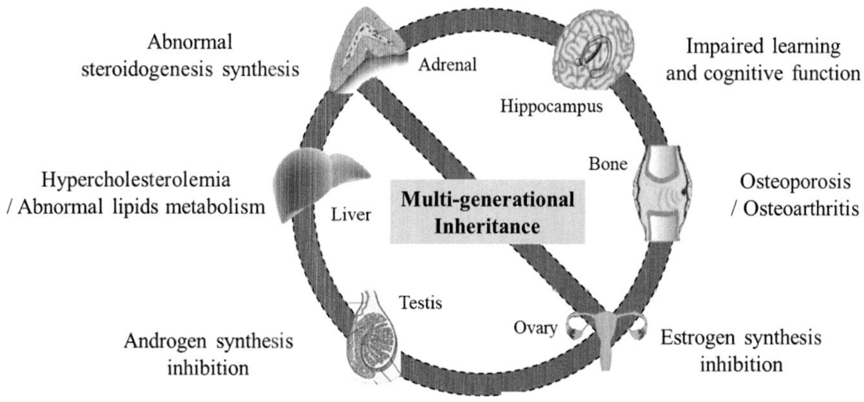

Fig. 8.2 Transgenerational genetic consequences of fetal-originated disease

4 Epigenetic Reprogramming Mechanisms Underlying Multigenerational Genetic Effects of Fetal-Originated Diseases

Epigenetic modifications represent stable genetic alterations encompassing DNA methylation, histone modification, and noncoding RNA activities. These epigenetic aberrations are increasingly recognized for their pivotal role in multigenerational genetic effects. They mediate not only the abnormal phenotypic changes in somatic cells but also facilitate transmission through germline cells.

4.1 Aberrations in Somatic Epigenetic Modifications

Methylation, a stable epigenetic modification, is generally inversely related to gene expression. Epidemiological studies have revealed that daughters of mothers who consumed synthetic estrogen diethylstilbestrol during pregnancy exhibit a twofold increased risk of breast cancer. This heightened risk is linked to altered DNA methylation levels in genes governing stem cell regulation and cellular differentiation (Hilakivi-Clarke 2014). Moreover, prenatal exposure to inflammation significantly elevates hypertension incidence in offspring, persisting into the F2 and F3 generations. This effect is associated with multigenerational alterations in DNA methylation in the thoracic aorta of offspring, influencing PI2K/Akt signaling activation and leading to vascular remodeling in the thoracic aorta and superior mesenteric artery, culminating in vasoconstrictive dysfunction (Guan et al. 2022). Histone modifications and noncoding RNAs also play critical roles in epigenetic regulation prior to transcription. The authors' team previous research indicates that these modifications often occur in interconnected networks. For instance, in PDE rat models, diminished levels of miR-466b-3p in F1, F2, and F3 testicular Leydig cells lead to decreased histone acetylation and expression in the StAR promoter region via targeted regulation of HDAC7 (Liu et al. 2023). Additionally, PDE impacts estrogen synthesis in offspring's ovarian granulosa cells by modulating miR-320-3p expression (Gong et al. 2021) and contributes to osteoarthritis susceptibility by affecting miR-98-3p expression in long bone cells (Han et al. 2022). The microRNA let-7 has been found to inhibit steroid synthesis in adrenal cortical cells (He et al. 2021). In conclusion, adverse environmental exposures during pregnancy can influence somatic cell functions through various epigenetic modifications, with these changes enduring across multiple generations.

4.2 *Epigenetic Reprogramming in Germ Cells*

In the realm of multigenerational genetic effects, germ cells serve as vital conduits for the transmission of genetic information from parents to offspring. Epigenetic modifications in these cells can be perpetuated into the somatic cells of progeny, giving rise to transgenerational genetic effects. Through extensive research, our team has discovered that rats from the F1, F2, and F3 generations, derived maternally post PDE, exhibit abnormal functions in multiple tissues and organs, as well as cross-generational genetic effects. These include altered gonadal hormone synthesis, impaired hippocampal cognitive functions, and disrupted bone development. Analysis of F1 and F2 generation female rat oocytes revealed significant changes in the expression of key microRNAs and DNA methylation levels in certain miRNA promoter regions. This suggests that critical epigenetic modifications in oocytes might bypass reprogramming processes, thereby mediating maternal multigenerational genetic effects. Furthermore, the impact of paternal transgenerational inheritance is gaining increased attention. For instance, chronic paternal stress results in notable alterations in the tsRNA expression profile in sperm. These aberrant tsRNA profiles can be transmitted to offspring, affecting liver glucose and lipid metabolism, and even influencing the dietary habits and lifestyles of the progeny (Chen et al. 2016). Other studies have demonstrated that chronic paternal stress can alter the DNA methylation levels of imprinted genes in the imprinting control regions of sperm. These changes can persist in offspring tissues and organs, leading to aberrant expression of imprinted genes and their downstream signaling pathways, ultimately inducing related diseases. Globally, research suggests that epigenetic modifications in germ cells are primary catalysts for multigenerational genetic effects, both maternally and paternally. However, due to the intricate proliferation and differentiation processes of germ cells, as well as existing technical constraints, the precise mechanisms of these epigenetic changes in germ cells require further elucidation.

5 Summary and Prospect

This chapter comprehensively examines the multigenerational genetic impacts of diseases originating in the fetal stage. It delves into the phenomenon where parental exposure to adverse environmental conditions can induce functional gene expression alterations and abnormalities in tissue and organ functions. This is achieved through the epigenetic modification of key functional genes in the offspring's somatic cells. Additionally, such environmental exposures can also influence epigenetic modifications in germ cells, including oocytes and sperm. These epigenetic changes can be passed down to progeny cells, thereby instigating multigenerational genetic effects. This review contributes to a deeper understanding of the etiology and enduring impacts of fetal-originated disease. It provides a theoretical foundation for the early diagnosis, proactive monitoring, and prevention of these diseases and offers guidance for improved eugenics and reproductive health planning.

References

Boskovic A, Rando O J. 2018. Annu Rev Genet, 52: 21–41.
Burton N O, Greer E L. 2022. Semin Cell Dev Biol, 127: 121–132.
Cao J, Chen Y, Xia X, et al. 2021. Toxicol Lett, 351: 78–88.
Cao N, Lan C, Chen C, et al. 2022. Circulation, 146(14): 1082–1095.
Chamorro-Garcia R, Sahu M, Abbey R J, et al. 2013. Environ Health Perspect, 121(3): 359–366.
Chen Q, Yan M, Cao Z, et al. 2016. Science, 351(6271): 397–400.
Constantinof A, Boureau L, Moisiadis V G, et al. 2019. Sci Rep, 9(1): 18211.
Estermann M A, Major A T, Smith C A. 2021. Genes (Basel), 12(9): 1459.
Gong X, Dai S, Wang T, et al. 2023. Cell Biol Toxicol, 39(3):867–883.
Gong X, Zhang J, Ge C, et al. 2021. Pharmacol Res, 165: 105435.
Guan X, Dan G R, Yang Y, et al. 2022. Acta Pharmacol Sin, 43(6): 1419–1429.
Guidotti G, Calabrese F, Auletta F, et al. 2012. Neuropsychopharmacology, 37(3): 746–758.
Han H, Xiao H, Wu Z, et al. 2022. Exp Mol Med, 54(3): 298–308.
He B, Wen Y, Hu S, et al. 2019. J Endocrinol, 242(3): 211–226.
He Z, Zhang J, Chen G, et al. 2021. Sci Total Environ, 792: 148440.
He Z, Zhang JZ, Chen YW, et al. 2023. Cell Commun Signal, 2023; 21(1): 294.
Hu W, Wang GH, He B, et al. 2020a. Toxicology, 432:152378.
Hu W, Yuan C, Luo HW, et al. 2020b. Toxicol Lett, 331:167–177.
Hilakivi-Clarke L. 2014. Breast Cancer Res, 16(2): 208.
Hojsager F D, Andersen M, Juul A, et al. 2022. Environ Int, 167: 107417.
Hu W, Wang G, He B, et al. 2020. Toxicology, 432: 152378.
Huang T, Zhang W, Lin T, et al. 2022. Food Chem Toxicol, 160: 112803.
Liu L, Li B, Li Q, et al. 2023. J Adv Res, 47:123–135.
Luo H, Deng Z, Liu L, et al. 2014. Toxicol Appl Pharmacol, 274(3): 383–392.
Luo M C, Yi Y W, et al. 2023. Acta Pharm Sin B, 3(9): 3708–3727.
Lv F, Wan Y, Chen Y, et al. 2018. Endocrinology, 159(3): 1401–1415.
Ma Z, Wang Y, Quan Y, et al. 2022. Hum Genomics, 16(1): 34.
Nicolella H D, de Assis S. 2022. Int J Mol Sci, 23(9).
Tie K, Zhao Z, Wu ZX et al. 2022. Biochem Pharmacol, 203: 115196.
Trefts E, Gannon M, Wasserman D H. 2017. Curr Biol, 27(21): R1147-R1151.

Wasterlain C G, Shirasaka Y. 1994. Brain Dev, 16(4): 279–295.
Yang Y, Jiang W, Yang S, et al. 2020. Mol Nutr Food Res, 64(8): e1900823.
Yohn N L, Bartolomei M S, Blendy J A. 2015. Prog Biophys Mol Biol, 118(1–2): 21–33

Chapter 9
Developmental Toxicity of Caffeine and Related Fetal-Originated Disease

Yinxian Wen and Kexin Liu

Abstract An increasing number of studies suggest that caffeine is detrimental to fetal development and can cause intrauterine dysplasia in offspring, leading to susceptibility to various diseases after birth, with potential multigenerational genetic effects. Prenatal caffeine exposure has been shown to have both direct toxic effects on the fetus by crossing the placenta and indirect effects by stimulating the maternal hypothalamic-pituitary-adrenal axis and opening the placental glucocorticoid barrier, which can result in overexposure of the fetus to maternal-originated glucocorticoids, leading to fetal multi-organ dysfunction. Therefore, the developmental toxicity of caffeine poses both immediate and long-term hazards. Intrauterine maternal-originated glucocorticoid overexposure caused by prenatal caffeine exposure can induce developmental programming changes in offspring, affecting the glucocorticoid-insulin-like growth factor 1 axis, hypothalamic-pituitary-adrenal axis, and the renin-angiotensin system, leading to postnatal susceptibility to various diseases. Epigenetics is involved in regulating the programming mechanisms of caffeine developmental toxicity. This chapter provides a comprehensive review of the short-term and long-term hazards and mechanisms of caffeine developmental toxicity, providing an experimental and theoretical basis for the analysis of the international hot theory of "Developmental Origins of Health and Disease".

1 Introduction

Caffeine, also known as 1,3,7-trimethylxanthine, is a xanthine derivative commonly found in coffee, tea, chocolate, and some caffeine-containing medications (Brent et al. 2011). Due to its stimulant effects on the central nervous system and its

Y. Wen (✉)
Division of Joint Surgery and Sports Medicine, Department of Orthopedic Surgery, Zhongnan Hospital of Wuhan University, Wuhan, Hubei, China

K. Liu
Department of Pharmacy, Zhongnan Hospital of Wuhan University, Wuhan, Hubei, China

H. Wang et al. (eds.), *Fetal Origin of Diseases*,
https://doi.org/10.1007/978-981-97-5730-5_9

potential to improve cardiovascular function, caffeine-containing beverages are widely consumed to reduce fatigue, increase alertness, and improve heart and digestive health (Heckman et al. 2010). Approximately 60% to 75% of pregnant women worldwide consume caffeinated beverages, making caffeine consumption during pregnancy a common practice (Maslova et al. 2010; Frary et al. 2008; Group 2008). However, the safety of caffeine consumption during pregnancy has been a concern. While caffeine intake in adulthood may prevent or alleviate metabolic syndrome (MS), epidemiological and laboratory studies suggest that prenatal caffeine exposure (PCE) can be reproductive and developmental toxicity (Fortier et al. 1993; Weng et al. 2008; Kuczkowski 2009). PCE may also be an important cause of a range of adult metabolic diseases, including MS, through the "intrauterine developmental origin" mechanism (Momoi 2006; Bakker et al. 2010). The World Health Organization recommends that pregnant women who consume more than 300 mg of caffeine per day, and their offspring, are at increased risk of intrauterine growth retardation (IUGR) (Guilbert 2003). The American Congress of Obstetricians and Gynecologists and the UK Food Standards Agency recommend that pregnant women consume less than 200 mg of caffeine per day (American College of and Gynecologists 2010; Crozier et al. 2012).

Over the past decade, the author's team has demonstrated that PCE can result in IUGR and increase the susceptibility of offspring to various metabolic diseases. We have proposed an innovative mechanism called "intrauterine neuroendocrine metabolism programming," which is related to the hypothalamic-pituitary-adrenal (HPA) axis (Xu et al. 2012b). The core of this mechanism is the "glucocorticoid-insulin-like growth factor 1 (GC-IGF1) axis programming change." In this chapter, we provide a comprehensive review of the intrauterine programming mechanism and research progress on the developmental toxicity and metabolic disease susceptibility induced by PCE.

2 Developmental Toxicity and Long-Term Harms of Caffeine

Developmental toxicity refers to the harm caused to the structure or function of the offspring before and after birth as a result of exposure to a poor living environment. The hazards of developmental toxicity can be both short-term and long-term (Setzer et al. 2001).

2.1 Adverse Pregnancy Outcomes in Offspring Caused by Prenatal Caffeine Exposure

In humans and mammals, caffeine is rapidly and completely absorbed from the stomach and small intestine following oral administration. The plasma concentration of caffeine reaches its peak within 30–120 min after administration, and an increase in dosage accelerates the absorption process without any obvious first-dose effect. Once absorbed, caffeine enters the bloodstream and is distributed throughout the body to achieve equilibrium. The main metabolic pathway of caffeine in humans is through Cytochrome P450 1A2 (CYP1A2) in the liver, resulting in the formation of metabolites such as paraxanthine, theobromine, and theophylline (Nehlig et al. 1992). The half-life of caffeine in the human body ($t_{1/2}$) varies significantly among individuals, depending on factors such as age, sex, pregnancy status, liver function intensity, the number of metabolic enzymes, and co-administration of other drugs. The $t_{1/2}$ of caffeine in healthy individuals ranges from 2 to 6 h, but it can be extended up to 18 h in late pregnancy due to the reduced activity of maternal liver CYP1A2 (Tsutsumi et al. 2001; Bakker et al. 2010). In pregnant women, caffeine is fully absorbed after oral administration and reaches peak concentration within 30–60 min (Newton et al. 1981). Furthermore, caffeine, being highly fat soluble, can easily cross the placenta and various fetal tissue barriers, such as the blood-brain barrier and blood-testosterone barrier, and distribute throughout the fetal body (Miners and Birkett 1996). Therefore, caffeine can have a direct effect on various fetal tissues, including the hippocampus (Xu et al. 2012c).

The adverse effects of PCE on offspring have been extensively studied, and previous research has reported that PCE is associated with spontaneous abortion (Greenwood et al. 2010), congenital malformation (Schmidt et al. 2009), growth retardation, and even death. Clinical studies have shown that maternal caffeine intake is associated with lower birth weight and smaller head circumference in newborns, and the severity of these adverse outcomes is dose-dependent (Watkinson and Fried 1985; Fernandes et al. 2020). Fortier et al. have also reported that excessive caffeine intake, such as more than 11 times the normal amount (3 cups of coffee per day), is associated with the occurrence of small for gestational age, which is characterized by reduced fetal growth and birth weight. Animal studies have shown that high doses of caffeine can induce congenital malformations, such as heart malformations, missing toes, and imperforation of the anus, and increase the rate of stillbirth, intrauterine growth retardation (IUGR), and abnormal bone development in offspring (Ikeda et al. 1982; Smith et al. 1987; Nehlig and Debry 1994). IUGR is a common developmental toxicity, and its clinical diagnostic criteria include a birth weight of less than 2500 g for infants born after 37 weeks of gestation or fetal weight less than 2 standard deviations of the mean weight for their gestational age (Bakker et al. 2010). In addition to genetic factors, adverse intrauterine environments, such as maternal and exogenous factors, can also contribute to the incidence of IUGR (Martin-Gronert and Ozanne 2007). PCE has been found to reduce birth weight in male monkeys (Gilbert and Rice 1991), and our recent series of animal experiments have confirmed that caffeine exposure during and after pregnancy can reduce body weight and length and increase the incidence of IUGR in fetal rats, including mice and rats, in a dose-dependent manner (20–180 mg/kg·d) (Liu et al. 2012; Xu et al. 2012b; Xu et al. 2012c). These findings suggest that caffeine has developmental toxicity and is one of the significant causes of IUGR, emphasizing the need to limit caffeine intake during pregnancy to prevent adverse outcomes in offspring.

2.2 Changes in Developmental Programming and Homeostasis of Multiple Organs in Offspring Induced by Prenatal Caffeine Exposure

Mounting evidence suggests that the predisposition to multiple diseases after birth caused by adverse intrauterine environments is linked to abnormal functional development of the hypothalamic-pituitary-adrenal (HPA) axis. The HPA axis is a crucial neuroendocrine axis of the stress response and a vulnerable target site for intrauterine injury. Exposure to high levels of GCs during pregnancy not only impairs the function of the neuroendocrine system in offspring after birth but also leads to metabolic syndrome, HPA axis-related diseases, and neurobehavioral changes. These permanent alterations to the endocrine system can have long-term and profound effects on health. Therefore, HPA axis programming changes are considered the most likely mechanism mediating the fetal origin of many adult diseases, such as metabolic syndrome (MS) (Reynolds 2010). Our previous studies have demonstrated that PCE can reduce placental 11 beta-hydroxysteroid dehydrogenase type 2 (11β-HSD2) expression, weaken the placental barrier to maternal GCs, and cause fetal rats to be overexposed to maternal GCs, resulting in impaired HPA axis development and IUGR (Xu et al. 2012c). These findings suggest that fetal HPA axis programming is a critical mechanism of fetal developmental toxicity and susceptibility to adult diseases caused by PCE.

The harmful effects of PCE on fetal growth and development have been extensively studied. PCE can create an unfavorable intrauterine environment that results in fetal growth retardation due to the loss of normal hormone regulation and changes in the programming of fetal neuroendocrine metabolism. The authors' team has reported that PCE can induce IUGR in rat offspring, which leads to multiple endocrine axis programming changes and resultant multiple organ abnormalities. For instance, PCE can cause a reduction in the synthetic function of the extracellular matrix of fetal chondrocytes, leading to delayed growth plate development, delayed differentiation of growth plate chondrocytes into calcified cells, and ultimately, bone growth retardation (Tan et al. 2012). Additionally, PCE can decrease the contents of corticosterone and aldosterone, as well as the expression levels of steroid synthase in the adrenal gland of fetal rats, which persists into adulthood (Chen et al. 2018). Furthermore, PCE can lead to abnormal morphological development of the fetal hypothalamus, such as increased or decreased neuronal generation, decreased endplate dendrite structure, and altered HPA axis activity (Lu et al. 2020). Moreover, PCE can cause hyperactivity of liver triglyceride and cholesterol synthesis in utero in offspring (Liu et al. 2021; Hu et al. 2019; Del et al. 2015), impaired postnatal glucose metabolism (Kou et al. 2017), and renal dysplasia (Zhu et al. 2019). Taken together, these studies suggest that PCE can cause changes in the developmental programming and homeostasis of multiple organs, which may persist after birth and predispose individuals to various diseases in adulthood.

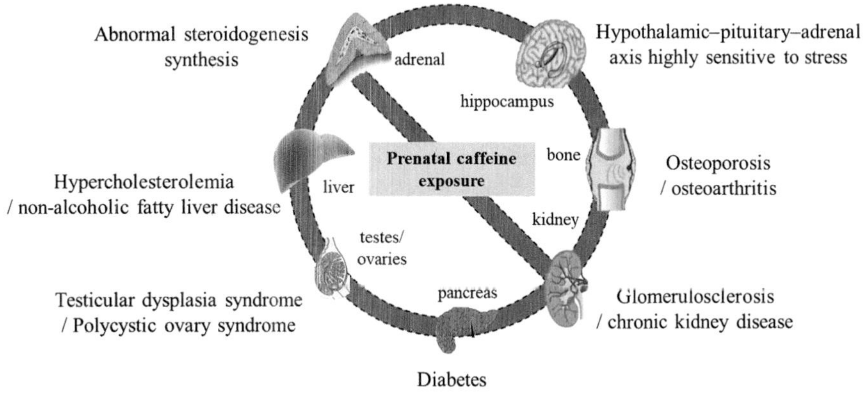

Fig. 9.1 Increased susceptibility to various diseases in adult offspring after prenatal caffeine exposure

2.3 Susceptibility to the Diseases in the Adult Offspring Due to Prenatal Caffeine Exposure

Numerous epidemiological studies have demonstrated that caffeine intake during adolescence can increase susceptibility to adult MS (Hino et al. 2007). The author team has conducted a series of animal experiments that confirm the adverse effects of PCE on the neuroendocrine and metabolic programming of offspring rats, which increase their susceptibility to MS and multiple organ diseases in adulthood (Xu et al. 2012b; Wang et al. 2014). These diseases include nonalcoholic fatty liver disease (NAFLD), hypercholesterolemia, hyper-stress sensitivity of the HPA axis, diabetes-like reaction, hyper-adrenal function, glomerulosclerosis, osteoporosis and osteoarthritis, and testicular and ovarian dysplasia (Xu et al. 2012b; Wang et al. 2014; Ao et al. 2015; Luo et al. 2015; Wu et al. 2015; Zhou et al. 2015; He et al. 2016; Zhang et al. 2020; Lv et al. 2021; Park et al. 2015) (Fig. 9.1).

3 Inducement and Mechanism of the Developmental Toxicity by Caffeine

In addition to its direct effects on fetal development, caffeine can also lead to overexposure of the offspring to high concentrations of maternal-originated GC through the opened-placental barrier, resulting in long-term effects on the offspring and even multigenerational effects.

3.1 Direct Effects and Mechanisms of the Developmental Toxicity by Caffeine

Caffeine has a high-fat solubility, easily through the placenta into the fetal tissue, so caffeine has a direct effect on the fetus. The study of the author's team showed that caffeine can increase the gene expression of the steroidogenic acute regulator (StAR) in adrenal cells with the demethylation of CpG in the promoter region (Ping et al. 2012). It also reduced 11 beta-hydroxysteroid dehydrogenase type 2 (11β-HSD2) expression and increased expression of 11β-HSD1 and glucocorticoid receptor (GR) is associated with increased methylation of the 11β-HSD2 promoter region (Xu et al. 2012c). This result may lead to increased local GC concentrations in the fetal hippocampus and excessive activation of GR, resulting in hippocampal toxicity, which is the main cause of postnatal hyper-stress sensitivity of the HPA axis (Xu et al. 2012c). Recent studies in our laboratory have shown that caffeine can open the placental GC barrier by inhibiting the placental cAMP/PKA signaling pathway and up-regulating the expression of EGR1, resulting in abnormal genetic modification and expression inhibition of the 11β-HSD2 promoter region (to be published).

Due to its high-fat solubility, caffeine can easily pass through the placenta and directly affect fetal tissues. Research conducted by the author's team has shown that caffeine exposure can increase the gene expression of the steroidogenic acute regulator (StAR) in adrenal cells by demethylating CpG in the promoter region (Ping et al. 2012). Additionally, caffeine exposure can decrease the expression of 11β-HSD2 and increase the expression of 11β-HSD1 and GR by increasing the methylation of the 11β-HSD2 promoter region (Xu et al. 2012c). These changes may lead to an increase in local GC concentrations in the fetal hippocampus and excessive activation of GR, resulting in hippocampal toxicity, which is the main cause of postnatal hyper-stress sensitivity of the HPA axis (Xu et al. 2012c). Recent studies conducted by our laboratory have also shown that caffeine exposure can open the placental GC barrier by inhibiting the placental cAMP/PKA signaling pathway and up-regulating the expression of EGR1, which leads to abnormal genetic modification and expression inhibition of the 11β-HSD2 promoter region (to be published). Furthermore, the increase of fetal blood caffeine levels caused by maternal caffeine exposure during pregnancy can also have a direct effect on various fetal tissues. For instance, caffeine exposure can upregulate the expression of HMGCR and FASN in the liver by activating the cAMP/PKA signaling pathway, leading to enhanced lipid synthesis function in the fetal liver (Hu et al. 2019). Moreover, caffeine exposure can activate the miR375/CTGF signal, which leads to H-type dysangiogenesis and inhibition of osteogenic differentiation during long bone development (He et al. 2021), among other effects (Fig. 9.2).

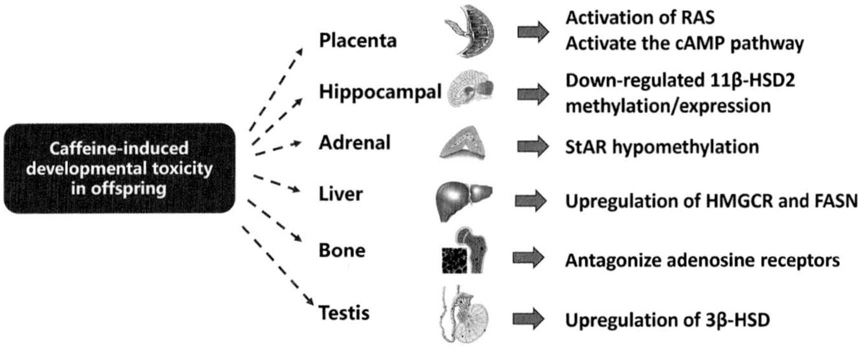

Fig. 9.2 Direct effect and mechanism of caffeine-induced developmental toxicity in offspring

3.2 *Indirect Effects and Mechanisms by Maternal-Derived Glucocorticoid*

Maintaining normal intrauterine GC levels is crucial for regulating fetal tissue morphology and functional maturity, as excessively high concentrations of GC exposure can lead to fetal developmental abnormalities (Fowden et al. 1998). During pregnancy, the circulating GC level of the mother is 5–10 times higher than that of the fetus. The maintenance of this concentration difference is mediated by the placental GC barrier, which is primarily composed of 11β-HSDs and P-glycoprotein (P-gp).

Within the GC metabolic enzyme system, 11β-HSD1 activates GC via its reductive activity, whereas 11β-HSD2 inactivates GC through its oxidative activity (Chapman et al. 2013). Both 11β-HSD1 and 11β-HSD2 are expressed in the placenta (Xu et al. 2012c), and the balance between the two enzymes constitutes the GC barrier of the placenta. Intrauterine adverse environments, such as hypoxia, ischemia, and maternal stress, have been shown to downregulate the expression of placental 11β-HSD2, leading to the opening of the placental GC barrier and excessive exposure of the fetus to maternally derived GC (Straley et al. 2014; Seth et al. 2015; Kosicka et al. 2016; Togher et al. 2017). Our team has recently demonstrated that PCE can directly inhibit placental 11β-HSD2 expression through the cAMP/PKA/EGR1 signaling pathway, resulting in increased concentrations of corticosterone in fetal serum and placenta (to be published). In addition, high concentrations of GC in maternal serum under PCE can activate the glucocorticoid receptor-CCAAT enhancer binding protein α-early growth response 1 (GR-C/EBPα-Egr1 pathway), leading to the inhibition of 11β-HSD2 expression in male placenta (Yu et al. 2022). These findings suggest that PCE can inhibit 11β-HSDs and open the placental GC barrier through caffeine or high concentrations of GC, leading to excessive exposure of the fetus to maternally derived GC.

P-gp, a member of the ATP-binding cassette (ABC) transporter family, utilizes ATP to efflux foreign substances from the maternal surface of placental trophoblasts back to the maternal circulation, providing a protective function against high maternal GC levels. ABCB1/MDR1, also known as P-gp, is the most abundant drug efflux transporter expressed on the apical membrane of the maternal surface of the placental syncytial trophoblast. P-gp constitutes an additional placental GC barrier that pumps GC outwards back to the maternal surface against the concentration gradient, thereby limiting the passage of maternally derived GCs across the placenta to the fetus. Previous studies have demonstrated that maternal exposure to cannabis, lipopolysaccharides, or malnutrition can inhibit placental P-gp expression (Soo et al. 2012; Bloise et al. 2013; Feinshtein et al. 2013). Recently, our team found that PCE can activate the placental ryanodine receptor/c-Jun N-terminal kinase (RYRs/JNK) pathway. The RYRs/JNK pathway inhibits the expression of Y-box binding protein-1 (YB-1), resulting in the epigenetic modification and downregulation of P-gp expression, leading to impaired placental GC efflux and restricted fetal development (Ge et al. 2021).

Furthermore, the authors' team also discovered a correlation between the expression of p-JNK, YB-1, and P300 in the placenta of IUGR children and the expression of P-gp, indicating that the placental p-JNK/YB-1/P300 pathway is a common mechanism for the decreased expression of P-gp in IUGR children. This common mechanism may induce the development of a fetal "thrifty phenotype" and increase susceptibility to multiple chronic diseases in adulthood under adverse intrauterine environments (Soo et al. 2012; Reynolds 2013). The thrifty phenotype hypothesis, initially proposed by David Barker and Hales, suggests that when a fetus experiences malnutrition in early life, it undergoes short-term compensatory adaptations to the adverse intrauterine environment, which become permanent "programming." Consequently, the fetal body size is smaller, and some essential internal organs undergo structural and functional changes. These individuals are more sensitive to the nutrient-rich environment after birth, which increases the likelihood of developing metabolic diseases.

4 Mechanisms of Multi-Organ Programming and Disease Susceptibility in Offspring Induced by Prenatal Caffeine Exposure

Intrauterine programming refers to the process of permanent changes in tissue morphology and function caused by early developmental damage, resulting in increased susceptibility to various diseases (Zhang et al. 2014). Based on numerous systematic studies (Liu et al. 2012; Tan et al. 2012; Xu et al. 2012c; Kou et al. 2014; Wang et al. 2014), the theory of "intrauterine neuroendocrine metabolic programming" was proposed to explain the susceptibility of PCE-induced IUGR offspring to various diseases after birth. This theory suggests that maternal-derived GC overexposure induced by PCE can alter the functional development of the offspring's HPA axis and induce intrauterine programming changes of the "GC-IGF1 axis," resulting in postnatal HPA axis hypersensitivity and GC-dependent lipid metabolic function changes, increasing susceptibility to various metabolic diseases. Various chronic stimuli after birth, such as overnutrition and high mental stress, can further disrupt this neuroendocrine metabolic programming, eventually inducing or exacerbating the occurrence of metabolic diseases.

4.1 Epigenetic Mechanisms and Multigenerational Genetic Effects of Developmental Toxicity by Caffeine

Despite the extensive reporting of offspring susceptibility to metabolic diseases and their heritable effects resulting from adverse environmental conditions during pregnancy, the intrauterine programming mechanism, particularly the epigenetic mechanism, has yet to be fully elucidated and remains a current research focus. Epigenetic modifications, including DNA methylation, histone modification, and noncoding RNAs, are present in the normal development of higher eukaryotic cells. A growing body of evidence suggests that epigenetic modifications play a key role in the intrauterine programming of various chronic diseases (Luo et al. 2015).

The author's team has discovered abnormal epigenetic modifications in PCE fetal rats related to multiple organ development-related genes. These modifications include high DNA methylation in the promoter region of 11β-HSD2 in the hippocampus (Xu et al. 2012a; Xia et al. 2014), DNA methylation and histone deacetylation in the promoter region of the adrenal SF-1 (Ping et al. 2014), decreased histone methylation in the IGF1 promoter P2 region, and reduced IGF1 expression in the liver (Tan et al. 2012; Fu et al. 2009), as well as histone acetylation in the promoter region of cholesterol anabolic functional genes (Xu et al. 2018; Hu et al. 2019). Histone deacetylase sirtuin1 (SIRT1) and acetylase p300 mediate the histone acetylation level and expression of functional genes of cholesterol metabolism in intrauterine and postnatal rats of PCE offspring (Xu et al. 2018; Hu et al. 2019). These epigenetic modifications can serve as biomarkers and early warning targets for PCE-induced offspring-related disease susceptibility. Recent studies by our team have demonstrated that PCE can induce multigenerational damage to articular cartilage in female offspring rats, which is related to the decreased H3K9ac level of the chondrocyte TGFβ signaling pathway induced by intrauterine high maternal-originated GC (Zhao et al. 2020). Abnormal synthesis of adrenal steroids in female adult offspring of PCE exhibits multigenerational inheritance and is related to GC-dependent programming of the imprinted gene H19/let-7 axis (He et al. 2021). These findings provide evidence for the multigenerational genetic effects of PCE on fetal-originated disease susceptibility in offspring.

4.2 The "Two-Programming" Mechanisms for Fetal-Originated Disease Associated with Developmental Toxicity by Caffeine

In the 1980s, British scholar David Barker and others proposed the "fetal origin theory" of adult metabolic diseases, suggesting that the occurrence of adult metabolic diseases is not only related to poor lifestyle but may also be related to abnormal birth weight caused by exposure to adverse factors during fetal development in utero. Building on previous studies of multiple organs and different developmental stages of progeny in PCE rat models, the authors' team have proposed a systematic mechanism theory of caffeine developmental toxicity.

During pregnancy, exposure of the mother to various substances, including caffeine, can lead to an increase in maternal endogenous GC levels (Ana et al. 2015). Additionally, prenatal exposure to polycyclic aromatic hydrocarbons (PAHs) can disrupt the placental GC barrier, resulting in excessive exposure of the fetus to maternal-derived GC, which can limit fetal development (Yu et al. 2022; Ge et al. 2021) (to be published). This excess exposure is also a significant cause of the increased risk of metabolic diseases in offspring later in life (Stroud et al. 2016). Moisiadis and Matthews (2014) reported that changes in maternal-derived GC levels during pregnancy mediated changes in fetal development programming due to adverse events during pregnancy, which played a crucial role in fetal-originated disease caused by xenobiotic exposure during pregnancy. Our team conducted animal experiments and found that maternal-derived GC overexposure of fetuses induced by prenatal PAHs exposure can activate the GR in various organs and recruit a variety of epigenetic enzymes, which can jointly regulate the expression of functional genes of multiple organs, leading to programming changes (Zhou et al. 2019; Xu et al. 2018). This programming change can persist after birth, eventually causing susceptibility to multiple organ-related diseases in the offspring into adulthood. This is the "first programming" mechanism of PAH-induced susceptibility to multiple diseases in offspring.

IGF1 is a growth factor synthesized primarily in the liver, which plays a vital role in regulating cellular proliferation, differentiation, and metabolism in various tissues (Netchine et al. 2009). Recent studies have shown that GC, a crucial factor in regulating fetal tissue's morphological and functional maturation, can inhibit the expression and secretion of IGF1 (Kou et al. 2014; Wang et al. 2014). Through a series of previous studies, the author team discovered that PCE-induced IUGR fetal mice exhibited maternal-derived GC overexposure and reduced IGF1 levels in fetal serum and multi-tissue organs. It was further observed that the serum corticosterone concentration of IUGR offspring decreased after birth, while the liver IGF1 expression increased, and catch-up growth and metabolic disorders were aggravated under a high-fat diet (Shen et al. 2014; Wang et al. 2014; Xu et al. 2015). This programming effect is the "GC-IGF1 axis" programming change caused by exposure to adverse environmental conditions (such as xenobiotics) during pregnancy, resulting in low IGF1 concentration in utero and high IGF1 concentration after birth, leading to fetal IUGR and catch-up growth after birth. This is an overall adaptation and compensatory change of IUGR individuals in response to different living environments after birth. The "GC-IGF1 axis" is the strongest physiological compensatory axis for body development and is the "second programming" mechanism of PCE-induced multi-disease susceptibility in offspring mediated by programming changes in the reduction of fetal metabolic rate in utero, catch-up growth after birth, and metabolic disorders in the later period.

4.3 The "Two-Strike" Mechanisms for Fetal-Originated Disease Associated with Developmental Toxicity by Caffeine

IUGR offspring exposed to PCE also exhibit hyperreactivity of the HPA axis under chronic stress after birth. When these offspring experience chronic stimulation, such as a high-fat diet after birth, it can result in accelerated catch-up growth and further increase susceptibility to multiple organ diseases (Xu et al. 2012c). The authors' team conducted an analysis and discovered that PCE-induced hypercholesterolemia in offspring rats was caused by xenobiotic exposure during pregnancy, a high-fat diet after birth, and the interaction between different genders (Guo et al. 2018; Qi et al. 2017). In recent years, based on the research progress in the intrauterine pathogenesis of various fetal-originated diseases, the authors' team proposed the "two-strike" mechanism of adult metabolic diseases induced by PCE in the rat IUGR model.

According to the proposed "two-strike" mechanisms, PCE acts as the "first strike" leading to intrauterine maternal-derived GC overexposure and "two-programming" mechanism changes in offspring's multi-organ. The "first programming" mechanism involves specific changes (enhancement or reduction) in various tissues and organs' function, which can continue from intrauterine to postnatal and even adult stages. This programming is related to the offspring's "thrifty phenotype" programming, mediated by abnormal epigenetic modification. The "second programming" mechanism is the programming change of the "GC-IGF1 axis" in various fetal tissues and organs. This programming can cause low IGF1 concentration in utero and high IGF1 concentration after birth, leading to IUGR and catch-up growth under conditions of hyper-nutrition after birth. PCE and resulting maternal high GCs, as the first strike, interact with these "two-programming" mechanism leading to multi-organ development programming, changes in functional homeostasis after birth, and susceptibility to various diseases in adulthood. However, environmental changes or adverse lifestyle habits after birth, such as a high-fat diet or chronic stress, may act as a "second strike" and accelerate or aggravate the occurrence of fetal-originated disease in offspring (Fig. 9.3).

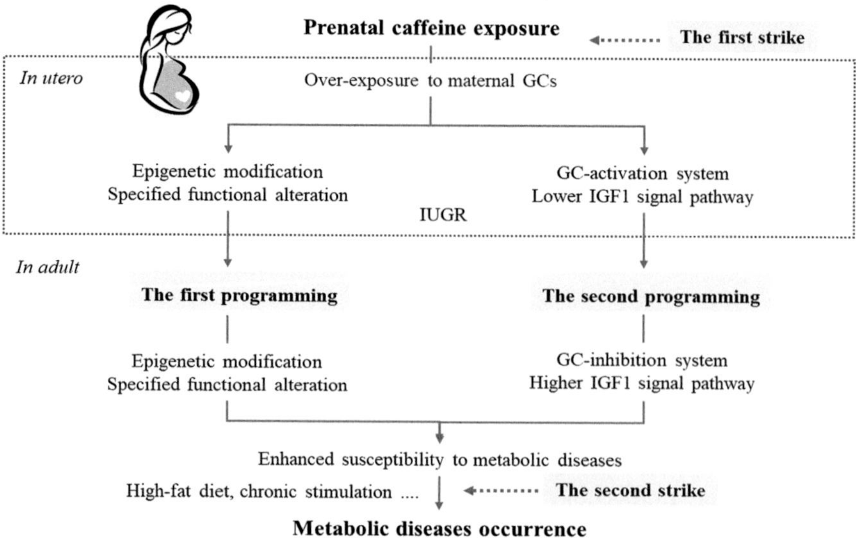

Fig. 9.3 Intrauterine mechanisms of susceptibility to metabolic diseases in offspring by prenatal caffeine exposure

5 Summary and Prospect

Given the increasing consumption of caffeinated foods and drinks, the safety implications of caffeine have become a significant concern. Although many studies have reported the reproductive and developmental toxicity of caffeine, few studies have investigated its long-term effects and mechanisms. Through a series of animal experiments combined with clinical studies, the author team discovered that adverse intrauterine environments caused by PCE can lead to maternal-originated high GC concentration overexposure of the fetus due to the open placental GC barrier. This can cause a variety of endocrine development programming changes, such as the changes in "GC-IGF1 axis," resulting in abnormal development of multiple organ functions, and the surviving fetus presents a "thrifty phenotype." Catch-up growth occurs after birth. Intrauterine dysplasia and catch-up growth after birth can induce and exacerbate the incongruity between the local functional development of various tissues and organs and the systemic development, eventually leading to increased susceptibility to various chronic diseases, represented by MS. During the study, the authors' team found that the PCE-induced rat IUGR model exhibited many similar toxic manifestations and common programming mechanisms with other IUGR models induced by exposure to xenobiotics and dietary restriction during pregnancy. An in-depth understanding of the intrauterine neuroendocrine metabolism programming mechanism and related epigenetic studies can provide new ideas for early prediction and intervention of "fetal-originated disease." With the progress of the theory of "fetal-originated disease," it is believed that there will be more methods to overcome the toxic effects of caffeine on pregnant women and children and eliminate various long-term diseases caused by caffeine exposure. This is of great significance for improving the population's quality of life.

References

American College of and Gynecologists. Committee Opinion, Number 462, August 2010.
Ana C G, Navarro-Fontestad C, Mangas-Sanjuan V, et al. 2015. Eur J Pharm Sci, 73: 57–63.
Ao Y, Sun Z, Hu S, et al. 2015. Toxicol Appl Pharmacol, 287(2): 128-138.
Bakker R, Steegers E A, Obradov A, et al. 2010. Am J Clin Nutr, 91(6): 1691–1698.
Bermejo M. 2015. Eur J Pharm Sci, 73: 57–63.
Bloise E, Bhuiyan M, Audette M C, et al. 2013. PLoS One, 8(6): e65728.
Brent R L, Christian M S, Diener R M. 2011. Reprod Toxicol, 92(2): 152–187.
Chapman K, Holmes M, Seckl J. 2013. Physiol Rev, 93(3): 1139-1206.
Chen G, Yuan C, Duan F, et al. 2018. Toxicol Appl Pharmacol, 341: 64–76.
Crozier T W, Stalmach A, Lean M E, et al. 2012. Food Funct, 3(1): 30–33.
Del Bas J M, Crescenti A, Arola-Arnal A, et al. 2015. J Nutr Biochem, 26(12): 1670–1677.
Feinshtein V, Erez O, Ben-Zvi Z, et al. 2013. PeerJ, 1: e153.
Fernandes P R, Almeida F B, da Cunha M V, et al. 2020. Physiol Behav, 223: 112966.
Fortier I, Marcoux S, Beaulac-Baillargeon L. 1993. Am J Epidemiol, 137(9): 931–940.
Fowden A L, Li J, Forhead A J. 1998. Proc Nutr Soc, 57(1): 113-122.
Frary C D, Johnson R K, Wang M Q. 2008. J Am Diet Assoc, 108(4): 110–113.
Fu Q, Yu X, Callaway C W, et al. 2009. FASEB J, 23(8): 2438–2449.
Ge C Y, Xu D, Yu P X, et al. 2021. BMC Med, 19(1): 311.
Gilbert S G, Rice D C. 1991. Fundam Appl Toxicol, 17(3): 454-465.
Greenwood D C, Alwan N, Boylan S, et al. 2010. Eur J Epidemiol, 25(4): 275-280.
Group C S. 2008. BMJ, 337: a2332.
Guilbert J J. 2003. Educ Health (Abingdon), 16(2): 230.
Guo Y, Luo H, Wu Y, et al. 2018. Reprod Toxicol, 79: 47–56.
He Z, Zhu C, Huang H, et al. 2016. Toxicology Research, 5(2): 388-398.
He Z, Zhang J, Chen G, et al. 2021. Sci Total Environ, 792: 148440.
Heckman M A, Weil J, Gonzalez de Mejia E. 2010. J Food Sci, 75(3): R77–87.
Hino A, Adachi H, Enomoto M, et al. 2007. Diabetes Res Clin Pract, 76(3): 383-389.
Hu S, Liu K, Luo H, et al. 2019. Toxicology, 418: 11-21.
Ikeda G J, Sapienza P, McGinnis M L, et al. 1982. J Appl Toxicol, 2(6): 307-314.
Kosicka K, Siemiatkowska A, Glowka F K. 2016. Int J Endocrinol, 2016: 5279462.
Kou H, Liu Y, Liang G, et al. 2014. Toxicol Appl Pharmacol, 275(2): 79-87.
Kou H, Wang G H, Pei L G, et al. 2017. Naturwissenschaften, 104(11–12): 89.
Kuczkowski K M. 2009. Ann Fr Anesth Reanim, 28(6): 605–607.
Liu Y, Xu D, Feng J, et al. 2012. Toxicol Appl Pharmacol, 262(2): 205-216.
Liu H Z, He B, Hu W. 2021. Biochem Pharmacol, 185: 114420.
Lu J, Li Q, Xu D. 2020. Toxicol Lett, 331: 31–41.
Luo H, Li J, Cao H, et al. 2015. Sci Rep, 5: 17746.
Lv F, Fan G, Wan Y, et al. 2021. Sci Total Environ, 789: 147691.
Martin-Gronert M S, Ozanne S E. 2007. J Intern Med, 261(5): 437-452.
Maslova E, Bhattacharya S, Lin SW, et al. 2010. Am J Clin Nut, 92: 1120-1132.
Miners J O, Birkett D J. 1996. Gen Pharmacol, 27(2): 245–249.
Moisiadis V G, Matthews S G. 2014. Nat Rev Endocrinol, 10(7): 391–402.
Momoi T. 2006. Curr Mol Med, 6(1): 111–118.
Nehlig A, Debry G. 1994. J Am Coll Nutr, 13(1): 6-21.

Nehlig A, Daval J L, Debry G. 1992. Brain Res Brain Res Rev, 17(2): 139–170.
Netchine I, Azzi S, Houang M, et al. 2009. J Clin Endocrinol Metab, 94(10): 3913-3921.
Newton R, Broughton L J, Lind M J, et al. 1981. Eur J Clin Pharmacol, 21(1): 45–52.
Park M, Choi Y, Choi H, et al. 2015. Int J Endocrinol, 2015: 368475.
Ping J, Wang J F, Liu L, et al. 2014. Toxicology, 321: 53-61.
Ping J, Lei Y Y, Liu L, et al. 2012. I Chem Biol Interact, 195(1): 68–75.
Qi Y, Luo H, Hu S, et al. 2017. Cell Physiol Biochem, 44(2): 657–670.
Reynolds R M. 2010. J Steroid Biochem Mol Biol, 122(1–3): 3–9.
Reynolds R M. 2013. Psychoneuroendocrinology, 38(1): 1–11.
Schmidt R J, Romitti P A, Burns T L, et al. 2009. Birth Defects Res A Clin Mol Teratol, 85(11): 879–889.
Seth S, Lewis A J, Saffery R, et al. 2015. Int J Mol Sci, 16(11): 27482–27496.
Setzer R W, Lau C, Mole M L, et al. 2001. Toxicol Sci, 59(1): 49–58.
Shen L, Liu Z, Gong J, et al. 2014. Toxicol Appl Pharmacol, 274(2): 263–273.
Smith S E, McElhatton P R, Sullivan F M. 1987. Food Chem Toxicol, 25(2): 125–133.
Soo P S, Hiscock J, Botting K J, et al. 2012. Reprod Toxicol, 33(3): 374–381.
Straley M E, Togher K L, Nolan A M, et al. 2014. Placenta, 35(8): 533–538.
Stroud L R, Papandonatos G D, Parade S H, et al. 2016. Psychosom Med, 78(9): 979–990.
Tan Y, Liu J, Deng Y, et al. 2012. Toxicol Lett, 214(3): 279–287.
Togher K L, Treacy E, O'Keeffe G W, et al. 2017. Psychiatry Res, 255: 17–26.
Tsutsumi K, Kotegawa T, Matsuki S, et al. 2001. Clin Pharmacol Ther, 70(2): 121–125.
Wang L, Shen L, Ping J, et al. 2014. Toxicol Lett, 224(3): 311–318.
Watkinson B, Fried P A. 1985. Toxicol Teratol, 7(1): 9–17.
Weng X, Odouli R, Li D K. 2008. Am J Obstet Gynecol, 198(3): 279. e1-8.
Wu D M, He Z, Ma L P, et al. 2015. Toxicol Appl Pharmacol, 285(2): 89–97.
Xia L P, Shen L, Kou H, et al. 2014. Toxicol Lett, 226(1): 98–105.
Xu D, Liang G, Yan Y E, et al. 2012a. Toxicol Lett, 209(3): 282–290.
Xu D, Wu Y, Liu F, et al. 2012b. Toxicol Appl Pharmacol, 264(3): 395–403.
Xu D, Zhang B, Liang G, et al. 2012c. PLoS One, 7(9): e44497.
Xu D, Bai J, Zhang L, et al. 2015. Toxicol Res, 4(1): 112–120.
Xu D, Luo H W, Hu W, et al. 2018. FASEB J, 32(10): 5563–5576.
Yu P X, Zhou J, Ge C Y, et al. 2022. Sci Total Environ, 827: 154396.
Zhang C, Xu D, Luo H, et al. 2014. Toxicology, 325: 74–84.
Zhang Q, Pei L G, Liu M, et al. 2020. Food Chem Toxicol, 135: 111057.
Zhao Z, Qin J, Pei L G, et al. 2020. Toxicology, 442: 152533.
Zhou J, Kou H, Guo Y, et al. 2015. Chin J Pharmacol Toxicol, 1(29): 8.
Zhou J, Zhu C Y, Luo H W, et al. 2019. FASEB J, 33(1): 1110–1123.
Zhu Y N, Chen H Y, Zhao X Q. 2019. Toxicol Lett, 314: 63–74.

Chapter 10
Developmental Toxicity of Dexamethasone and Related Fetal-Originated Disease

Hao Xiao and Yi Liu

Abstract Dexamethasone, a synthetic glucocorticoid, is widely used to treat preterm labor and related diseases during pregnancy. However, the prenatal application of dexamethasone has a "double-edged sword" effect. Based on the latest research, this chapter summarizes the fetal developmental toxicity of dexamethasone, its long-term susceptibility to disease, and proposes a systematic intrauterine programming mechanism of fetal developmental toxicity and susceptibility to adult disease induced by dexamethasone. The mechanism of developmental toxicity is related to its direct effect on dexamethasone and the indirect effect of endogenous glucocorticoid reduction in vivo, both of which cause abnormal epigenetic modification (i.e., two-programming mechanisms). Intrauterine dexamethasone exposure can also lead to the multigenerational inheritance of altered programming in offspring's multiple organs by changing the epigenetic reprogramming in germ cells. This chapter contributes to a comprehensive understanding of the developmental toxicity of dexamethasone and provides a theoretical and experimental basis for guiding the rational application of dexamethasone during pregnancy and early prevention and treatment of related fetal-originated diseases.

1 Introduction

Dexamethasone, a synthetic glucocorticoid, is widely used to treat preterm birth-related diseases, such as threatened preterm birth, placenta previa, and multiple pregnancies. Its ability to penetrate the placenta and promote fetal lung and other

H. Xiao (✉)
Division of Joint Surgery and Sports Medicine, Department of Orthopedic Surgery, Zhongnan Hospital of Wuhan University, Wuhan, Hubei, China
e-mail: HaoXiao@whu.edu.cn

Y. Liu
Department of Pharmacology, School of Basic Medical Science, Wuhan University, Wuhan, Hubei, China

H. Wang et al. (eds.), *Fetal Origin of Diseases*,
https://doi.org/10.1007/978-981-97-5730-5_10

important tissue and organ maturation has effectively reduced neonatal mortality (Jobe and Goldenberg 2018). According to the WHO's maternal and child health survey of 359 institutions in 29 countries, the prophylactic treatment rate of dexamethasone for preterm infants at 22–36 weeks of gestation was 54% on average, with some countries reaching as high as 91% (Chawanpaiboon et al. 2019). However, recent studies have suggested that the use of dexamethasone during pregnancy has a "double-edged sword" effect. Prenatal exposure to dexamethasone has been shown to induce changes in multiple-organ programming, functional homeostasis, and susceptibility to related diseases in adulthood (Murphy et al. 2008). Nonetheless, the mechanisms underlying these effects have not been systematically elucidated. In this chapter, the author's team presents a comprehensive review of the latest international research progress and their own series of studies, proposing a mechanism for the developmental toxicity of dexamethasone. The mechanism is not only related to the direct effect of dexamethasone but also to the indirect effect of endogenous glucocorticoid reduction in vivo. Both of these effects are associated with abnormal epigenetic modifications, which can be referred to as "two-programming." Additionally, dexamethasone can cause multigenerational inheritance effects on the programming of multiple-organ development in offspring through epigenetic reprogramming in germ cells. This review offers insights into the double-edged sword effect and developmental toxicity of dexamethasone, providing both experimental and theoretical basis for rational drug use during pregnancy, exploring early prevention and treatment strategies, preventing, and controlling dexamethasone-related fetal-originated disease, and analyzing the international hot theory of "the developmental origin of health and diseases."

2 Developmental Toxicity and Long-Term Hazards of Dexamethasone

Despite the widely recognized effectiveness of prenatal dexamethasone application in promoting lung maturation and reducing neonatal mortality, the developmental toxicity of dexamethasone in offspring cannot be overlooked. A plethora of epidemiological, clinical, and laboratory studies have provided evidence that prenatal dexamethasone use can lead to structural and functional abnormalities in the multi-tissue organs of the fetus, including but not limited to the brain, liver, adrenal gland, bone, gonad, heart, pancreas, and kidney. These abnormalities can result in changes in multiple-organ development programming after birth, functional homeostasis, and increased susceptibility to adult chronic diseases, such as nonalcoholic fatty liver disease (NAFLD) and osteoarthritis (Fig. 10.1).

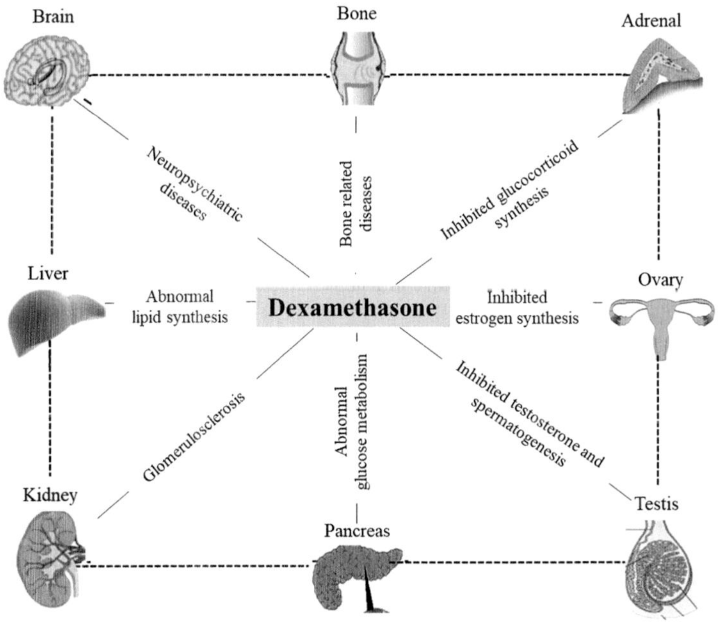

Fig. 10.1 Dexamethasone exposure during pregnancy can cause multiple-organ dysplasia and multiple disease susceptibility in adults

2.1 Prenatal Clinical Application of Dexamethasone

Globally, approximately 15 million premature infants are born each year, accounting for approximately one in 10 live births. Among these infants, respiratory distress syndrome due to pulmonary hypoplasia is the leading cause of premature infant death (Roberts and Dalziel 2006). The World Health Organization survey results indicate that the average preventive treatment rate of synthetic glucocorticoids, including dexamethasone and betamethasone, was 54%, and in some cases, it was as high as 91% (Brownfoot et al. 2013; Norton 2015; Chawanpaiboon et al. 2019). As a result, dexamethasone is widely used in the treatment of premature birth.

The classic clinical approach for prenatal dexamethasone involves administering an intramuscular injection of 6 mg dexamethasone every 12 h for a total of four doses. The optimal time for a single course of prenatal dexamethasone treatment for fetal lung maturation is between 24 h and 7 d after treatment. In cases where women have a tendency toward preterm birth or are at high risk of preterm birth, and there is no delivery within 7 d after a single course of dexamethasone treatment, multiple courses of treatment may be administered to reduce perinatal mortality. Multicourse treatment of dexamethasone involves either repeated medication every week until delivery in cases where the woman is still at risk of premature delivery, or repeated medication within 1 week before the inevitable occurrence of preterm labor after a single course of treatment for an extended period (Asztalos et al. 2013). There exists a clear "time window" during which dexamethasone can promote lung maturation in premature infants. The 2016 European Consensus Guidelines for the Management of Respiratory Distress Syndrome recommend that prenatal glucocorticoid therapy should be administered before 34 weeks gestation for women with threatened preterm birth. The limit of glucocorticoid use is 23 weeks of gestation, during which fetuses have few original alveoli, a low survival rate, and high long-term complications. Therefore, the optimal time for dexamethasone administration should be between 23 and 34 weeks of gestation (Sweet et al. 2019).

2.2 *Multiple-Organ Developmental Toxicity and Disease Susceptibility in Offspring Induced by Prenatal Dexamethasone Exposure*

The fetal period is a crucial stage of body development, during which multiple organs complete their morphological and functional development. For instance, during the fetal period, the testes form spermatogenic tubules and acquire the function of testosterone synthesis. Similarly, the adrenal glands develop cortical and medullary areas and acquire the function of glucocorticoid synthesis. The prenatal developmental toxic effects of dexamethasone exposure (PDE) have been reported numerous times and include the following multiple-organ developmental toxic effects:

As the main component of the central nervous system, the brain plays a critical role in regulating various life activities. Epidemiological studies have established a link between fetal exposure to high levels of glucocorticoids and an increased risk of hyperactivity and cognitive deficits (e.g., attention deficit and hyperactivity disorder) after birth (Rodriguez and Bohlin 2005; Talge et al. 2007). Our research team administered 0.2 mg/kg.d dexamethasone to Wistar rats from 9 to 20 days of gestation and observed that the proliferation of hippocampal neural progenitor cells was inhibited, and the hippocampal function was impaired in female offspring, who were also more prone to depression after birth (Huang et al. 2019). These findings suggest that PDE can lead to hippocampal damage, resulting in long-term cognitive and emotional disorders in offspring.

As the largest metabolic organ in mammals, the liver plays a crucial role in fetal development during the intrauterine period. Epidemiological studies have reported that prenatal dexamethasone application can result in abnormal postnatal lipid metabolism and increased susceptibility to adult liver-related diseases, such as NAFLD (Azadbakht et al. 2014; Carpinello et al. 2018). Clinical studies have also found an increased incidence of progenitor cirrhosis in women who received prenatal glucocorticoid treatment (Cotton and Heap 1976). Our research team discovered that PDE can lead to liver cell proliferation inhibition and liver dysplasia (Zhang et al. 2021), as well as an increase in adult liver triglyceride synthesis function (Liu et al. 2021). Regarding cholesterol metabolism, we observed that the expression of low-density lipoprotein receptor (LDLR) in the fetal liver of male offspring exposed to PDE is increased and persists after birth, inducing hypercholesterolemia in adult offspring (Li et al. 2020). In summary, dexamethasone has developmental toxicity on the liver, which can cause changes in progeny liver development programming, lipid metabolic function homeostasis, and susceptibility to related diseases (such as NAFLD and hypercholesterolemia).

The prenatal application of dexamethasone significantly affects the development of the motor system, particularly the long bones and articular cartilage of offspring. Studies have reported that dexamethasone use can lead to inhibited bone formation and decreased bone mass, resulting in an increased incidence of osteoporosis (Rubin and Bilezikian 2002). After administering dexamethasone during pregnancy, the level of serum osteocalcin in male pig offspring after birth was significantly reduced, the thickness of the proliferating, resting, and hypertrophic regions of the growth plates of the femur and tibia were decreased, and the thickness of the bone trabeculae was thinned (Swolin-Eide et al. 2002; Tomaszewska et al. 2013). Our research team discovered that bone developmental toxicity induced by PDE can extend from intrauterine to postnatal, leading to susceptibility to adult osteoporosis (Xiao et al. 2018). Furthermore, intrauterine glucocorticoid exposure can directly inhibit the expression of vascular endothelial factors in growth plate chondrocytes, inhibiting angiogenesis and leading to bone retardation (Koedam et al. 2002; Shangguan et al. 2021). In summary, dexamethasone has developmental toxicity on bone, leading to the susceptibility of offspring to osteoarticular-related diseases (such as osteoporosis) in adulthood. The mechanism is related to the abnormal differentiation of growth plate cartilage, osteogenesis, osteoclast, and inhibition of angiogenesis.

The gonads comprise the ovaries and testes. Animal studies have demonstrated that dexamethasone exposure in early pregnancy can lead to abnormal expression of fetal testicular sex differentiation genes and affect testicular development in offspring mice (Yun et al. 2016). Furthermore, the study confirmed that dexamethasone exposure during the middle and late trimester can reduce the testicular area and the number of Leydig cells and inhibit steroidal synthesis function in fetal offspring mice. Furthermore, the weight of the testes is reduced, the interstitial area is expanded, and the area and thickness of the spermatogenic tubules and epithelium are shortened, resulting in decreased sperm count and motility. These alterations in sperm count and morphology have been observed to persist into the F2 generation (Jeje and Raji 2017). The research by the authors' team has discovered that female offspring exposed to dexamethasone during the middle and late stages of gestation exhibit abnormal follicular development in adulthood, as well as changes in reproductive function such as earlier vaginal opening, an increased proportion of secondary follicles, and a prolonged estrous cycle. Additionally, PDE female offspring display a decreased ovarian estrogen synthesis function with transgenerational genetic effects (Lv et al. 2018). These findings suggest that PDE has the capacity to induce the programming of gonadal development and homeostasis in offspring, with multi-generational or even cross-generational effects.

Regarding the cardiovascular system, epidemiological studies have revealed a significantly increased incidence of multiple cardiovascular diseases in adulthood among low birth weight infants (Barker 2004). Animal experiments have also demonstrated that adult mice offspring exposed to dexamethasone during early or late pregnancy exhibit hypertension-related manifestations such as increased basic blood pressure, increased pulse pressure difference, and reduced blood vessel compliance (O'Sullivan et al. 2013). In Wistar fetal rats exposed to dexamethasone during late pregnancy, the expression of calc-binding protein and calponectin in the heart was twice that of the control group, which increased to four times in adulthood. This abnormal increase was closely associated with the occurrence of cardiovascular diseases such as hypertension and arrhythmia (Langdown et al. 2003). These findings suggest that dexamethasone has developmental toxicity on the cardiovascular system, which can lead to the onset of various cardiovascular diseases in adulthood.

3 Inducement and Mechanism of the Developmental Toxicity by Dexamethasone

The developmental toxicity of dexamethasone is associated with its direct cytotoxic effects, which regulate cell proliferation, differentiation, apoptosis, endoplasmic reticulum stress, and autophagy. Conversely, PDE can downregulate the level of endogenous GCs in the offspring, indirectly regulating their development and programming their susceptibility to multiple diseases through endogenous low GCs.

3.1 Direct Effects and Mechanisms of Developmental Toxicity by Dexamethasone

The developmental toxicity of dexamethasone is associated with its direct cytotoxic effects, which involve diverse mechanisms such as cell proliferation, differentiation, apoptosis, endoplasmic reticulum stress, and autophagy. Zuloaga et al. (2012) found that the use of dexamethasone during pregnancy increased the expression of genes related to apoptosis (e.g., caspase-3, Bad) in the hippocampus and paraventricular nucleus of the hypothalamus in offspring. In female fetal mice exposed to dexamethasone during the middle and late trimester, the number of cells labeled with cell proliferation antigen (PCNA) in the ovary was significantly lower than that of the control group, while the number of cells labeled with caspase-3 was increased by 2–3 times. These results suggest that dexamethasone can cause decreased proliferation but increased apoptosis in ovarian cells (Ristic et al. 2019). Exposure to dexamethasone during late pregnancy activates the MAPK signaling pathway in fetal neuronal precursor cells, resulting in inhibited cell proliferation, shortened S phase of the cell cycle, and accelerated termination (Samarasinghe et al. 2011). Furthermore, the research team led by the authors' team confirmed that the cell proliferation ability of fetal testis and kidneys in PDE offspring was weakened compared to the control group, as demonstrated by Ki67 staining (Liu et al. 2018; Li et al. 2019). These findings suggest that abnormal cell proliferation and apoptosis play a crucial role in the developmental toxicity of dexamethasone. Endoplasmic reticulum stress (ERS) is a cellular response to misfolded/unfolded protein aggregation and calcium ion balance disorder in the endoplasmic reticulum lumen (Oakes and Papa 2015; Lebeaupin et al. 2018). Previous studies by the authors' team have revealed that PDE can increase the expression of ryanodine receptor 1 (RYR1) and calcium ion content in fetal cartilage, leading to an increase in the expression of ERS-related genes. This, in turn, results in the inhibition of TGF-β signaling pathway, decreased matrix content, and chondrodysplasia (Wang et al. 2020). These findings suggest that PDE-induced fetal tissue and organ dysplasia may be related to ERS. Autophagy, a lysosomal dependent degradation mechanism, is closely associated with the occurrence of various metabolic disorders (Jimenez-Sanchez et al. 2012). In the ovarian primordial follicles of progenitor rats exposed to dexamethasone during late pregnancy, the expression of the autophagy marker gene Lamp1 was significantly increased, and autophagy was enhanced. It was proposed that dexamethasone could cause cytotoxicity by inducing autophagy or apoptosis, leading to abnormal ovarian function in progenitor rats (Hulas-Stasiak et al. 2016). The research by the authors' team found that PDE could activate GR/FOXO1 signals in the liver of male fetal mice, inducing autophagy and inhibiting cell proliferation, which led to fetal liver dysplasia (Zhang et al. 2021).

3.2 *Indirect Effects and Mechanisms by Endogenous Glucocorticoid*

Insulin-like growth factor 1 (IGF1) is a type of cell growth factor with insulin-like biological activity, and the IGF signaling pathway is central to the endocrine regulatory system of the body (Piecewicz et al. 2012). It is involved in the processes of cell proliferation, differentiation, and metabolism before and after birth (Randhawa and Cohen 2005). During the intrauterine period, the synthesis and secretion of IGF1 mainly come from the fetal liver and regulate the development of various organs through autocrine and paracrine pathways. In other tissues, the IGF1 autocrine and paracrine regulatory system is initially established during the middle and late stages of embryonic development, and gradually improves after birth (Halmos and Suba 2019). The authors' team have recently discovered that the level of endogenous corticosterone in fetal blood of PDE offspring is significantly decreased, the acetylation level and expression of the IGF1 promoter region in fetal adrenal tissue are inhibited, and steroid synthesis function is decreased.

After birth, the acetylation level and expression of the IGF1 promoter region in adrenal tissue are continuously inhibited due to the continuous decrease of serum corticosterone level. After chronic stress, the levels of IGF1 in blood and adrenal tissues also increase with the increased blood corticosterone level, demonstrating a clear positive programming effect of the "GC-IGF1 axis" (Chen et al. 2021). The authors' team also observed the same programming effect in the testes, long bones, and liver. These findings suggest that, in the PDE rat model, IGF1-related signaling pathways play a crucial role in offspring low birth weight, postnatal multi-tissue organ functional homeostasis, and susceptibility to multiple adult diseases. The positive programming effect of the "GC-IGF1 axis" in offspring multi-organs implies weakened self-compensatory function, which is distinct from the negative programming of the "GC-IGF1 axis" in the exposure to endogenous high glucocorticoids during pregnancy. This represents a novel concept in the pathogenesis of fetal-originated disease.

4 Mechanisms of Multi-Organ Programming and Disease Susceptibility in Offspring Induced by Prenatal Dexamethasone Exposure

The mechanisms of multi-organ programming and related disease susceptibility in offspring induced by PDE are complex. Based on epigenetic programming theory, the authors' team has recently proposed intrauterine "two-programming" mechanisms for fetal-originated disease associated with dexamethasone developmental toxicity.

4.1 Epigenetic Mechanisms and Multigenerational Genetic Effects of Developmental Toxicity by Dexamethasone

The physiological development of fetal organs is a precise and coordinated process that is sensitive to and easily affected by the external environment (Goyal et al. 2019). Epigenetic modification, as a heritable form of gene regulation that does not alter DNA sequence, mainly includes DNA methylation, histone modification, non-coding RNA, etc. It is characterized by accuracy, sustainability, and networking (Cavalli and Heard 2019). An increasing number of studies have confirmed that epigenetic modifications play a crucial role in the regulation of dexamethasone-induced offspring developmental toxicity. Relevant examples are summarized in Table 10.1.

DNA methylation is a crucial epigenetic modification that has been demonstrated to play a significant role in gene regulation. Studies have revealed that dexamethasone, a synthetic glucocorticoid, can directly interact with the glucocorticoid response element (GRE) in the promoter region of the target gene, and in cooperation with DNA methyltransferases (DNMTs), it can alter the methylation level of the gene promoter region, thereby modulating the expression of relevant functional

Table 10.1 Epigenetic modification of multiple-organ functional genes in offspring caused by prenatal dexamethasone exposure

Gene	Function	Epigenetic modification	Animal model (species, dose, time)	References
DNA methylation				
Reelin /GAD67	Regulates neuronal migration, dendritic growth and branching	DNA methyltransferase1 / MeCP2 meditates hypermethylation of Reelin promoter in brain	SD rats, 0.1 mg/kg·d GD14–21	Lui et al. (2015)
BKα/ BKβ1	As an important vascular medium, it plays a crucial role in determining vascular tension	Increased DNA methylation in the vascular BKα /BKβ1 promoter region	SD rats, 200 μg/kg·d GD14–21	Xu et al. (2020)
Leptin	A key regulator of acute and long-term metabolic health and energy expenditure	Methyltransferases/ leptin meditates hypermethylation in adipose tissue	SD rats, 0.1 mg/kg·d GD14–21	Tsai et al. (2018)
BMP4	Plays an important role in embryonic heart development	Hypermethylation of BMP4 promoter in heart	SD rats, 0.1 mg/kg·d GD14.5–21	Peng et al. (2018)
ATGL/ HSL	Catalyze the hydrolysis of triglycerides to free fatty acids	Hypormethylation of ATGL and HSL promoter in liver	SD rats, 0.1 mg/kg·d GD0–21	Zhao et al. (2020)
Histone modification				
BDNF	Widely distributed in the central nervous system and involved in neurons survival, differentiation, growth, and development	Glucocorticoid receptor/ histone deacetylase 2 meditates inhibited BNDF H3K14ac in hippocampal	Wistar rats, 0.2 mg/kg·d GD9–20	Huang et al. (2019)
TNF-α	A class of inflammatory response factors, involved in a variety of diseases	Decreased H3ac, H3K4me1/3, and H3K36me3 of TNF-α in the spleen	SD rats, 0.1 mg/kg·d GD14–21	Yu et al. (2014)
IFN-γ	A class of important immunomodulatory factors with antiviral, antitumor, and immunomodulatory effects	Decreased H3K9 ac and H3K36me3 of IFN-γ in the spleen	SD rats, 0.1 mg/kg·d GD14–21	Yu et al. (2016)
ACE	An important part of the RAS and participates in the mediation of various life activities	GR-C/EBPα/p300 meditates decreased/ increased H3K27ac of ACE in bone and hippocampal	Wistar rats, 0.2 mg/kg·d GD9–20	Xiao et al. (2018), Hu et al. (2020)
StAR	The rate-limiting enzyme of steroid hormone synthesis system	Glucocorticoid receptor/ histone deacetylase 7 meditates inhibited StAR H3K9ac in testis	Wistar rats, 0.2 mg/kg·d GD9–20	Liu et al. (2018)

(continued)

Table 10.1 (continued)

Gene	Function	Epigenetic modification	Animal model (species, dose, time)	References
AT2R	Highly expressed in fetuses and neonates, plays an important role in fetal tissue development	Reduced H3K9/27 ac of AT2R in kidney Reduced H3K14/27 ac of AT2R in pancreas	Wistar rats, 0.2 mg/kg·d GD9–20	Li et al. (2019), Kou et al. (2020)
Noncoding RNAs				
LDLR	Participate in liver cholesterol metabolism, responsible for cholesterol reversal	Glucocorticoid receptor/ DGCR8 meditates inhibited miR-148a/ LDLR expression in liver	Wistar rats, 0.2 mg/kg·d GD9–20	Li et al. (2020)
YY1	Zinc finger protein family of proteins, plays an important role in liver lipids accumulation	Glucocorticoid receptor/ miR-122 meditates YY1/ACE2-MAS1activiation in liver	Wistar rats, 0.2 mg/kg·d GD9–20	Liu et al. (2021)
ECM proteins	Closely related to lung related cell maturation and extracellular matrix synthesis	Glucocorticoid receptor/ miR-29 meditates inhibited ECM expression in lungs	Fetal pulmonary fibrocytes treats with dexamethasone For 10 days	Chuang et al. (2018)
RUNX2	Highly expressed in fetal ovary and related to luteinosis and estrogen synthesis	Glucocorticoid receptor/ miRNA320a-3p meditates inhibited RUNX2/P450arom expression in ovary	Wistar rats, 0.2 mg/kg·d GD9–20	Gong et al. (2021)
SOX2	Important for the development of central nervous system	Glucocorticoid receptor/ SIRT1meditates inhibited miR-134-5p/ SOX2 expression in hippocampal	Wistar rats, 0.2 mg/kg·d GD9–20	Jiang et al. (2021)

GAD67 glutamic acid decarboxylase 67, *BK* high-conductance Ca^{2+}-activated K^{+} channels, *BMP4* bone morphogenetic protein-4, *ATGL* adipose triglyceride lipase, *HSL* hormone-sensitive lipase, *GD* gestational day, *BDNF* brain-derived neurotrophic factor, *TNF-α* tumor necrosis factor-α, *IFN-γ* interferon γ, *ACE* angiotensin-converting enzyme, *StAR* steroidogenic acute regulatory protein, *AT2R* angiotensin II receptor type 2, *H3K9ac* histone 3 lysine 9 acetylation, *H3K4me* histone 3 lysine 4 methylation, *GR* glucocorticoid receptor, *LDLR* low-density lipoprotein receptor, *DGCR8* DiGeorge syndrome critical region 8 gene, *YY1* Yin Yang-1, *ACE2* angiotensin-converting enzyme 2, *MAS1* Mas receptor, *ECM* extracellular matrix, *RUNX2* RUNX family transcription factor 2, *P450arom* cytochrome P450 aromatase, *SOX2* sex-determining region Y-box 2, *SIRT1* sirtuin 1

genes (Crudo et al. 2013). Notably, cohort studies have shown that prenatal exposure to dexamethasone in patients with congenital adrenal insufficiency leads to abnormal brain structure and function in their adult offspring, which is closely related to changes in DNA methylation of multiple functional genes in T cells (Van't

Westeinde et al. 2020). Furthermore, various animal studies have demonstrated that PDE can result in the abnormal development of multiple organs such as the hippocampus, liver, and blood vessels, which is associated with abnormal DNA methylation modifications of crucial genes. Histone modification, which mainly encompasses acetylation and methylation, plays a critical role in regulating gene transcriptional modification (Jenuwein and Allis 2001; Lawrence et al. 2016; Zhang et al. 2021). Numerous studies have demonstrated that PDE can alter the histone modification of key target genes in some organs by modulating histone-modifying enzymes such as acetyltransferases and deacetylases, leading to dysplasia of organs and susceptibility to multiple diseases after birth. Noncoding RNAs, a class of short-stranded RNAs that lack coding ability, have yet to be fully understood in terms of biosynthesis and regulation of target genes (Cech and Steitz 2014). Currently, the role of noncoding RNA in fetal-originated diseases has garnered increasing attention. For instance, PDE may act on fetal lung fibroblasts via miR-29, contributing to the development of adult pulmonary fibrosis (Sahebjami and Domino 1989; Chuang et al. 2018). The various types of epigenetic modifications observed under the PDE model suggest that PDE can affect organ development through multiple pathways.

Multigenerational and intergenerational genetic effects refer to the phenomenon whereby exposure of an individual to an adverse external environment during the sensitive stage of development leads to the transmission of the same abnormal phenotype to their offspring, which is often closely associated with epigenetic modifications in germline cells (Boskovic and Rando 2018). Several studies have demonstrated that glucocorticoids can program offspring to develop multigenerational organ dysfunction after chronic maternal/paternal stress. The impact of PDE on multigenerational changes in offspring through maternal inheritance and its underlying mechanism have been extensively investigated. For instance, in rats exposed to dexamethasone, the offspring exhibit hyperinsulinemia, reduced glucose tolerance, and an increased incidence of cardiovascular disease, and this abnormal phenotype can result in transgenerational genetic effects, whether inherited through the paternal or maternal line (Drake et al. 2005). It has been suggested that PDE may further influence the development of the grandchild by affecting epigenetic inheritance in the offspring's sperm or oocytes. A series of studies by the author and their team have revealed that exposure to dexamethasone during the middle and late trimester can lead to multiple-organ function impairments in the offspring and have multigenerational genetic effects. The author found that PDE can cause abnormal ovarian development in female offspring rats, including inhibited estrogen synthesis function and reduced numbers of primal follicles, which is associated with the inhibition of miR-320-3p/RUNX2/P450arom signaling pathway expression. The abnormal ovarian phenotype and related pathway changes persist in the F1–F3 generation. Additionally, the expression changes of miR-320-3p were also observed in F1 and F2 oocytes (Gong et al. 2021). The aforementioned studies indicate that exposure to dexamethasone in utero can result in multiple-organ dysfunction in the offspring and have transgenerational inheritance effects, which are all linked to oocyte epigenetic reprogramming.

4.2 The "Two-Programming" Mechanisms for Fetal-Originated Disease Associated with Developmental Toxicity by Dexamethasone

The renin-angiotensin system (RAS) is a cascade reaction system in which angiotensin is converted into angiotensin I (AngI) by renin, and AngI is degraded into AngII by ACE. The latter plays a role mainly through AT1R and AT2R (Chen et al. 2009). Studies have shown that PDE can alter the vascular local RAS via AT2R, leading to an increased risk of hypertension in male adult offspring rats (Tain et al. 2014). Previous research by our team has revealed that PDE can decrease the expression of AT1R and AT2R in the fetal kidney, inhibit the expression of glial cell-derived neurotrophic factor and its receptor, and cause fetal kidney dysplasia and susceptibility to adult glomerulosclerosis (Li et al. 2019). Additionally, increased acetylation of the local ACE promoter region in PDE offspring can lead to sustained activation of local bone RAS, resulting in inhibited osteogenic differentiation and decreased peak bone mass after birth (Xiao et al. 2018). Interestingly, this phenomenon can be reversed when ACE is silenced (Xiao et al. 2020). Furthermore, PDE can program offspring hippocampal glutamate/GABA function through RAS, leading to susceptibility to temporal lobe epilepsy in adulthood (Hu et al. 2020). Epigenetic modification of pancreatic beta cells AT2R can induce glucose intolerance and abnormal glucose tolerance in offspring (Kou et al. 2020). These findings suggest that PDE can induce changes in offspring RAS programming, which increases the risk of multiple chronic diseases in adulthood, known as the "first programming" mechanism.

Our previous study found that xenobiotics such as caffeine and ethanol during pregnancy can cause fetal rats to be overexposed to maternal glucocorticoids. Additionally, the blood IGF1 level and IGF1 expression in multiple tissues, such as the liver, adrenal gland, and bone, were decreased. After birth, due to the continuous inhibition of offspring adrenal development, the glucocorticoid level continues to decrease, and the IGF1 level in blood and tissue gradually increases. However, after chronic stress, increased glucocorticoid levels were associated with decreased IGF1 levels in adults. Therefore, we propose that maternal glucocorticoid exposure during pregnancy can negatively regulate the blood glucocorticoid and blood/tissue IGF1 level in the progeny (i.e., the "GC-IGF1 axis"), while the intrauterine GC-IGF1 axis is programmed as a self-compensatory mechanism. This mechanism is involved in the inhibited intrauterine growth of offspring, catch-up growth after birth, and susceptibility to multiple diseases in adults (Shen et al. 2014; Wang et al. 2014). Does intrauterine GC-IGF1 axis programming occur under the influence of dexamethasone exposure during pregnancy? Our findings suggest that PDE can cause a decrease in maternal glucocorticoid levels, while inhibiting the acetylation level and expression of the IGF1 promoter region in the fetal adrenal gland, leading to decreased steroid synthesis function. After birth, the level of acetylation and expression of the IGF1 promoter region in the adrenal gland and the function of steroid synthesis were continuously inhibited, while the level of glucocorticoid in the blood was continuously decreased. After chronic stress in adults, blood and adrenal IGF1 levels increase simultaneously with the increase of glucocorticoid levels, demonstrating an obvious positive-programming effect of the "GC-IGF1 axis" (Chen et al. 2021). Additionally, a similar phenomenon of positive-programming of the "GC-IGF1 axis" was also observed in the testicle, liver, and bone in the PDE model (Liu et al. 2021). The positive-programming of the offspring "GC-IGF1 axis" implies that the self-compensatory function of the body is weakened. It is suggested that the endogenous hypoglucocorticoid-mediated positive programming "GC-IGF1 axis" under PDE also plays a crucial role in multi-organ development and abnormal functional homeostasis, which can be considered as the "second programming."

5 Summary and Prospect

The application of dexamethasone during pregnancy has a "double-edged sword" effect, as it can effectively treat preterm birth and related diseases, but also cause developmental toxicity in the offspring and susceptibility to adult diseases. The severity of this toxicity depends on the dose of dexamethasone, the course of treatment, and the specific time window for intrauterine organ development. The developmental toxicity caused by dexamethasone during pregnancy has the characteristics of "multi-pathway, multiorgan, and multi-target," leading to abnormal multi-organ function and susceptibility to a series of diseases in adulthood. The present research study posits that the administration of dexamethasone could potentially result in the increased susceptibility of adult offspring to multiple diseases. This can occur through both direct and indirect effects of endogenous glucocorticoids. Additionally, the study suggests that abnormal epigenetic modifications of key genes, induced by PDE, may evade oocyte reprogramming modification and mediate maternal multigenerational inheritance effects. These findings have significant implications for the identification of early warning markers or intervention targets of fetal diseases arising from dexamethasone developmental toxicity. Furthermore, this study provides valuable insights for the safe use of drugs in clinical settings.

References

Asztalos E V, Murphy K E, Willan A R, et al. 2013. JAMA Pediatr, 167(12): 1102–1110.
Azadbakht L, Kelishadi R, Saraf-Bank S, et al. 2014. Nutrition, 30(2): 150–158.
Barker D J. 2004. J Am Coll Nutr, 23(6 Suppl): 588S–595S.
Boskovic A, Rando O J. 2018. Annu Rev Genet, 52: 21–41.
Brownfoot F C, Gagliardi D I, Bain E, et al. 2013. Cochrane Database Syst Rev(8): CD006764.
Carpinello O J, DeCherney A H, Hill M J. 2018. Semin Reprod Med, 36(3-04): 177–182.
Cavalli G, Heard E. 2019. Nature, 571(7766): 489–499.
Cech T R, Steitz J A. 2014. Cell, 157(1): 77–94.
Chawanpaiboon S, Vogel J P, Moller A B, et al. 2019. Lancet Glob Health, 7(1): e37-e46.

Chen X F, Wang J A, Hou J, et al. 2009. Clin Sci (Lond), 116(11): 819–826.
Chen Y, Xia X, Fang M, et al. 2021. Sci Total Environ, 797: 149084.
Chuang T D, Sakurai R, Gong M, et al. 2018. Am J Physiol Regul Integr Comp Physiol, 315(5): R1017-R1026.
Cotton P B, Heap T R. 1976. Ann R Coll Surg Engl, 58(4): 324.
Crudo A, Petropoulos S, Suderman M, et al. 2013. Endocrinology, 154(11): 4170–4181.
Drake A J, Walker B R, Seckl J R. 2005. Am J Physiol Regul Integr Comp Physiol, 288(1): R34- R 38.
Gong X, Zhang J, Ge C, et al. 2021. Pharmacol Res, 165: 105435.
Goyal D, Limesand S W, Goyal R. 2019. J Endocrinol, 242(1): T105-T119.
Halmos T, Suba I. 2019. Orv Hetil, 160(45): 1774–1783.
Hu S, Yi Y, Jiang T, et al. 2020. Arch Toxicol, 94(9): 3201–3215.
Huang S, Dong W, Jiao Z, et al. 2019. Toxicol Sci, 171(2): 369–384.
Hulas-Stasiak M, Dobrowolski P, Tomaszewska E. 2016. Reprod Fertil Dev, 28(7): 1038–1048.
Jeje S O, Raji Y. 2017. J Dev Orig Health Dis, 8(1): 101–112.
Jenuwein T, Allis C D. 2001. Science, 293(5532): 1074–1080.
Jiang T, Hu S, Dai S, et al. 2021. Cell Biol Toxicol, 38(1): 69–86.
Jimenez-Sanchez M, Thomson F, Zavodszky E, et al. 2012. Prog Neurobiol, 97(2): 67–82.
Jobe A H, Goldenberg R L. 2018. Am J Obstet Gynecol, 219(1): 62–74.
Koedam J A, Smink J J, van Buul-Offers S C. 2002. Mol Cell Endocrinol, 197(1–2): 35–44.
Kou H, Gui S, Dai Y, et al. 2020. Toxicol Appl Pharmacol, 404: 115187.
Langdown M L, Holness M J, Sugden M C. 2003. Biochem J, 371(Pt 1): 61–69.
Lawrence M, Daujat S, Schneider R. 2016. Trends Genet, 32(1): 42–56.
Lebeaupin C, Vallee D, Hazari Y, et al. 2018. J Hepatol, 69(4): 927–947.
Li B, Zhu Y, Chen H, et al. 2019. Toxicology, 411: 32–42.
Li L, Hu W, Liu K, et al. 2020. Toxicol Appl Pharmacol, 395: 114979.
Liu H, He B, Hu W, et al. 2021. Biochem Pharmacol, 185: 114420.
Liu M, Chen B, Pei L, et al. 2018. Toxicology, 408: 1–10.
Lui C C, Hsu M H, Kuo H C, et al. 2015. Dev Neurosci, 37(2): 105–114.
Lv F, Wan Y, Chen Y, et al. 2018. Endocrinology, 159(3): 1401–1415.
Murphy K E, Hannah M E, Willan A R, et al. 2008. Lancet, 372(9656): 2143–2151.
Norton M E. 2015. Obstet Gynecol Surv, 70(2): 79–81.
O'Sullivan L, Cuffe J S, Paravicini T M, et al. 2013. PLoS One, 8(7): e69149.
Oakes S A, Papa F R. 2015. Annu Rev Pathol, 10: 173–194.
Peng J, Zhou Y, Zhang Z, et al. 2018. Cell Death Dis, 9(8): 834.
Piecewicz S M, Pandey A, Roy B, et al. 2012. PLoS One, 7(2): e32191.
Randhawa R, Cohen P. 2005. Mol Genet Metab, 86(1–2): 84–90.
Ristic N, Nestorovic N, Manojlovic-Stojanoski M, et al. 2019. Fundam Clin Pharmacol, 33(2): 199–207.
Roberts D, Dalziel S. 2006. Cochrane Database Syst Rev (3): CD004454.
Rodriguez A, Bohlin G. 2005. J Child Psychol Psychiatry, 46(3): 246–254.
Rubin M R, Bilezikian J P. 2002. J Clin Endocrinol Metab, 87(9): 4033–4041.
Sahebjami H, Domino M. 1989. Exp Lung Res, 15(6): 961–973.
Samarasinghe R A, Di Maio R, Volonte D, et al. 2011. Proc Natl Acad Sci U S A, 108(40): 16657–16662.
Shangguan Y, Wu Z, Xie X, et al. 2021. Biochem Pharmacol, 185: 114414.
Shen L, Liu Z, Gong J, et al. 2014. Toxicol Appl Pharmacol, 274(2): 263–273.
Sweet D G, Carnielli V, Greisen G, et al. 2019. Neonatology, 115(4): 432–450.
Swolin-Eide D, Dahlgren J, Nilsson C, et al. 2002. J Endocrinol, 174(3): 411–418.
Tain Y L, Sheen J M, Chen C C, et al. 2014. Free Radic Res, 48(5): 580–586.
Talge N M, Neal C, Glover V, et al. 2007. J Child Psychol Psychiatry, 48(3–4): 245–261.
Tomaszewska E, Dobrowolski P, Puzio I. 2013. J Anim Physiol Anim Nutr (Berl), 97(4): 785–796.
Tsai C C, Lin Y J, Yu H R, et al. 2018. Int J Mol Sci, 19(11).

Van't Westeinde A, Karlsson L, Nordenstrom A, et al. 2020. J Clin Endocrinol Metab, 105(8):2575–2586.
Wang L, Shen L, Ping J, et al. 2014. Toxicol Lett, 224(3): 311–318.
Wang Y Z, Li Q X, Zhang D M, et al. 2020. Biochim Biophys Acta Mol Cell Res, 1867(10): 118791.
Xiao H, Wen Y, Pan Z, et al. 2018. Cell Death Dis, 9(6): 638.
Xiao H, Wen Y, Wu Z, et al. 2020. Bone, 115578.
Xu T, Zhao M, Li H, et al. 2020. Arterioscler Thromb Vasc Biol, 40(11): e284-e295.
Yu H R, Kuo H C, Chen C C, et al. 2014. Immunology, 143(4): 651–660.
Yu H R, Tain Y L, Sheen J M, et al. 2016. Int J Mol Sci, 17(10): 1610.
Yun H J, Lee J Y, Kim M H. 2016. Biochem Biophys Res Commun, 471(1): 149–155.
Zhang D, Liu K, Hu W, et al. 2021. Toxicology, 449: 152664.
Zhang Y, Sun Z, Jia J, et al. 2021. Adv Exp Med Biol, 1283: 1–16.
Zhao N, Yang S, Sun B, et al. 2020. Eur J Nutr, 59(4): 1707–1716.
Zuloaga D G, Carbone D L, Quihuis A, et al. 2012. J Neurosci Res, 90(7): 1403–1412.

Chapter 11
Fetal Origin of Metabolic Syndrome

Dan Xu and Wen Hu

Abstract Metabolic syndrome (MS) is a complex of metabolic disorders that includes obesity, hypertension, and glucose-lipid metabolism abnormalities. MS is a known risk factor for cardiovascular disease. While there is growing epidemiological and laboratory evidence supporting the fetal origin of MS, the underlying mechanisms remain unclear. This chapter aims to elucidate the intrauterine programming mechanisms underlying the main characteristics of fetal-originated MS caused by various adverse environmental conditions during pregnancy, including abnormal glucose and lipid metabolism, hypertension, and obesity. The changes in hypothalamic-pituitary-adrenal axis-related neuroendocrine metabolic programming and its "two-programming and two-strike" mechanism play a crucial regulatory role in the occurrence of fetal-originated MS. This chapter reviews the research progress of the occurrence and intrauterine programming mechanism of fetal-originated MS, providing a theoretical and experimental basis for its clinical early warning, diagnosis, and treatment.

1 Introduction

Metabolic syndrome (MS) is a cluster of metabolic risk factors that coexist in the same individual, including obesity, hypertension, and glucose and lipid metabolism disorders. These factors may increase the risk of developing metabolic diseases such as diabetes, coronary heart disease, stroke, and fatty liver disease. The prevalence of MS has been increasing due to the improvement of people's living standards, but its etiology and pathogenesis have not been fully elucidated. There is a

D. Xu (✉)
Department of Pharmacology, School of Pharmaceutical Sciences, Wuhan University, Wuhan, Hubei, China
e-mail: xuyidan70188@whu.edu.cn

W. Hu
Department of Pharmacy, Zhongnan Hospital of Wuhan University, Wuhan, Hubei, China

H. Wang et al. (eds.), *Fetal Origin of Diseases*,
https://doi.org/10.1007/978-981-97-5730-5_11

growing understanding that MS is the result of the complex interaction between multiple genes and environmental factors, including genetics and immunity. While MS and related diseases in adulthood have clear disease states, they can often be traced back to the origin, including childhood, adolescence, and even the fetal period. The adverse intrauterine environment during pregnancy is closely associated with fetal development and susceptibility to adult chronic metabolic diseases. The adverse intrauterine environment during pregnancy has been established as a significant factor in fetal development and susceptibility to adult chronic metabolic diseases. Intrauterine growth retardation (IUGR) is a common manifestation of fetal growth restriction, which refers to the retardation of the growth potential of embryos (fetus) during intrauterine development, resulting in low birth weight. The diagnostic criteria for IUGR are based on the birth weight of a fetus greater than 37 weeks of gestation, which is less than 2500 g, or the fetal weight is less than two standard deviations from the average weight for the corresponding gestational age. Children with IUGR are at an increased risk of developing MS and related metabolic diseases in adulthood. Therefore, research on the programming mechanism of MS from the perspective of intrauterine development has become a hot topic and international frontier.

2 Current Research Status of Metabolic Syndrome

With the continuous development of the social economy and lifestyle changes, the incidence rate of adult MS has increased sharply worldwide, leading to a significant health and economic burden on society. As a result, MS has become a new focus of attention in the field of chronic diseases and public health.

2.1 Current Incidence of Metabolic Syndrome

MS is a metabolic disorder that affects protein, fat, carbohydrate, and other substances and is recognized as a significant risk factor for cardiovascular disease. In 1988, Reaven coined the term "syndrome X" to describe the occurrence of obesity, hyperglycemia, hypertension, abnormal lipid metabolism, hyperinsulinemia, and other high-risk factors for cardiovascular disease in the same patient. In 1999, the World Health Organization recommended the unified name "MS" for this disease and provided the first definition. In 2005, the International Diabetes Alliance introduced a new definition of MS, which emphasized that central obesity, determined by waist circumference, was the basic condition. The diagnostic criteria include the presence of any two of the following four indicators: (1) increased triglyceride levels, (2) decreased high-density lipoprotein-cholesterol levels, (3) increased blood pressure, and (4) increased fasting blood glucose levels. While the oral glucose tolerance test is recommended for determining the presence of diabetes, it is not necessary to diagnose MS.

In recent years, the prevalence of MS has been increasing worldwide due to the improvement of people's living standards, the acceleration of pace of life, the rise of high-fat and high-calorie diets, and the reduction of daily exercise. Epidemiological surveys have shown (Park et al. 2007) that in Asia, the incidence rate of MS in Chinese and Korean men ranges between 8% and 13%, while the prevalence in women ranges between 2% and 18%. In Europe and Latin America, the prevalence rate is 25%. MS significantly increases the risk of developing type 2 diabetes and is also a high-risk factor for cardiovascular disease, which can increase the incidence rate of coronary heart disease and stroke.

2.2 *Main Mechanisms of MS*

The pathophysiology of MS is complex and not yet fully understood. Recent studies have shown that MS is primarily associated with insulin resistance (IR), obesity, abnormal fat distribution, inflammation, changes in the hypothalamic-pituitary-adrenal (HPA) axis, as well as genetic and environmental factors. IR, overweight, hyperinsulinemia, and central obesity are considered the basis of dyslipidemia, hypertension, and hyperglycemia, which are also important factors in the incidence of MS. Of these, central obesity resulting from visceral fat accumulation and IR play crucial roles in the pathophysiological process of MS. IR is considered the core of MS, and chronic subclinical inflammation is believed to promote the progression of atherosclerosis, leading to the impaired function of blood vessel walls, which is an important pathogenesis of MS. The inflammatory factors in individuals with obesity, IR, and dyslipidemia are significantly higher than those in the normal population and can cause coagulation abnormalities and blood hypercoagulability, leading to vascular damage. Prolonged exposure to inflammatory factors can also impair islet cell function, disrupt normal endocrine function of adipocytes, and cause the formation of atherosclerosis, leading to hyperglycemia, dyslipidemia, and hypertension (Fig. 11.1).

3 Evidence for Intrauterine Developmental Origins of MS

Although the clinical manifestations of MS are typically discovered in late adulthood, current research on MS is primarily focused on the onset and development stages of adult diseases. However, there is increasing epidemiological and laboratory evidence that the causes of MS can be traced back to the intrauterine period. Therefore, exploring the intrauterine developmental origins of MS is of great significance for the effective prevention and control of MS.

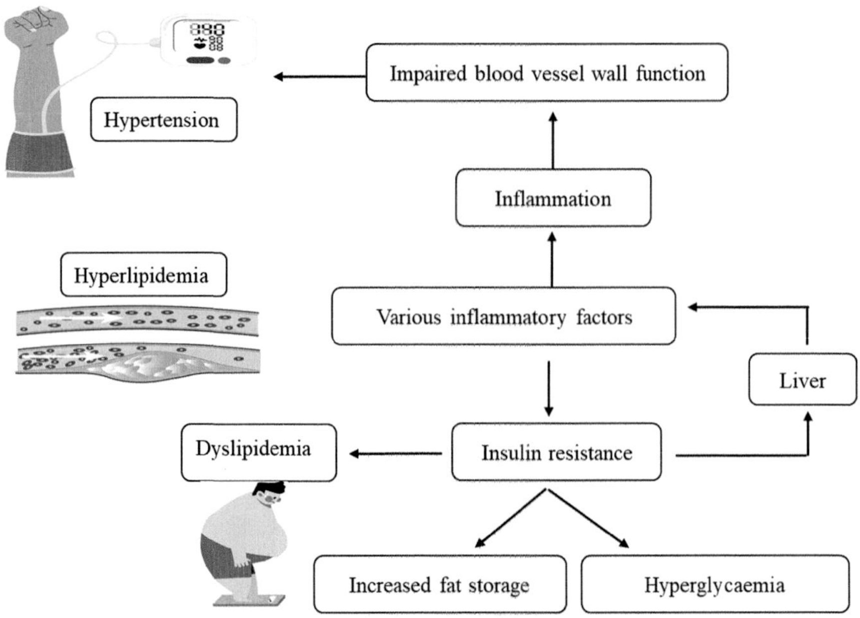

Fig. 11.1 Relationship between insulin resistance and metabolic syndrome

3.1 Epidemiological Evidence

Barker et al. (1989) conducted a study on the relationship between birth weight and adult blood pressure or cardiovascular disease, which revealed the risk of low birth weight. Subsequent studies further confirmed that IUGR is closely associated with the occurrence and development of MS-related symptoms. According to statistics, the incidence of IUGR in European and American countries is 2.5%, while in China, it is between 3% and 5%. A large systematic study involving 444,000 participants concluded that there is a negative correlation between birth weight and systolic blood pressure, with the effect size being approximately 2 mm Hg/kg (Huxley et al. 2000). Furthermore, compared to white adolescents, black adolescents exhibit lower birth weight but higher blood pressure (Cruickshank et al. 2005). The study findings reveal that the likelihood of MS in fetuses with IUGR is 2.53 times greater compared to normal fetuses. Infants with IUGR may exhibit hazardous indications of MS in the early postnatal phase, including hyperinsulinemia and IR (Silveira and Horta 2008). Furthermore, epidemiological investigations have confirmed that adverse prenatal conditions can also elevate the vulnerability of offspring to MS in later life. A cohort study discovered that adults who experienced prenatal exposure to the 1983–1985 Ethiopian famine were 2.94 times more prone to developing MS, and a greater proportion of MS high-risk anthropometric parameters and dyslipidemia parameters were observed in the exposed group (Arage et al. 2020). Another cohort study examining the health and development of children born between 1959 and 1967 found that offspring of 1801 mothers aged 44 to 54 who smoked during pregnancy had an increased risk of diabetes (La Merrill et al. 2015). These epidemiological findings provide support for the fetal origin hypothesis of MS.

3.2 *Clinical and Basic Laboratory Research*

As research into the fetal origin of MS continues to advance, numerous animal experiments have confirmed this hypothesis. Exposure of the fetus to harmful substances due to maternal intake during pregnancy can lead to an adverse intrauterine environment, resulting in IUGR and long-term changes in the liver metabolic function of the offspring (Ayres et al. 2011). In addition to the prenatal malnutrition model, prenatal overnutrition can also alter the intrauterine metabolic environment's homeostasis. Studies have shown that increasing dietary fat intake during pregnancy and lactation predisposes offspring to an MS-like phenotype in adulthood. Common adverse environmental models during pregnancy include exposure to xenobiotics, uterine artery ligation, and low-calorie and low-protein diet models. The influence of postnatal adverse factors such as intrauterine dysplasia and high-fat diet can further increase the susceptibility to MS (Rueda-Clausen et al. 2011).

3.2.1 Abnormal Blood Glucose and Lipid with Fetal Origin

Studies have found that newborn rats subjected to uterine artery ligation had impaired insulin secretion after glucose stimulation, which increased with age (Simmons et al. 2001). A similar phenomenon of insufficient insulin secretion occurs in the low-calorie model of late pregnancy. The offspring of pregnant rats fed a low-protein diet had a lower birth weight, and the male offspring experienced age-dependent impaired glucose tolerance, developing symptoms similar to human type 2 diabetes at the age of 17 months (Buckley et al. 2005). The authors' team also conducted research that found that rat offspring exposed to xenobiotics, such as caffeine and ethanol, or with dietary restriction during pregnancy, had impaired glucose metabolism and were susceptible to diabetes after birth (Kou et al. 2017; Xiao et al. 2017, 2019).

Population surveys have shown that smoking during pregnancy can cause hyperlipidemia, hypertension, and high waist circumference in children and dyslipidemia in adults (Seal et al. 2013). Similarly, maternal alcohol consumption or obesity during pregnancy can lead to abnormal lipid metabolism in the offspring. Animal experiments have confirmed that the adverse intrauterine environment results in abnormal lipid metabolism in IUGR fetal rats, and this effect could cause MS susceptibility in adulthood (Papoutsi et al. 2007). The IUGR offspring rats induced by dietary restriction during pregnancy showed an increase and accumulation of triglyceride synthesis in the liver, with gender differences after birth, which was prominent in males but unchanged in females (Choi et al. 2007). However, Carbone's study (2012) demonstrated more severe liver fat accumulation under a high-fat diet in female rats with IUGR. The author's team conducted research that found that an adverse environment during pregnancy, such as alcohol and caffeine exposure, causes significant lipid accumulation in the liver of the offspring, leading to nonalcoholic fatty liver disease (NAFLD) under excessive nutrition, such as a high-fat diet, after birth (Shen et al. 2014; Wang et al. 2014). Moreover, a high-fat diet could exacerbate the hypercholesterolemia of rats exposed to caffeine during pregnancy, with female rats being more susceptible (Guo et al. 2018). These findings indicate that the intrauterine origin of abnormal glucose and lipid metabolism is widely recognized as one of the principal phenotypes of MS (Table 11.1).

3.2.2 Abnormal Blood Pressure with Fetal Origin

A US study found that low birth weight babies had a significantly increased risk of high blood pressure as adults (Hughson et al. 2006). Further studies have shown that a loss of birth weight increases the risk of hypertension (Bergvall et al. 2007). In Sweden, a retrospective epidemiological study found that blood pressure is negatively correlated with birth weight in children aged 5–13 (Leon et al. 2005). Similar results have been found in various ethnic and regional populations, suggesting that low birth weight is a risk factor for adult hypertension (Hemachandra et al. 2006). Animal studies have shown that prenatal exposure to lipopolysaccharide (0.79 mg/kg) leads to hypertension in offspring rats (Hao et al. 2010). However, prenatal exposure to lipopolysaccharide combined with prenatal and postnatal high-fat diets led to decreased blood pressure and IR development in 3-month-old rat offspring (Hao et al. 2014). A low-protein diet reduces the number of nephrons in F1 offspring, leading to hypertension (Langley-Evans et al. 1999). Exposure to high salt during pregnancy increases the susceptibility to hypertension in adult rat offspring (Liu et al. 2018). Dexamethasone exposure during pregnancy significantly increases angiotensin II type 1 receptor expression in the brain stem and arterial blood pressure of adult sheep offspring (Dodic et al. 2006). Furthermore, the number of nephrons in offspring exposed to dexamethasone during pregnancy is significantly reduced, and the expression of renin and renin precursors is significantly upregulated, leading to nephrogenic hypertension (Singh et al. 2007). In conclusion, the occurrence of adult hypertension is closely related to the adverse environment during pregnancy.

3.2.3 Obesity with Fetal Origin

Obesity has long been attributed to genetic factors, an unhealthy diet, and a sedentary lifestyle. However, it is now widely recognized that obesity is a complex disease caused by both genetic and environmental factors. Studies have shown that maternal overnutrition can increase the risk of obesity in offspring, and IUGR

Table 11.1 Abnormal glucose and lipid metabolism in offspring caused by the adverse environment during pregnancy

Type of study	Adverse environment during pregnancy	Species	Phenomenon	Main references
Clinical research	Smoking	Human	Hyperlipidemia, hypertension, and high waist circumference in childhood and disordered blood total cholesterol levels in adulthood	Godfrey et al. (1996)
	Smoking	Human	MS onset in adolescence, blood pressure, blood total cholesterol, low-density lipoprotein cholesterol, and triglyceride levels increased	Efstathiou et al. (2012)
Basic research	Obesity	Human	Increased liver fat content in newborns	Modi et al. (2011)
	Polyfluoroalkyl substances	Human	Lipid accumulation	Mora et al. (2018)
	Psychostimulants	Mouse	Increased blood glucose and glucose intolerance	Korchynska et al. (2020)
	Dexamethasone	Rat	Abnormal glucose tolerance	Kou et al. (2020)
	Arsenic	Rat	Hyperglycemia	Ohara et al. (1993)
	Ethanol	Rat	Hyperglycemia and hyperlipidemia after a high-fat diet	Xia et al. (2014)
	Dietary restrictions	Rat	Blood cholesterol increases in adulthood	Qu et al. (2007)
	Dietary restrictions	Rat	Enhancement and accumulation of hepatic triglyceride synthesis	Choi et al. (2007)
	Polystyrene	Rat	Postnatal lipid disorder in serum and the liver of offspring rats	Luo et al. (2019)
	Caffeine	Rat	Lipid accumulation in the liver after high-fat diet	Wang et al. (2014)
	Ethanol	Rat	Increased susceptibility to nonalcoholic fatty liver disease	Shen et al. (2014)
	Perfluoro compound	Rat	Increased susceptibility to nonalcoholic fatty liver disease	Lv et al. (2013)
	Perfluoro octanoic acid	Rat	Increased triglyceride	Quist et al. (2015)

animals with low maternal protein models are more likely to develop obesity during postnatal development (Bol et al. 2009). Moreover, the offspring of uterine artery ligation rats exhibit weight catch-up growth and reach obesity at 26 weeks of age (Simmons et al. 2005). Exposure to bisphenol A during the early development period can significantly increase body weight and lead to IR in female mice. The decrease in serum leptin level caused by bisphenol A exposure was one of the causes of obesity and IR occurrence (Angle et al. 2013). Conversely, studies have found that a high-protein diet given in early pregnancy can reduce the occurrence of adult obesity by programmatically increasing the expression of uncoupling protein of offspring (Dong et al. 2008). The fetal-originated obesity has been found in a growing number of animal models, indicating that the adverse environment during pregnancy is closely related to the occurrence of offspring obesity.

4 Intrauterine Programming Mechanism of Fetal-Originated MS

Intrauterine programming refers to the permanent changes in tissue morphology and function caused by damage in the initial stage of development. These changes lead to alterations in the expression patterns of functional genes in tissues and organs, which usually last from adolescence to adulthood and even throughout life. The pathogenesis of fetal-originated MS is highly complex, with current studies mainly focusing on pathogenic factors, including abnormal blood glucose, dyslipidemia, and obesity. Among them, programming changes in glucose and lipid metabolism, HPA axis-related neuroendocrine metabolic programming, and the "two-programming and two-strike" mechanism are the main mechanisms of fetal-originated MS. Understanding the mechanism of intrauterine programming of fetal-originated MS under adverse environmental exposure during pregnancy will provide an essential theoretical basis for exploring early prevention and treatment strategies of fetal-originated MS.

4.1 Changes in Metabolic Programming of Glucose, Lipid, and Adipose Tissue

A study conducted by Dumortier et al. (2007) revealed that the low-calorie mouse model exhibited a decrease in neurogenin 3 and pancreatic and duodenal homeobox-1 (Pdx-1) positive cell numbers at the 15th day of gestation. This reduction in pancreatic progenitor cell conversion to endocrine pancreas resulted in impaired β-cell differentiation. Valtat et al. (2011) reported that inactivation of the glucocorticoid receptor (GR) prevented β-cell depletion induced by intrauterine low calories during mouse pancreatic development. Gesina et al. (2006) demonstrated that glucocorticoid (GC) can interfere with normal β-cell differentiation and reduce Pdx-1 expression in in vitro pancreatic development experiments. Ma et al. (2013) found that exposure to bisphenol A during pregnancy and lactation led to global DNA hypomethylation and increased DNA methylation of glucokinase gene promoter, as well as decreased glucokinase expression in offspring. Kou et al. (2020) recently reported that prenatal dexamethasone exposure reduced the levels of H3K27ac in the angiotensin receptor II receptor 2 promoters of female offspring before and after birth, resulting in β-cell dysfunction and abnormal glucose tolerance. Furthermore, Kou et al. (2017) found that increased glucose tolerance in offspring exposed to caffeine during pregnancy was associated with decreased pancreatic β-cell mass. These findings suggest that adverse conditions during pregnancy can contribute to abnormal glucose metabolism by affecting the development of pancreatic β-cells.

Sohi et al. (2011) reported that maternal protein restriction in adult rat offspring resulted in increased expression of 3-hydroxy-3-methylglutarate monoacyl-CoA reductase and cholesterol 7-hydroxylase in the liver. This increase was associated with increased levels of histone H3 acetylation and H3K4 trimethylation (H3K4me3) and decreased levels of H3K9me1 and H3K27me3, affecting cholesterol metabolism. Quist et al. (2015) found that prenatal exposure to perfluorooctanoic acid in mice fed a high-fat diet resulted in lipid accumulation in offspring, leading to hepatocyte injury and mitochondrial abnormalities. The authors' team conducted a series of studies and found that offspring rats exposed to caffeine during pregnancy had increased blood cholesterol levels after birth, accompanied by increased cholesterol synthesis and decreased cholesterol reversal transport in the liver of offspring both in utero and after birth (Xia et al. 2014). These findings suggest that changes in liver cholesterol metabolism genes are involved in the occurrence of fetal dyslipidemia (Xu et al. 2018). The authors' team also discovered that caffeine can directly increase liver cholesterol synthesis function through the A2AR/cAMP/PKA signaling pathway, leading to postnatal hypercholesterolemia (Hu et al. 2019). Berry et al. (2013) reported that dexamethasone exposure in late pregnancy can cause fat accumulation in offspring. Our team confirmed that prenatal dexamethasone exposure can lead to postnatal hypercholesterolemia by activating GR in the liver of rat offspring and enhancing the expression of miR-148a, thereby reducing the expression of low-density lipoprotein receptor, a key gene for liver cholesterol reversing transport (Li et al. 2020). Therefore, epigenetic changes in liver lipid metabolism-related genes caused by adverse environments during pregnancy are closely related to dyslipidemia after birth.

In the adipose tissue of IUGR offspring, Desai et al. (2008) reported that upregulation of peroxisomal proliferator-activated receptor γ expression could promote glucose and lipid uptake, leading to accelerated obesity. Zhang et al. (2011) found that fat deposition in the offspring of malnourished ewes during the perinatal period was related to the expression of insulin-like growth factor (IGF)2 in the adrenal gland and the reduction of DNA methylation in the proximal imprinting control region. Our team discovered that caffeine exposure during pregnancy increases liver triglyceride synthesis and susceptibility to NAFLD in female rat offspring by activating the GR-CCAAT enhancer binding proteins alpha (C/EBp-α)-SIRT1 pathway (Hu et al. 2019). Guan et al. (2005) observed that the expression of a series of genes related to lipid synthesis was increased in the adipose tissue of children with IUGR. In adults with IUGR, downregulation of hormone-sensitive lipase promoted fat synthesis and reduced lipolysis. Changes in these lipid-metabolizing enzymes also play a role in fetal-derived obesity.

4.2 Hypothalamic-Pituitary-Adrenal Axis-Related Neuroendocrine Metabolic Programming Alteration

Exposure to maternal GC is a common occurrence in adverse maternal environments during pregnancy. Numerous studies have demonstrated that glucocorticoids, as the main effector product of the HPA axis, can regulate IGF1 expression in various tissues and cells. This regulation is mainly observed as increased IGF1 secretion at low glucocorticoid levels (Binoux et al. 1982) and inhibited IGF1 secretion at high glucocorticoid levels (Robson et al. 2002; Inder et al. 2010). In conclusion, adverse environmental conditions during pregnancy that increase maternal glucocorticoid levels may inhibit IGF1 secretion, thereby affecting fetal development. Recently, based on the results of our and other laboratory studies, we have proposed an "intrauterine neuroendocrine metabolic programming" mechanism to explain the susceptibility of offspring to chronic diseases in adulthood due to adverse environmental conditions during pregnancy. This mechanism suggests that the increase in maternal GC, caused by various adverse environmental conditions during pregnancy, inhibits the development of the HPA axis in fetal rats and leads to characteristic glucocorticoid-dependent glycolipid metabolism changes in peripheral tissues such as the liver and skeletal muscle (Liang et al. 2011; Liu et al. 2012; Xu et al. 2012). These neuroendocrine metabolic changes could persist after birth, leading to abnormal function of the HPA axis, including low basal activity and high stress sensitivity, as well as GC-IGF1-dependent changes in glycolipid metabolism phenotype. Specifically, under low basal activity of the HPA axis (low GC level), glucose metabolism was accelerated but lipid metabolism was slowed down, while under high stress sensitivity of the HPA axis (high GC level), glucose metabolism was slowed down but lipid metabolism was accelerated. These changes in glucose and lipid metabolism phenotype were similar to those observed under overexposure to maternal glucocorticoids in utero (Liang et al. 2011; Liu et al. 2012; Xu et al. 2012). Based on these findings, we have proposed, for the first time, an "intrauterine neuroendocrine metabolic programming mechanism" to explain the susceptibility of adult MS and related metabolic diseases in IUGR offspring induced by prenatal xenobiotic exposure (Fig. 11.2). This intrauterine programming change could increase the susceptibility of offspring to multiple chronic diseases after birth (Shen et al. 2014; Wang et al. 2014).

4.3 The "Two-Programming and Two-Strike" Mechanisms of Fetal-Originated Metabolic Syndrome

Drawing from previous systematic research on glucolipid metabolic programming and HPA axis-related neuroendocrine metabolic programming, the authors' team proposed a "two-programming and two-strike" mechanisms for fetal-originated

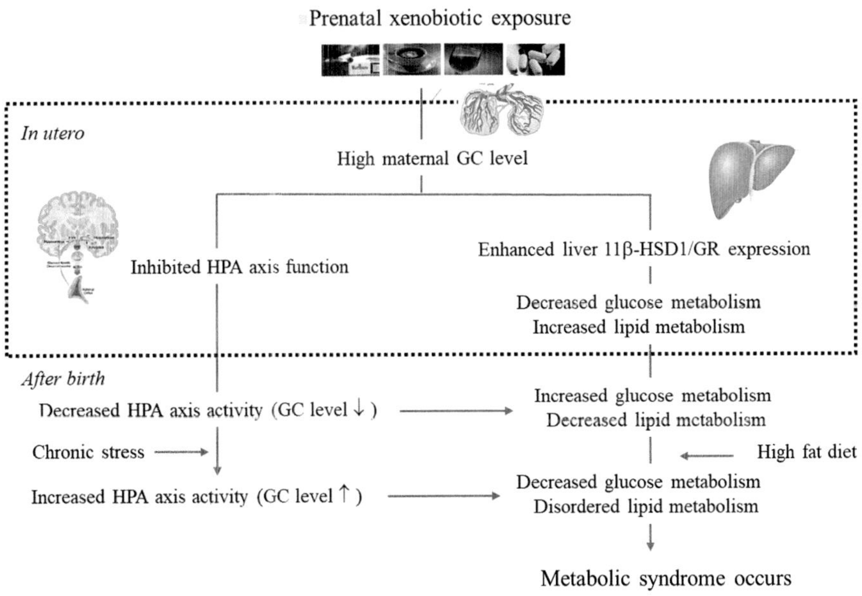

Fig. 11.2 Hypothalamic-pituitary-adrenal axis-related neuroendocrine metabolic programming alteration in fetal-originated metabolic syndrome. *GR* glucocorticoid receptor, *GC* glucocorticoid

MS, as depicted in Fig. 11.3. In a rat model, exposure to caffeine and nicotine during pregnancy resulted in epigenetic modifications mediated by xenobiotics or maternal GC, leading to the upregulation of liver cholesterol synthesis genes in offspring. This enhanced phenotype persisted postnatally, resulting in continued enhanced function of cholesterol synthesis in adult offspring, and increased susceptibility to MS and related metabolic diseases (Xu et al. 2018). The first programming mechanism involves changes that persist before and after birth. The second programming mechanism is the "GC-IGF1 axis" programming in multi-organs. The authors' team investigation revealed that offspring exposed to xenobiotics during pregnancy exhibited elevated levels of intrauterine maternal glucocorticoids, leading to the inhibition of fetal liver IGF1 expression and IUGR. Following chronic stress after birth, fetal liver IGF1 expression was further inhibited due to the prior exposure to intrauterine high levels of maternal glucocorticoids, resulting in compensatory development of liver function. Additionally, the enhanced liver fat metabolism was further exacerbated after a high-fat diet was introduced (Shen et al. 2014; Wang et al. 2014). The long-term effects of nicotine and ethanol exposure during pregnancy were observed in F1 and F2 generation rats, primarily due to epigenetic programming changes of the "GC-IGF1 axis" mediated by the GC activation system, including 11β-hydroxysteroid dehydrogenase type (11β-HSD)1/11β-HSD2, and GR-C/EBp-α (Hu et al. 2020; Hu et al. 2020).

In the case of the "first strike," such as xenobiotic exposure during pregnancy, the liver regulates its metabolic function through two forms of programming.

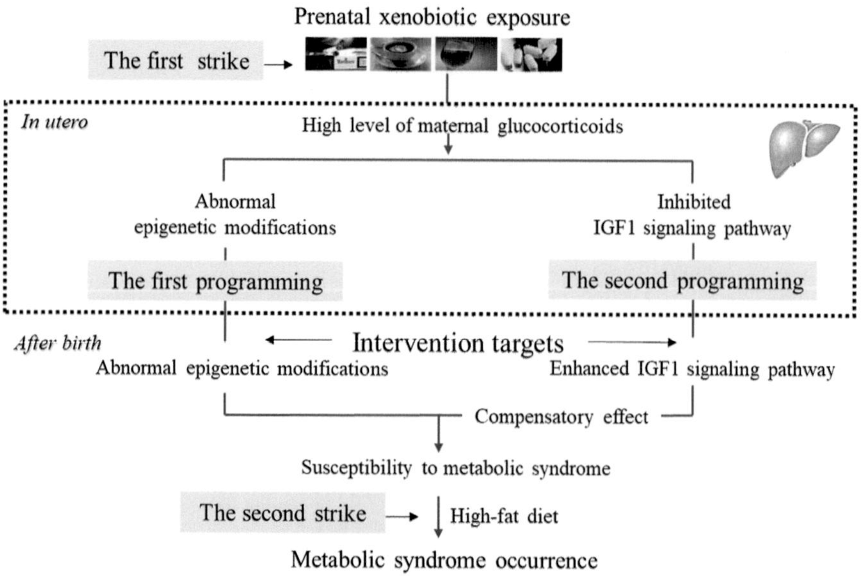

Fig. 11.3 The "two-programming and two-strike" mechanisms for fetal-originated metabolic syndrome

Subsequently, environmental changes such as chronic stress, high-fat diet, and mental stress act as the "second strike," which may ultimately lead to the development of MS and related metabolic diseases. The authors' team research confirmed that exposure to xenobiotics, including caffeine, nicotine, or ethanol, or dietary restriction during pregnancy as the "first strike," can inhibit the development of the HPA axis-related neuroendocrine and metabolic changes in fetal rats. In adult offspring rats, blood ACTH and corticosterone levels decreased in the basal state but increased significantly after chronic stress (as a "second strike"), leading to increased susceptibility to MS and related metabolic diseases (Xu et al. 2015). The authors' team proposed that exposure to xenobiotics during pregnancy and dietary restriction as the "first strike" can result in increased GC levels and decreased IGF1 levels in IUGR fetuses. Conversely, adult blood exhibited decreased GC levels and increased IGF1 levels, accompanied by catch-up growth. Additionally, a high-fat diet and chronic stress as a "second strike" could exacerbate liver metabolic damage. The study suggests that various tissues and organs of IUGR offspring exposed to xenobiotics during pregnancy undergo adaptive changes in endogenous GC and IGF1 to adapt to the adverse intrauterine environment, ensuring fetal growth and development. The postnatal catch-up growth, glucolipid metabolism disorder, and susceptibility to MS in IUGR offspring are associated with postnatal changes in the liver "GC-IGF1 axis" programming.

The elucidation of the "two-programming and two-strike" mechanism of fetal-originated MS provides a robust foundation for identifying early warning biomarkers and evaluating harmful factors during pregnancy. Additionally, early postnatal interventions targeting epigenetic modifications of related functional genes may be an effective means of preventing fetal-originated MS, thus offering a new approach for preventing fetal-originated disease and their associated complications.

5 Summary and Prospect

The development of MS is attributed to various factors, including genetic, dietary, physical activity, and environmental factors. Available epidemiological and laboratory evidence strongly supports the fetal origin of MS. Fetal programming induced by adverse intrauterine environments, including exposure to xenobiotics and maternal malnutrition, is the primary factor leading to offspring susceptibility to MS after birth. With the improvement of economic levels and quality of life, the incidence of MS has been increasing annually. Implementing effective measures to prevent fetal-originated MS can delay the development of metabolism-related diseases and prevent and control related complications. However, the lack of effective prevention and treatment of MS may be due to an incomplete understanding of its etiology and pathophysiological mechanisms. The occurrence and development of fetal-originated MS are the result of multiple factors. The "two-programming and two-strike" theory provides a theoretical basis for further analysis of fetal-originated MS. It is believed that with the deepening of research on fetal-originated MS, the understanding of the pathophysiological processes, diagnostic criteria, and clinical significance of fetal-originated MS will continue to deepen, providing new ideas for developing early warning and prevention technologies related to fetal-originated MS-related diseases.

References

Dong Y M, Sun C S, Cui L L. 2008. Chin J Prev Med, 42(001): 16-20.
Arage G, Belachew T, Hassen H, et al. 2020. Br J Nutr, 124(10): 1052-1060.
Angle B M, Do R P, Ponzi D, et al. 2013. Reprod Toxicol, 42: 256-268.
Ayres C, Silveira P P, Barbieri M A, et al. 2011. J Dev Orig Health Dis, 2(3): 162–167.
Buckley A J, Keserü B, Briody J, Thompson M, et al. 2005. Metabolism, 54(4): 500–507.
Barker D J, Osmond C, Golding J, et al. 1989. BMJ, 298(6673): 564–567.
Bergvall N, Iliadou A, Johansson S, et al. 2007. Circulation, 115(23): 2931–2938.
Berry M J, Jaquiery A L, Oliver M H, et al.2013 . Acta Obstet Gynecol Scand, 92(7): 862–865.
Binoux M, Lassarre C, Hardouin N. 1982. Eur J Endocrinol, 99(3): 422–30.
Bol V V, Delattre A I, Reusens B, et al. 2009. Am J Physiol Regul Integr Comp Physiol, 297(2): R291–299.
Carbone D L, Zuloaga D G, Hiroi R, et al. 2012. Endocrinology, 153(1): 295–306.
Choi G Y, Tosh D N, Garg A, et al. 2007. Am J Obstet Gynecol, 196(5): 471–477.

Cruickshank J K, Mzayek F, Liu L, et al. 2005 Circulation, 111(15):1932–1937.
Desai M, Guang H, Ferelli M, et al. 2008. Reprod Sci, 15(8): 785–796.
Dodic M, McAlinden A T, Jefferies A J, et al. 2006. J Physiol, 571(Pt 3): 651–660.
Dumortier O, Blondeau B, Duvillie B, et al. 2007. Diabetologia, 50(12): 2495–2503.
Efstathiou S P, Skeva II, Zorbala E, et al. 2012. Circulation, 125(7): 902–910.
Gesina E, Blondeau B, Milet A, et al. 2006. Diabetologia, 49(12): 2939–2947.
Godfrey K, Robinson S, Barker D J, et al. 1996. BMJ, 312(7028): 410–414.
Guan H, Arany E, van Beek J P, et al. 2005. Am J Physiol Endocrinol Metab, 288(4): E663-E673.
Guo Y, Luo H, Wu Y, et al. 2018. Reprod Toxicol, 79: 47–56.
Hao X Q, Zhang H G, Yuan Z B, et al. 2010. Hypertens Res, 33(1): 76–82.
Hao X Q, Du J X, Li Y, et al. 2014. PLoS One, 9(2): e88127.
Hemachandra A H, Klebanoff M A, Furth S L. 2006. J Am Soc Nephrol, 17(9): 2576–2581.
Hughson M D, Douglas-Denton R, Bertram J F, et al. 2006. Kidney Int, 69(4): 671–678.
Huxley R R, Shiell A W, Law C M, et al. 2000. J Hypertens, 18(7): 815–831.
Hu S W, Liu K X, Luo H W, et al. 2019. Toxicology, 418: 11–21.
Hu W, Wang G H, He B, et al. 2020. Toxicology, 432: 152378.
Hu W, Yuan C, Luo H W, et al. 2020. Toxicol Lett, 331: 167–177.
Inder W J, Jang C, Obeyesekere V R, et al. 2010. Clin Endocrinol (Oxf), 73(1): 126–132.
Kou H, Wang G H, Pei L G, et al. 2017. Naturwissenschaften, 104(11–12): 89.
Kou H, Gui S X, Dai Y G, et al. 2020. Toxicol Appl Pharmacol, 404: 115187.
Korchynska S, Krassnitzer M, Malenczyk K, et al. 2020. EMBO J, 39(1): e100882.
La Merrill M A, Cirillo P M, Krigbaum N Y, et al. 2015. J Dev Orig Health Dis, 6(3): 242–249.
Langley-Evans S C, Welham S J, Jackson A, et al.1999. Life Sci, 64(11): 965–974.
Leon D A, Koupil I, Mann V, et al. 2005. Circulation, 112(22): 3478–3485.
Liang G, Chen M, Pan X L, et al. 2011. Exp Toxicol Pathol, 63(7–8): 607–611.
Liu Y, Qi L, Wu J, et al.2018. J Hypertens, Dec;36(12): 2369–2379.
Liu L, Liu F, Kou H, et al. 2012. Toxicol Lett, 214(3): 307–313.
Li L, Hu W, Liu K X, et al. 2020. Toxicol Appl Pharmacol, 395: 114979.
Luo T, Zhang Y, Wang C, et al. 2019. Environ Pollut, 255(Pt 1): 113122.
Lv Z, Li G, Li Y, et al. 2013. Environ Toxicol, 28(9): 532–542.
Ma Y, Xia W, Wang D Q, et al. 2013. Diabetologia, 56(9): 2059–2067.
Modi N, Murgasova D, Ruager-Martin R, et al. 2011. Pediatr Res, 70(3): 287–291.
Mora AM, Fleisch AF, Rifas-Shiman SL, et al.2018. Environ Int, 111: 1–13.
Ohara Y, Peterson T E, Harrison D G. 1993. J Clin Invest, 91(6): 2546–2551.
Papoutsi M, Dudas J, Becker J, et al. 2007. Cell Tissue Res, 330(2): 209–220.
Park H S, Kim S M, Lee J S, et al. 2007. Diabetes Obes Metab, 9(1): 50–58.
Qu X, Lam E, Doughman Y Q, et al. 2007. Embo Journal, 26(21): 4445–4456.
Quist E M, Filgo A J, Cummings C A, et al.2015. Toxicol Pathol, 43(4): 546–557.
Robson H, Siebler T, Shalet SM. et al.2002. Pediatr Res, 52(2): 137–147.
Rueda-Clausen C F, Dolinsky V W, Morton J S, et al. 2011. Diabetes, 60(2): 507–516.
Seal N, Krakower G, Seal J. 2013. Jnp-J Nurse Pract, 9(10): 695–705.
Shen L, Liu Z, Gong J, et al. 2014. Toxicol Appl Pharmacol, 274(2): 263–273.
Silveira V M, Horta B L. 2008. Rev Saude Publica. 42(1): 10–18.
Sohi G, Marchand K, Revesz A, et al. 2011. Molecular endocrinology, 25(5): 785–798.
Simmons R A, Suponitsky-Kroyter I, Selak M A. 2005. J Biol Chem, 280(31): 28785–28791.
Simmons R A, Templeton L J, Gertz S J. 2001. Diabetes, 50(10): 2279–2286.
Singh R R, Moritz K M, Bertram J F, et al. 2007. Am J Physiol Renal Physiol, 293(2): F548-F554.
Valtat B, Dupuis C, Zenaty D, et al. 2011. Diabetologia, 54(2): 350–359.
Wang L, Shen L, Ping J, et al. 2014. Toxicol Lett, 224(3): 311–318.
Xia L P, Shen L, Kou H, et al. 2014. Toxicol Lett, 226(1): 98–105.
Xiao D, Kou H, Zhang L, et al. 2017. Arch Med Res, 48(1): 35–45.

Xiao D, Kou H, Gui S, et al. 2019. Front Endocrinol (Lausanne), 10: 34.
Xu D, Wu Y, Liu F, et al. 2012. Toxicol Appl Pharmacol, 264(3): 395–403.
Xu D, Bai J, Zhang L, et al. 2015. Toxicol Res, 4(1): 112–120.
Xu D, Luo H W, Hu W, et al. 2018. FASEB J, 32(10): 5563–5576.
Zhang S, Rattanatray L, McMillen IC, et al. 2011. Prog Biophys Mol Biol, 106(1): 307–314.

Chapter 12
Fetal Origin of Diabetes

Hao Kou and Yu Guo

Abstract Diabetes mellitus is a chronic metabolic disease characterized by elevated blood glucose levels resulting from defective insulin secretion and/or action. While genetics and human lifestyle are known factors contributing to the development of diabetes, emerging evidence suggests that an adverse environment during early development may also play a role. Intrauterine growth retardation, caused by adverse factors during pregnancy (such as maternal diseases, nutritional imbalances, and exposure to foreign substances), may lead to abnormal development and function of fetal pancreatic β-cells and peripheral glucose-metabolizing organs such as the liver and fat. These changes may result in continuous adaptive changes in glucose-insulin metabolism, including decreased insulin secretion ability and insulin resistance, predisposing individuals to diabetes later in life. The occurrence and development of fetal-originated diabetes may be related to the "two-programming and two-strike" mechanism, in which epigenetic modifications play a significant role.

1 Introduction

Diabetes mellitus (DM) is a significant global health concern that has become an epidemic due to population aging, urbanization, and lifestyle changes. The prevalence of DM has increased rapidly worldwide over the past few decades, and DM has become a major public health issue (Jaacks et al. 2016). The latest data from the International Diabetes Federation (IDF) Global Diabetes Overview 2019 reports that 463 million adults currently have DM, with an estimated 578 million adults

H. Kou (✉)
Department of Pharmacy, Zhongnan Hospital of Wuhan University, Wuhan, Hubei, China
e-mail: kouhao007@whu.edu.cn

Y. Guo
Department of Pharmacology, School of Basic Medical Science, Wuhan University, Wuhan, Hubei, China

H. Wang et al. (eds.), *Fetal Origin of Diseases*,
https://doi.org/10.1007/978-981-97-5730-5_12

projected to have DM by 2030 and 700 million by 2045. Studies on the etiology and pathogenesis of DM suggest that both genetic and environmental factors contribute to its occurrence, with genetic susceptibility playing a decisive role in its development. While genetic factors play a role in the development of DM, the sharp rise in diabetes rates over the past few decades cannot be attributed to genetic factors alone. Rapid lifestyle changes around the world are also contributing to the increasing prevalence of DM (Wu et al. 2014). Emerging evidence suggests that the risk of DM depends not only on lifestyle factors in adulthood but also on the living environment in early life (Jiang et al. 2013; Berends and Ozanne 2012; Vaiserman 2017; Estampador and Franks 2018). Unfortunately, our understanding of how these early factors trigger the development of DM in the body is limited. This chapter aims to review and discuss the characteristics and intrauterine programming mechanism of fetal-originated DM.

2 Research Status of Diabetes Mellitus

DM is a chronic metabolic disorder characterized by hyperglycemia resulting from impaired insulin secretion and/or insulin resistance. This disease can be broadly classified into two types—type 1 diabetes mellitus (T1DM) and type 2 diabetes mellitus (T2DM).

2.1 Current Research Situation of Diabetes

The initial global projection of diabetes incidence was based on a series of population studies conducted between 1994 and 1997, which approximated that between 100 and 124 million individuals worldwide were affected by diabetes (Mccarty and Zimmet 1997; Amos et al. 1997). However, the latest estimations based on the IDF Global Diabetes Overview 2021 tenth Edition indicate that the number of adults aged 20–79 years with diabetes has increased to 537 million (10.5%). Notably, the prevalence of diabetes is higher in men than in women (10.8% vs 10.2%). The prevalence of diabetes also tends to rise with age, and there are significant regional differences in its prevalence.

2.2 Main Pathogenesis of Diabetes

The pathogenesis of diabetes is a complex and multifactorial process that varies depending on the specific type of diabetes. Even within the same type, there can be differences in the underlying causes. Therefore, the pathogenesis of diabetes can be broadly classified based on the different types of diabetes.

In the case of T1DM, the pathogenesis typically involves autoimmune diseases, genetic factors, and environmental factors such as viral infections, chemical poisons, and dietary factors. These external factors can trigger autoimmune reactions mediated by T lymphocytes in genetically susceptible individuals, leading to selective islet β-cell destruction and functional failure. This exacerbates the insufficient insulin secretion in the body, ultimately resulting in diabetes. It has also been confirmed that insulin resistance plays a role in the onset and/or accelerated deterioration of T1DM.

T2DM is a complex polygenic disease caused by both genetic and environmental factors. The characteristics of its genetic factors are as follows: (1) there are many genes involved in the pathogenesis; (2) the degree of each gene involved in the disease was not equal; (3) each gene only confers a certain degree of susceptibility to the individual; and (4) the total effect of polygenic abnormalities forms genetic susceptibility. Environmental factors affecting the onset of T2DM include aging, modern lifestyle, overnutrition, insufficient physical activity, intrauterine environment and stress, and chemical toxicants. Under the combined action of genetic factors and the above environmental factors, obesity (especially central obesity), which is closely related to insulin resistance and T2DM, can induce insulin resistance and T2DM (Kahn et al. 2006). Recent studies suggest that the progression of diabetes is still largely due to insulin secretion defects and β-cell apoptosis, which further lead to compensatory dysregulation of insulin resistance (Helman et al. 2016). β-cell dysfunction leading to varying degrees of insulin deficiency and tissue (especially skeletal muscle and liver) insulin resistance are the two main stages of T2DM pathogenesis. In the presence of insulin resistance, normal blood glucose can be maintained if the β-cells can compensate for increased insulin secretion, whereas T2DM occurs when the function of the β-cells cannot compensate for insulin resistance (Cnop et al. 2005).

3 Evidence of Intrauterine Origin of Diabetes Mellitus

Over the past two decades, there has been increasing awareness of the intrauterine origins of diabetes, although the impact of environmental factors in adulthood on the development and progression of DM is well-established. Both epidemiological and laboratory studies have demonstrated that the intrauterine environment can have a significant influence on the development of DM in adulthood.

3.1 Epidemiological Investigation

There is mounting evidence that the intrauterine environment, in which the fetus grows and develops, can have long-term effects on health and survival. For instance, individuals who experienced intrauterine growth retardation (IUGR) or low birth weight have significantly higher rates of diabetes in adulthood. In a seminal cohort study of 300,000 men, Ravelli et al. (1976) found that maternal exposure to the Dutch famine of 1944–1945 during pregnancy resulted in significantly higher rates of obesity in their offspring by age 19. Subsequent studies have demonstrated a correlation between low-birth-weight infants in England and subsequent impaired glucose tolerance (Hales et al. 1991).

IUGR is a common manifestation of the adverse effects of pregnancy complications on fetal development. Studies conducted across different populations worldwide have demonstrated a significant correlation between IUGR and the subsequent development of T2DM (Hales and Barker 2001; Curhan et al. 1996; Valdez et al. 1994; Lithell et al. 1996; McKeigue et al. 1998; Leger et al. 1997; Jaquet et al. 2000; Egeland et al. 2000). Population-based studies have also indicated an increased risk of glucose intolerance and T2DM in offspring born to mothers with pre-existing diabetes (Pettitt et al. 1988). Moreover, some epidemiological studies have reported an increased incidence of diabetes in the offspring of mothers who used certain drugs, such as glucocorticoids, during pregnancy (Greene et al. 2013). Collectively, these studies suggest that DM has an intrauterine developmental origin.

3.2 Clinical and Basic Laboratory Research

To further investigate the intrauterine origin of DM and the underlying mechanisms, various animal models including rodents, sheep, guinea pigs, and nonhuman primates have been utilized in research, supported by epidemiological evidence. These animal models include models of maternal nutritional change, exposure to xenobiotics during pregnancy, and abnormal maternal metabolism. These models have provided valuable insights into the developmental origins of DM and have helped to elucidate the mechanisms underlying this phenomenon.

Maternal malnutrition is a significant contributor to diabetes in offspring. Fetal development relies on maternal nutrition, and inadequate maternal nutrient intake may result in IUGR in the offspring (Campos et al. 2012). The end of pregnancy and the early postnatal period are critical periods for the growth and development of pancreatic β-cells, and the amount of nutrients during this period can have long-term and even lifelong effects on the structure and function of the pancreas. To investigate the impact of maternal nutrition on offspring, researchers often employ maternal nutrient deficiency (such as dietary or protein restriction) and excess (such as high-fat diet) models in basic experimental studies. Studies have demonstrated that IUGR can occur in the offspring of pregnant rats when their food intake is restricted to 50% (i.e., feeding restriction) during the last trimester of pregnancy, and the quality of islet β-cells in the offspring decreases by 40% at birth. If the duration of dietary restriction is prolonged throughout pregnancy or even into lactation, the quality of islet β-cells after birth is further diminished (Dumortier et al. 2007). In rat models, selective protein restriction (40%–50% of the normal amount), even with normal energy intake during pregnancy, resulted in IUGR, reduced islet size and islet β-cell mass, and impaired glucose tolerance in adult offspring (Garofano et al. 1997). Dumortier et al. (2007) also demonstrated that low-protein diets during late pregnancy reduced the quality of islet β-cells by 53% in fetal rats at 21 days of gestation. Using a 60% high-fat diet-induced rodent obesity model, maternal obesity was found to impair the function of offspring islet β-cells, leading to abnormal glucose-insulin metabolism and impaired glucose tolerance (Graus-Nunes et al. 2015; Zambrano et al. 2016). Additionally, offspring mice on a vitamin D-restricted diet during pregnancy exhibited decreased islet β-cell quality at postnatal week (PW) 12 (Maia-Ceciliano et al. 2016).

Exposure to certain xenobiotics during pregnancy has been identified as a potential risk factor for the development of diabetes in offspring. Pregnant women often require pharmacological intervention for various medical conditions, but some drugs may exert deleterious effects on the developing fetus. For instance, dexamethasone, a commonly prescribed corticosteroid for preterm birth, has been shown to cause a reduction in pancreatic β-cell count and insulin content during both prenatal and postnatal stages in rodent models (De et al. 2007; Kou et al. 2020; Chen et al. 2017; de Vries et al. 2007). This has been associated with the onset of glucose intolerance, hyperglycemia, and hyperinsulinemia in adulthood. These findings have been corroborated by several independent laboratory studies, including those conducted by Roberts et al. (2017). In addition to dexamethasone, other drugs and environmental factors have been identified as potential risk factors for the development of diabetes in offspring following prenatal exposure. For example, metformin, folic acid, and psychostimulants such as amphetamine, cocaine, methamphetamine, and δ9-THC have been shown to affect glucose and insulin metabolism in the offspring, leading to the onset of DM (Salomaki et al. 2013; Kintaka et al. 2020; Korchynska et al. 2020). Furthermore, exposure to caffeine, nicotine, ethanol, PM2.5, bisphenol A (BPA), and phthalates during pregnancy has been associated with a decrease in the number of islet β-cells and glucose tolerance in the offspring in adulthood, which manifests as a "diabetes-like symptom" (Chen et al. 2018; Alonso-Magdalena et al. 2010; Zhao et al. 2019). These findings have been reported by both the authors' team and other laboratories.

3.3 *Changes of Glucose Metabolism and Insulin Sensitivity in Offspring*

Exposure to adverse factors during pregnancy can induce significant structural and functional alterations in critical organs involved in glucose metabolism in the fetus. This can result in the development of abnormal glucose-insulin metabolism in the offspring during the fetal period, and these changes may persist throughout their lifespan.

During the intrauterine period, fetal growth is primarily sustained by glucose, which is supplied by the maternal circulation since fetal gluconeogenesis is minimal

(Kalhan and Parimi 2000). To ensure proper fetal development, the mother must provide nutrients such as glucose, which are transported to the fetus *via* the placenta. This transport occurs through members of the glucose transporter family, which diffuse along a concentration gradient. The efficiency of nutrient transport across the placenta is dependent on various factors, including placental size, morphology, nutrient transport capacity, and uterine and fetal placental blood flow (Fowden et al. 2006; Brett et al. 2014). Placental dysfunction can lead to inadequate nutrient supply to the fetus, despite an adequate maternal supply during pregnancy, often resulting in IUGR. In different animal models, basal blood glucose and serum insulin levels of IUGR fetuses have been observed to decrease. In rat models with unilateral uterine artery ligation, IUGR fetal mice exhibited significantly lower blood glucose levels. Conversely, in ewe models of IUGR induced by high temperatures during pregnancy, Limesand et al. (2007) observed no difference in glucose utilization between IUGR and normal fetuses, despite significant reductions in plasma insulin (~70%) and glucose (~50%) concentrations in IUGR fetuses. These findings suggest increased insulin sensitivity in IUGR fetuses. In our study, male IUGR fetal mice exposed to alcohol during pregnancy (Xiao et al. 2019) and female IUGR fetal mice exposed to dexamethasone during pregnancy (Kou et al. 2020) exhibited decreased blood glucose, blood insulin, and proinsulin levels on gestational day 20. These changes may be attributed to the impaired coupling function of IUGR fetal islet β-cells during development and stimulation secretion, resulting in reduced glucose oxidation, insulin synthesis, and total amounts under glucose stimulation (Ding et al. 2012).

The adverse prenatal environment can negatively impact fetal growth, leading to IUGR and severe dysfunction of fetal pancreatic β-cells and adipose tissue. During this process, the fetus adapts to malnutrition by reducing insulin production and developing insulin resistance. This adaptive change may result in "developmental programming" that affects appetite and feeding behavior in adulthood (Estampador and Franks 2018). Such metabolic adaptations may provide short-term benefits for survival in adverse environmental conditions after birth by enhancing the body's ability to store energy. However, when adequate nutrition is available after birth, the implementation of this adaptive strategy may lead to glucose metabolism disorders and diabetes. Previous research has shown that a low-protein diet during pregnancy increased skeletal muscle insulin sensitivity at postnatal day 38, but this effect was reversed at 4 months of age (Zheng et al. 2012; Blesson et al. 2017). In our study (Xiao et al. 2019), we observed that male offspring exposed to prenatal ethanol exhibited decreased levels of basal blood glucose and insulin in utero, and transient insulin resistance occurred at PW6, as evidenced by increased basal blood glucose and insulin levels and an increased insulin resistance index. These findings are consistent with age-dependent insulin resistance observed in child cohort studies (Raab et al. 2013). Furthermore, we found that as male progeny exposed to alcohol during pregnancy aged, serum basal glucose levels significantly decreased at PW12, while basal insulin levels remained unchanged, and insulin sensitivity decreased. Although basal levels of serum glucose and insulin did not change at PW24, insulin sensitivity increased. These results suggest that IUGR offspring may exhibit age-dependent changes in glucose metabolism homeostasis and insulin sensitivity.

4 Intrauterine Programming Mechanisms and Multigenerational Effects of Fetal-Originated Diabetes

The pancreas is the primary organ responsible for insulin synthesis and secretion, which is the only hypoglycemic hormone in the body. Its development, differentiation, and functional establishment primarily occur during the fetal period, while islet differentiation in adults is essentially stagnant. Due to the limited developmental plasticity of β-cells after the perinatal period, changes in pancreatic developmental programming caused by adverse intrauterine environments often persist into adulthood, resulting in lifelong effects on individual glucose metabolism in offspring. Since diabetes and metabolic syndrome share common pathophysiological basis, fetal-originated diabetes may also involve the mechanism of "two-programming and two-strike".

4.1 *Epigenetic Mechanisms*

Epigenetics plays a critical role in the pathogenesis of fetal-originated diabetes. Abnormal maternal metabolic status during pregnancy, such as gestational diabetes or obesity, has been associated with changes in DNA methylation and histone modification of metabolic genes (El Hajj et al. 2013; Finer et al. 2015; Petropoulos et al. 2015; Rong et al. 2015). Maternal protein restriction during pregnancy can induce DNA hypomethylation of specific genes in the pup's liver, including glucocorticoid receptor (GR) and peroxisome proliferators-activated receptors (PPAR) α (Lillycrop et al. 2005, 2008). Reduced DNA methylation of these genes can lead to increased mRNA abundance, which may contribute to metabolic disorders later in life. Exposure to BPA throughout pregnancy and during lactation has been shown to result in decreased total DNA methylation, increased DNA methyltransferase (DNMT) 3b gene expression, and increased methylation of the glucokinase gene promoter in 3-week-old male offspring (Ma et al. 2013). These changes have been associated with insulin resistance in adult offspring.

Park et al. (2008) reported that Pdx1 expression was permanently reduced in β-cells of IUGR offspring induced by bilateral uterine artery ligation. This reduction was primarily due to the loss of upstream stimulatory factor-1 binding, recruitment of histone deacetylase 1 (HDAC1), and co-inhibitor Sin3A, and deacetylation of histones H3 and H4 at the proximal promoter of Pdx1 during the fetal stage. After birth, histone H3 lysine 4 trimethylation (H3K4me3) significantly decreased, while H3K9me2 significantly increased. In adult-onset diabetes, CpG islands in the Pdx1 proximal promoter were found to be methylated. We also found that persistent low expression of angiotensin receptor type 2 (AT2R) induced by H3K27 deacetylation mediated by HDAC2 programs the development and function of pancreatic β-cells, resulting in glucose intolerance in adult offspring exposed to dexamethasone prenatally (Kou et al. 2020).

An increasing body of research has demonstrated the involvement of noncoding RNAs in the programming of fetal-originated diabetes, with a particular emphasis on miRNAs. A study conducted by Alejandro et al. (2014) revealed that maternal low-protein diets can alter a variety of miRNAs in offspring, thereby affecting insulin secretion and glucose homeostasis. Additionally, Dumortier et al. (2014) observed increased expression of miR-375 in both fetal pancreas and 3-month-old pancreas in the low-protein diet group, which could regulate proliferation and insulin secretion of in vitro islet cells and mediate changes in the quality and function of offspring islet β-cells. Furthermore, maternal high-fat diet during pregnancy and lactation has been found to lead to impaired liver glucose and lipid homeostasis in the offspring and to be associated with the upregulation of PPAR signaling pathway-related genes early in the offspring's life (Zheng et al. 2014). This metabolic disorder is linked to the abnormal expression of several liver miRNAs, including upregulation of miR-143 and downregulation of miR-615-5p, miR-3079-5p, miR-124, and miR-101b (Zheng et al. 2016).

4.2 *Multigenerational Effects*

The programming of glucose metabolism induced by adverse intrauterine environments can be passed on to future generations. Pregnant women who are exposed to energy restriction and environmental stress are at a higher risk of giving birth to granddaughters with decreased birth weight and insulin resistance (Painter et al. 2005). In a rat model, protein-restricted dietary intake during F0 gestation resulted in altered glucose and insulin metabolism in both female and male F2 generations (Zambrano et al. 2005). Moreover, the effects of protein restriction during pregnancy on glucose homeostasis persisted until the F3 generation (Benyshek et al. 2006). Additionally, dexamethasone administration during late pregnancy induced increased expression of GR and its target gene phosphoenolpyruvate carboxykinase in the liver of both F1 and F2 generations (Drake et al. 2005). Although epigenetic changes are typically erased and re-established in each generation, some epigenetic states on alleles can also be passed on between generations. In a protein-restricted diet model, liver GR and PPARα gene promoters were demethylated and continued from F1 to F2 generations (Burdge et al. 2007). These changes are likely to be inherited into the F3 generation, as suggested by other studies (Hoile et al. 2011). These findings suggest that epigenetic modification may play a role in the multigenerational effects of fetal-originated diabetes.

4.3 The "Two-Programming and Two-Strike" Mechanisms

Fetal overexposure to maternal or exogenous high glucocorticoid caused by adverse environment during pregnancy (such as exposure to xenobiotics) can further activate fetal pancreatic GR and cause epigenetic modification of important genes in pancreatic development and insulin synthesis with the participation of epigenetic regulators. It leads to dysplasia of β-cells and decreased insulin synthesis. This epigenetic modification and abnormal insulin synthesis function can continue after birth, leading to abnormal glucose tolerance and susceptibility to diabetes in adults. We previously found that dexamethasone exposure during pregnancy can activate pancreatic local GR, recruit HDAC2, and cause the reduction of histone acetylation level and expression inhibition in the promoter region of AT2R in offspring rats. In addition, the number of β-cells and insulin synthesis function of female fetal mice were decreased. This AT2R epigenetic inhibition and low functional programming of the pancreas can persist after birth, eventually leading to elevated basal blood glucose and impaired glucose tolerance in female progeny (Kou et al. 2020). Coincidentally, for male offspring, prenatal dexamethasone exposure also induces epigenetic inhibition of angiotensin-converting enzyme 2 (ACE2) and low functional programming of islet β-cells by a similar GR/HDAC3 signal, which induces impaired glucose tolerance in the early postnatal period (Dai et al. 2022). This continuous change in the pre - and postnatal consistency is what we call "first programming."

The insulin-like growth factor 1 (IGF1) signaling pathway plays a crucial role in fetal pancreatic β-cell development (van Haeften and Twickler 2004). While IGF1 in the early stages of embryonic development is primarily synthesized in the liver and transported through the bloodstream, IGF1 synthesis within the pancreas itself occurs in the later stages of development and plays a role in both autocrine and paracrine functions (Le et al. 2001). As a potent mitogen, IGF1 interacts with the insulin-like growth factor 1 receptor (IGF1R) on the β-cell membrane, promoting the rapid division and proliferation of β-cells and inhibiting the apoptosis of mature β-cells, thereby maintaining the relative stability of β-cell numbers and insulin synthesis function (Lingohr et al. 2002). It is well-known that high levels of glucocorticoids can suppress the expression of IGF1 in various tissues and cells. Our previous research has confirmed that maternal corticosterone overexposure and fetal blood corticosterone level increase, caused by exogenous substances (such as caffeine and ethanol) during pregnancy, are associated with decreased blood and pancreatic IGF1 levels. After birth, the expression of IGF1 in the pancreas increased with a decrease in serum corticosterone levels under normal dietary conditions. However, chronic stress (characterized by high serum corticosterone levels) can lead to the inhibition of IGF1 expression in the pancreas that was previously induced by intrauterine maternal hypercorticosterone. This negative correlation between serum corticosterone levels and IGF1 expression in pancreatic tissue suggests the presence of an "GC-IGF1 axis." This type of programming can work in conjunction with first programming to further inhibit the morphological development and functional differentiation of pancreatic β-cells in utero, leading to decreased islet mass and fetal blood insulin levels. In the early postnatal period, this can induce the expression of low serum corticosterone and high pancreatic IGF1, promoting compensatory β-cell proliferation to meet the demands of catch-up growth. This process is characterized by age-dependent changes in the number and function of pancreatic β-cells and glucose metabolism, ultimately increasing susceptibility to diabetes in adults (Kou et al. 2020; Xiao et al. 2019).

In the presence of an adverse intrauterine environment, which act as a "first strike," individuals can adapt to the needs of fetal growth in utero through two forms of programming that regulate insulin-dependent glucose metabolism. However, they remain susceptible to diabetes after birth due to environmental changes, such as exposure to a high-fat diet, chronic stress, or advanced age, which can act as a "second strike" and ultimately lead to impaired glucose tolerance and diabetes in adulthood. Animal experiments have shown that postnatal induction factors, such as stress and cold exposure, can further exacerbate the abnormal glucose metabolism of IUGR offspring caused by the adverse intrauterine environment (Wei et al. 2011). In a study conducted by Gui et al. (2019), both male and female IUGR offspring induced by caffeine exposure during pregnancy were fed a high-fat diet after birth, and their adult steady-state metabolic phenotypes showed a diabetes-like response characterized by "hyperglycemia and hypoinsulinemia." Although the quantitative insulin sensitivity check index did not decrease significantly, and the expression of the liver insulin/IGF1 signaling pathway was still upregulated, the offspring in the caffeine group exhibited higher absolute blood glucose levels after glucose loading, especially in females. Additionally, there was no clear peak of blood insulin release following glucose stimulation in the caffeine group, and insulin levels were consistently lower than those in the control group. These findings suggest that a high-fat diet can further exacerbate the pancreatic dysplasia and insulin synthesis inhibition programmed by caffeine exposure during pregnancy in the offspring. While the increased expression of the insulin/IGF1 signaling pathway in the liver after birth can improve insulin sensitivity to a certain extent and lead to enhanced relative glucose tolerance, it is unable to fully reverse or compensate for the absolute deficiency of insulin secretion, ultimately inducing hyperglycemia.

The authors' team has confirmed that the occurrence of abnormal glucose metabolism and susceptibility to diabetes in adult rats exposed to xenobiotics during pregnancy involves a "two-programming and two-strike" mechanism (Fig. 12.1). The "first programming" involves continuous epigenetic modification of important regulatory factors of pancreatic development and insulin synthesis caused by high levels of GC in utero. The "second programming" is the postnatal catch-up growth of intrauterine islet β-cells caused by pancreatic GC-IGF1 axis programming. However, this type of programming cannot fully compensate for the postnatal islet dysplasia caused by suboptimal intrauterine environments. These "two-programming" events constitute the "first strike" on the islets that underlie the predisposition of IUGR

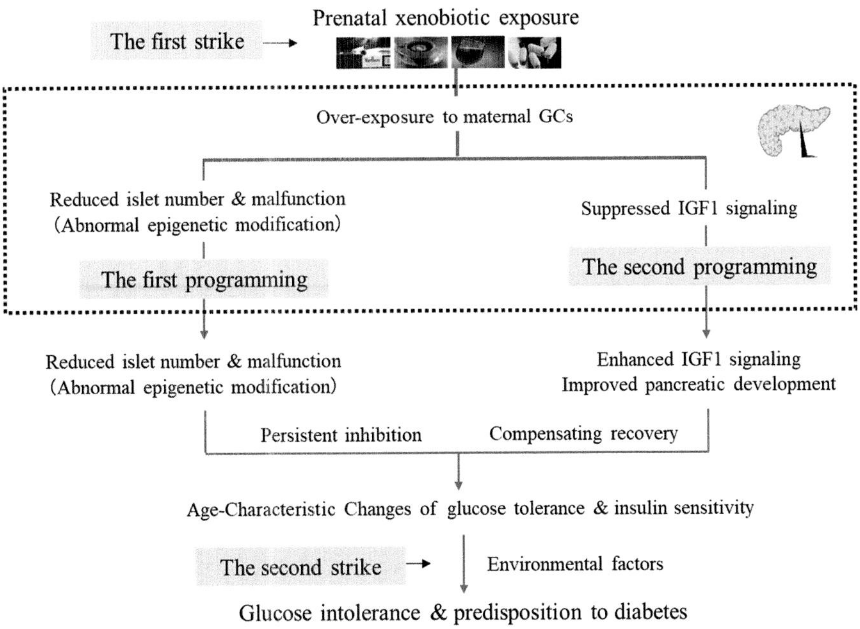

Fig. 12.1 The "two-programming and two-strike" mechanism of fetal-originated diabetes

offspring to diabetes after birth. A "second strike" in adulthood, such as a high-fat diet or chronic stress, can further induce the development of glucose intolerance.

5 Summary and Prospect

The deleterious effects of the prenatal environment can exert a profound impact on the development of the pancreas in offspring, as well as on the glucose metabolic function in adulthood. The underlying pathogenesis appears to be primarily associated with exposure to intrauterine glucocorticoids, aberrant epigenetic modifications of genes involved in pancreatic and peripheral organ development and function, and perturbations in the GC-IGF1 axis. Nevertheless, the characteristics, specific mechanisms, sexual dimorphisms, and key aspects of multigenerational inheritance of glucose metabolism dysfunction mediated by pancreatic dysplasia during the intrauterine and postnatal periods remain to be fully elucidated. As research on adverse prenatal environments and fetal-originated disease advances, so too does translational research on fetal-originated disease, which holds great significance for comprehending the fetal origins of adult diseases and for early detection and prevention of DM.

References

Alejandro E U, Gregg B, Wallen T, et al. 2014. J Clin Invest, 124(10): 4395–4410.
Alonso-Magdalena P, Vieira E, Soriano S, et al. 2010. Environ Health Perspect, 118(9): 1243–1250.
Amos A F, McCarty D J, Zimmet. 1997. Diabet Med, 14 Suppl 5: S1–85.
Benyshek D C, Johnston C S and Martin J F. 2006. Diabetologia, 49(5): 1117–1119.
Berends L M, Ozanne S E. 2012. Best Pract Res Clin Endocrinol Metab, 26(5): 569–580.
Blesson C S,Chinnathambi V,Kumar S, et al. 2017. Endocrinology, 158(4): 756–767.
Brett K E, Ferraro Z M, Yockell-Lelievre J, et al. 2014. Int J Mol Sci, 15(9): 16153–16185.
Burdge G C, Slaterjefferies J, Torrens C, et al. 2007. Br J Nutr, 97(3): 435–439.
Campos P H, Silva B A N, Donzele J L, et al. 2012. Animal, 6(5): 797–806.
Chen M, Liang S, Qin X B, et al. 2018. Am J Physiol Endocrinol Metab, 315(1): E72-E80.
Chen Y C, Huang Y H, Sheen J M, et al. 2017. Pediatr Neonatol, 58(2): 135–144.
Curhan G C, Willett W C, Rimm E B, et al. 1996. Circulation, 94(12): 3246–3250.
Cnop M, Welsh N, Jonas J C, et al. 2005. Diabetes, 54 Suppl 2(Suppl): S97-S107.
Dai Y, Kou H, Gui S, et al. 2022. Sci Total Environ, 826: 154095.
De V A, Holmes M C, Heijnis A, et al. 2007. J Clin Invest, 117(4): 1058–1067.
Ding G L, Wang F F, Shu J, et al. 2012. Diabetes, 61(5): 1133–1142.
Drake A J, Walker B R and Seckl J R. 2005. Am J Physiol-Reg I, 288(1): R34-R38.
Dumortier O, Blondeau B, Duvillié B, et al. 2007. Diabetologia, 50(12): 2495–2503.
Dumortier O, Hinault C, Gautier N, et al. 2014. Diabetes, 63(10): 3416–3427.
El Hajj N, Pliushch G, Schneider E, et al. 2013. Diabetes, 62(4): 1320–1328.
Estampador A C, Franks P W. 2018. Clin Chem, 64(1): 130–141.
Egeland G M, Skjaerven R, Irgens L M, et al. 2000. BMJ, 321(7260): 546–547.
Finer S, Mathews C, Lowe R, et al. 2015. Hum Mol Genet, 24(11): 3021–3029.
Fowden A L, Ward J W, Wooding F B P, et al. 2006. J Physiol, 572(Pt 1): 5–15.
Garofano A, Czernichow P, Breant B. 1997. Diabetologia, 40(10): 1231–1234.
Graus-Nunes F, Corte Frantz E D, Lannes W R, et al. 2015. Nutrition, 31(2): 380–387.
Greene N H, Pedersen L H, Liu S, et al. 2013. Int J Epidemiol, 42(1): 186–193.
Gui S X, Kou H, Guo Y, et al., 2019. Chin J Hosp Pharm, 39(9): 37–42.
Hales C N, Barker D J, Clark P M, et al. 1991. BMJ, 303(6809): 1019–1022.
Hales C N, Barker D J. 2001. Br Med Bull, 60: 5–20.
Helman A, Avrahami D, Klochendler A, et al. 2016. Diabetes Obes Metab, 18 Suppl 1: 58–62.
Hoile S P, Lillycrop K A, Thomas N A, et al. 2011. PLoS One, 6(7): e21668.
Jaacks L M, Karen R S, Unjali P G, et al. 2016. Best Pract Res Clin Endocrinol Metab, 30(3): 331–343.
Jaquet D, Gaboriau A, Czernichow P, et al. 2000. J Clin Endocrinol Metab, 85(4): 1401–1406.
Jiang X, Ma H, Wang Y, et al. 2013. J Diabetes Res, 2013: 485082.
Kalhan S, Parimi P, 2000. Semin Perinatol, 24(2): 94–106.
Kahn S E, Hull R L and Utzschneider K M. 2006. Nature, 444(7121): 840–846.
Kintaka Y, Wada N, Shioda S, et al. 2020. Heliyon, 6(4): e03597.
Korchynska S, Krassnitzer M, Malenczyk K, et al. 2020. EMBO J, 39(1): e100882.
Kou H, Gui S, Dai Y, et al. 2020. Toxicol Appl Pharmacol, 404, 115187.
Le Roith D, Scavo L, Butler A. 2001. Trends Endocrinol Metab, 12(2): 48–52.
Leger J, Levy-Marchal C, Bloch J, et al. 1997. BMJ, 315(7104): 341–347.
Lithell H O, McKeigue P M, Berglund L, et al. 1996. BMJ, 312(7028): 406–410.
Lillycrop K A, Phillips E S, Jackson A A, et al. 2005. J Nutr, 135(6): 1382–1386.
Lillycrop K A, Phillips E S, Torrens C, et al., 2008. Br J Nutr, 100(2): 278–282.
Limesand S W, Rozance P J, Smith D, et al. 2007. Am J Physiol-Endoc M, 293(6): 1716–1725.
Lingohr MK, Dickson LM, et al. 2002, Diabetes, 51(4): 966–976.
Maia-Ceciliano T C, Barreto-Vianna A R C, Barbosa-Da-Silva S, et al. 2016. Endocrine, 54(1): 1–10.
Ma Y, Xia W, Wang D Q, et al. 2013. Diabetologia, 56(9): 2059–2067.

Mccarty P, Zimmet P. 1997. Diabetes Care, 20: 1785–1790.
McKeigue P M, Lithell H O, Leon D A. 1998. Diabetologia, 41(10): 1133–1138.
Painter R C, Roseboom T J, Bleker O. 2005. Reprod Toxicol, 20(3): 345–352.
Park J H, Stoffers R D, et al. 2008. J Clin Invest, 118(6): 2316–2324.
Pettitt D J, Aleck K A, Baird H R, et al. 1988. Diabetes, 37(5): 622–628.
Petropoulos S, Guillemin C, Ergaz Z, et al. 2015. Endocrinology, 156(6): 2222–2238.
Raab J, Haupt F, Kordonouri O, et al. 2013. Diabetes metab Res Rev, 29(8): 631–635.
Ravelli G P, Stein Z A, Susser M W. 1976. N Engl J Med, 295(7): 349–353.
Roberts D, Brown J, Medley N, et al. 2017. Cochrane Database Syst Rev, 3: CD004454.
Rong C, Cui X, Chen J, et al. 2015. Exp Clin Endocrinol Diabetes, 123(5): 282–288.
Salomaki H, Vähätalo L H, Laurila K, et al. 2013. PLoS One, 8(2): e56594.
Vaiserman A M. 2017. Nutrients, 9 (3): 236.
Valdez R, Athens M A, Thompson G H, et al. 1994. Diabetologia, 37(6): 624–631.
van Haeften T W, Twickler T B. 2004. Eur J Clin Invest, 34(4): 249–255.
Wei J, Lin Y, Li Y, et al. 2011. Endocrinology, 152(8): 3049–3061.
Wu Y, Ding Y P, Yoshimasa T, et al. 2014. Int J Med Sci, 11(11): 1185–1200.
Xiao D, Kou H, Gui S, et al. 2019. Front Endocrinol (Lausanne), 10: 34.
Zambrano E, Martínez-Samayoa P M, Bautista C J, et al. 2005. J Physiol, 566(1): 225–236.
Zambrano E, Sosa-Larios T, Calzada L, et al., 2016. J Endocrinol, 231(1): 49–57.
Zhao W H, Wen X, Qu W, et al. 2019. Toxicol Lett, 315: 39–46.
Zheng S, Rollet M, Pan Y X. 2012. J Nutr Biochem, 23(9): 1064–1071.
Zheng J, Xiao X, Zhang Q, et al. 2014. Int J Mol Sci, 15(9): 14967–14983.
Zheng J, Zhang Q, Mul J D, et al. 2016. Endocrine, 54(1): 70–80.

Chapter 13
Fetal Origin of Nonalcoholic Fatty Liver Disease

Wen Hu and Ze Chen

Abstract With the emergence of the "Developmental Origins of Health and Disease" theory, it has been established that adverse intrauterine conditions can increase the susceptibility of offspring to nonalcoholic fatty liver disease (NAFLD) in cases of intrauterine growth retardation. However, a comprehensive theoretical framework underlying the occurrence of fetal-originated NAFLD is still lacking. Recent studies have revealed that various adverse prenatal environments lead to fetal overexposure to maternal glucocorticoids, which results in epigenetic programming changes in genes related to the glucocorticoid-insulin-like growth factor 1 axis, hypothalamic-pituitary-adrenal axis, and hepatic lipid metabolic function. Following birth, offspring exhibit catch-up growth and ectopic fat deposition based on the intrauterine programming changes under adequate nutritional conditions, thereby accelerating the onset of NAFLD. This chapter summarizes the research progress on fetal-originated NAFLD and its intrauterine programming mechanism and proposes the "two-programming and two-strike" mechanism of NAFLD. These mechanisms provide a theoretical and experimental basis for early clinical warning, diagnosis, and treatment of NAFLD.

1 Introduction

With the fast-paced lifestyle of modern society, work and life pressures are increasing, and high-calorie diets and excessive nutrient intake are becoming more prevalent. As a result, the incidence of metabolic syndrome and its related metabolic diseases is also increasing year by year. Nonalcoholic fatty liver disease (NAFLD), as a localized manifestation of metabolic syndrome in the liver, often coexists with obesity, lipid metabolism disorders, and abnormal glucose metabolism. While the

W. Hu (✉)
Department of Pharmacy, Zhongnan Hospital of Wuhan University, Wuhan, Hubei, China

Z. Chen
Department of Cardiology, Zhongnan Hospital of Wuhan University, Wuhan, Hubei, China

H. Wang et al. (eds.), *Fetal Origin of Diseases*,
https://doi.org/10.1007/978-981-97-5730-5_13

traditional view attributes NAFLD to environmental factors, gender, age, and lifestyle, recent studies have found that abnormal hepatic lipid metabolism caused by adverse intrauterine environments can also promote the development and progression of NAFLD. In the past two decades, both domestic and foreign scholars have put forward the new concept of the "Developmental Origins of Health and Disease (DOHaD)" theory, which posits that changes in the intrauterine environment can affect multiple aspects of fetal development, including the development of multiple organs, cell differentiation, and glucose and lipid metabolism, thereby increasing the risk of metabolic syndrome and a series of metabolic diseases in adulthood. This chapter reviews the intrauterine origin, development, and progression of NAFLD, providing an important theoretical and experimental basis for guiding a reasonable lifestyle during pregnancy.

2 Current Research Status of Nonalcoholic Fatty Liver Disease

NAFLD, a clinical-pathological syndrome characterized by the accumulation of triglycerides (TG) in liver parenchyma cells due to various pathogenic factors, excluding alcohol, is a potential risk factor for the development and progression of hepatitis, cirrhosis, and liver cancer. Additionally, there is substantial evidence indicating that NAFLD is a multisystem disease with systemic involvement that is closely associated with the risk of extrahepatic complications such as cardiovascular disease. The prevalence of NAFLD has increased dramatically in recent years, posing a significant burden on public health. As a large number of clinical and basic research studies are underway, the pathogenesis of NAFLD is gradually being understood.

2.1 Epidemiological Status of Nonalcoholic Fatty Liver Disease

NAFLD has emerged as a prevalent chronic liver disease that poses a significant threat to global health. The incidence and prevalence of NAFLD have increased rapidly worldwide over the past few decades. Epidemiological studies indicate that the prevalence of NAFLD is approximately 25% globally, ranging from 20% to 30% in Western countries and 5% to 18% in Asian countries. Nearly one billion individuals worldwide are afflicted with NAFLD (Sayiner et al., 2016). In China, with the rapid development of the economy, improvements in living conditions, and the influence of unhealthy lifestyles, NAFLD has become the second most prevalent liver disease following viral hepatitis. Its incidence is increasing year by year, and patients are becoming younger. In the United States, end-stage liver disease due to NAFLD is expected to be the leading cause of liver transplantation treatment within the next decade. The steep rise in the incidence of NAFLD has made it a significant medical and social concern for multiple fields, including public health, gastroenterology, endocrinology, and cardio-cerebrovascular diseases.

2.2 *Major Pathogenic Mechanisms of Nonalcoholic Fatty Liver Disease*

The liver is the primary organ responsible for lipid metabolism, and sources of TG include food intake and de novo synthesis of hepatic fatty acids. Under normal circumstances, TG uptake, synthesis, lipolysis, oxidation, and transport are maintained in a relative equilibrium. However, when one of these processes becomes abnormal, such as increased food intake or TG synthesis, decreased lipolysis or oxidation, or disordered lipid transport, a favorable environment for lipid peroxidation is created, which forms the basis for hepatocyte inflammatory reactions. Furthermore, lipid metabolism disorders and hepatic lipid accumulation lead to abnormal membrane structure and function of liver cells and a reduction in the number of insulin receptors on the surface of liver cells, resulting in decreased insulin sensitivity and exacerbating the severity of insulin resistance and NAFLD. Although ongoing research on NAFLD has yielded significant insights, our understanding of its pathogenesis remains incomplete. The prevailing view is that NAFLD is not an independent disease but rather a developmental and evolving liver lesion. Presently, the most widely accepted pathogenesis of adult NAFLD is the "two-strike" theory. The first strike refers to the excessive accumulation of lipids in liver cells, in which insulin resistance plays a critical role. Insulin resistance leads to disorders in liver triglyceride synthesis and transport, exacerbating hepatocyte lipid deposition. Moreover, the increase in hepatic lipid accumulation further aggravates insulin resistance, creating a vicious cycle that induces decreased liver tolerance and increased susceptibility to toxic substances, ischemia, and hypoxia (Salamone and Bugianesi, 2010). Based on this, oxidative stress, mitochondrial injury, and inflammatory response cause hepatocyte degeneration and necrosis, further promoting the development of liver fibrosis and cirrhosis, constituting the second strike. Therefore, the pathogenesis of NAFLD primarily involves abnormal hepatic lipid metabolism, altered insulin signaling pathways, and insulin resistance.

3 Evidence of the Intrauterine Origin of Nonalcoholic Fatty Liver Disease

The conventional belief is that the development of NAFLD is primarily associated with unhealthy lifestyle practices after birth. However, recent evidence from large-scale epidemiological studies suggests that offspring with low birth weight are predisposed to metabolic disorders in adulthood, thereby supporting the "fetal origin" hypothesis of adult diseases. This hypothesis posits that intrauterine malnutrition induces metabolic adaptive changes in the fetus that optimize the utilization of limited nutrients, thereby increasing the likelihood of survival in the face of adverse postnatal conditions. However, these adaptive changes also enhance the risk of metabolic disorders in the offspring in a nutrient-rich postnatal environment. Epidemiological inquiries and animal experiments have demonstrated that various adverse environmental exposures during pregnancy, such as malnutrition and xenobiotic exposure, may elevate the risk of developing NAFLD after birth, underscoring the intrauterine origin of NAFLD (Table 13.1).

3.1 Epidemiological Investigation and Clinical Study

Intrauterine growth retardation (IUGR) is a condition characterized by the impaired growth of the embryo/fetus due to various unfavorable factors during pregnancy, resulting in multi-organ developmental dysfunction, growth retardation, and low birth weight, which is a common developmental toxicity outcome. Epidemiological investigations have revealed that the incidence of adult metabolic syndrome in IUGR newborns (2.3%) is 5.75 times higher than that in normal fetuses (0.4%) (Alisi et al., 2011). Children with NAFLD have a fourfold higher incidence of IUGR than the general population. Low birth weight infants exhibit rapid catch-up growth within 6–12 months of life compared to normal birth weight infants. The incidence of NAFLD in childhood can be as high as 34.8%, accompanied by elevated insulin resistance and visceral fat deposition. Proper maternal diet during pregnancy and in the first year of life may prevent the onset of NAFLD (Faienza et al., 2013). The aforementioned studies suggest that low birth weight is a risk factor for NAFLD, and this association is likely to persist into adulthood.

Moreover, maternal metabolic disorders, such as gestational diabetes mellitus and/or obesity, can also elevate the liver fat content of the offspring. Magnetic resonance imaging and magnetic resonance spectroscopy studies have revealed that infants born to obese mothers with gestational diabetes exhibited a 70% increase in liver fat content, without a significant increase in abdominal or subcutaneous fat (Brumbaugh et al., 2013). In newborns from obese mothers without gestational

Table 13.1 Adult nonalcoholic fatty liver disease induced adverse intrauterine environment

Research type	Adverse exposure	Outcomes in offspring	Main references
Epidemiological studies	Intrauterine growth retardation	5.75-fold increase in metabolic syndrome incidence	Alisi et al. (2011)
	Low birth weight	Nonalcoholic fatty liver disease (NAFLD) incidence up to 34.8%	Efstathiou et al. (2015)
Clinical studies	Gestational diabetes mellitus and obesity	70% increase in baby's liver fat	Brumbaugh et al. (2013)
	Obesity	High correlation between neonatal liver fat content and maternal body mass index (BMI)	Modi et al. (2011)
	Obesity	Maternal BMI can predict offspring's liver fat content	Modi et al. (2011), Gale et al. (2014)
Basic laboratory studies	High-fat diet	Increased liver fat in offspring	McCurdy et al. (2009)
	High-fat diet	Increased liver fat content and NASH susceptibility	Bruce et al. (2009), Oben et al. (2010)
	Food restriction	Increased NAFLD susceptibility	Zhang et al. (2016)
	Bisphenol A	Increased NAFLD susceptibility	Jiang et al. (2014), Wei et al. (2014)
	Tributyltin	Increased NAFLD susceptibility	Grun et al. (2006)
	Perfluorinated compounds	Increased NAFLD susceptibility	Lv et al. (2013), Quist et al. (2015)
	Phthalate	Increased NAFLD susceptibility	Ortiz et al. (2013), Maranghi et al. (2010)
	Arsenic	Increased NAFLD susceptibility	Nohara et al. (2012)
	Caffeine	Increased NAFLD susceptibility	Wang et al. (2014)
	Ethanol	Increased NAFLD susceptibility	Shen et al. (2014) Xia et al. (2020)
	Dexamethasone	Increased NAFLD susceptibility	Liu et al. (2021)
	Nicotine	Increased NAFLD susceptibility	Xu et al. (2015)

diabetes, the liver fat content detected by magnetic resonance imaging was strongly associated with maternal body mass index (BMI) (Modi et al., 2011). Additionally, maternal BMI could predict changes in liver fat content in offspring 10–12 weeks after birth (Modi et al., 2011; Gale et al., 2014).

3.2 Basic Laboratory Research

The unfavorable intrauterine environment can induce abnormal lipid metabolism in offspring, leading to an increased susceptibility to metabolic disorders, including NAFLD, in adulthood. In the Japanese macaque model, chronic maternal high-fat diet caused a rise in hepatic TG and oxidative stress in fetal primates (McCurdy et al., 2009). Similarly, in a mouse model, maternal high-fat diet feeding during pregnancy and lactation resulted in hepatic lipid accumulation in offspring at 3 months, which progressively developed into NAFLD after weaning (Bruce et al., 2009; Oben et al., 2010). In a rat model of maternal feeding restriction, IUGR offspring exhibited heightened susceptibility to NAFLD (Zhang et al., 2016). Furthermore, exposure to endocrine-disrupting substances during early life, such as bisphenol A (Jiang et al., 2014; Wei et al., 2014), perfluoride (Lv et al., 2013; Quist et al., 2015), and arsenic (Nohara et al., 2012), may promote the development and progression of NAFLD. The authors' team also discovered that prenatal exposure to various xenobiotics, such as ethanol (Shen et al., 2014; Xia et al., 2020), caffeine (Wang et al., 2014), nicotine (Xu et al., 2015), and dexamethasone (Liu et al., 2021), increased the susceptibility to NAFLD in IUGR offspring, while postnatal high-fat diet could exacerbate or induce NAFLD. Therefore, fetal origin plays a role in adult NAFLD due to an adverse intrauterine environment. However, the underlying pathogenesis has yet to be fully elucidated.

4 Intrauterine Programming Mechanisms of Fetal-Originated Nonalcoholic Fatty Liver Disease

Intrauterine programming refers to alterations in the normal developmental process caused by an unfavorable intrauterine environment during fetal development, resulting in long-term or permanent changes in bodily functions. Although the intrauterine mechanism underlying the susceptibility to adult NAFLD in offspring with IUGR has not been systematically reported, several hypotheses have been proposed, including "thrifty phenotype", "nutritional programming" and "metabolic programming".

4.1 Epigenetic Mechanisms and Multigenerational Genetic Effects of Triglyceride Metabolic Programming

A mounting body of evidence suggests that epigenetic modifications play a crucial role in intrauterine programming, primarily encompassing DNA methylation, histone modifications, and noncoding RNA. Abnormal epigenetic modifications can adversely impact fetal growth and development caused by an unfavorable intrauterine environment and permanently induce phenotypic changes after birth. Cytosine methylation of CpG islands located in the promoter region of genes is frequently linked to gene expression suppression. Conversely, the effect of histone methylation on gene transcription activation or repression depends on the number of amino acid residues and methylomes present in the histone family. Histone acetylation is typically associated with transcriptional activation. In animal models, studies have shown that feeding restriction during pregnancy led to reduced histone methylation levels of insulin-like growth factor 1 (IGF1) in IUGR offspring, accompanied by catch-up growth (Tosh et al., 2010). Feeding restriction during pregnancy led to an increase in H3K9ac and H3K14ac levels in the promoter region of the hepatic lipoprotein lipase gene in offspring, which was associated with hepatic TG accumulation (Zhu et al., 2016). Furthermore, under cellular hypoxia, acetate induces high histone H3 acetylation and enhances the de novo pathway of fatty acid synthesis by increasing acetylation levels in the gene promoter regions of acetyl-CoA carboxylase (ACC) and fatty acid synthase (FASN) (Gao et al., 2016). The authors' team discovered that prenatal caffeine exposure increased expression levels and H3K14ac and H3K27ac levels in the gene promoter region of sterol regulatory element-binding protein 1c (SREBP1c) and FASN in female IUGR offspring, thereby increasing lipogenesis in the fetal livers of female offspring. These changes could persist after birth and even into adulthood, resulting in susceptibility to NAFLD (Hu et al., 2019).

Furthermore, microRNAs (miRNAs) have been implicated in the metabolic programming of liver TG. Among these miRNAs, miRNA-103, miRNA-107, and miRNA-122 constitute 70% of total miRNAs in the liver, all of which have been shown to upregulate the expression of SREBP1c protein. Conversely, overexpression of miRNA-107-3p has been shown to downregulate the expression of carnitine palmitoyltransferase 1α protein. The team of authors discovered that exposure to dexamethasone during pregnancy activated the GR and inhibited the expression of miR-122, thereby suppressing the angiotensin-converting enzyme 2 -Mas receptor 1 signaling pathway, which resulted in increased hepatic TG synthesis and decreased oxidative function. The decreased expression of miR-122 induced by dexamethasone was observed after birth, leading to increased TG synthesis and reduced oxidative function in hepatocytes, ultimately resulting in the development of NAFLD in adult rat offspring (Liu et al., 2021). In summary, the aforementioned studies suggest that epigenetic modifications of genes with critical roles in lipid metabolism, caused by adverse prenatal conditions, play a role in the pathogenesis and progression of NAFLD.

Numerous epidemiological studies have demonstrated that an adverse prenatal environment can increase the susceptibility of offspring to various chronic diseases in adulthood and may even have multigenerational genetic effects. The underlying mechanism of multigenerational inheritance resulting from intrauterine hepatic programming in IUGR offspring induced by an adverse prenatal environment remains unclear, and epigenetic modifications may play a crucial role. In a guinea pig model with prenatal betamethasone exposure, alterations in the total methylation rate of genomic DNA and related gene expression in multiple organs, including the liver of F1 fetuses and newborns, persisted into adulthood and were stably transmitted to F2 offspring. In a uteroplacental insufficiency model caused by bilateral uterine artery ligation, IUGR in F1 and F2 offspring resulted in abnormal lipid metabolism and metabolic syndrome, which could be largely reversed by supplementing their diet with essential nutrients that impact IGF1 methylation through the one-carbon metabolism pathway, indicating that epigenetic modifications can reverse or prevent fetal-originated metabolic diseases (Goodspeed et al., 2015). The authors' team also discovered that altered neuroendocrine metabolic programming induced by caffeine exposure during pregnancy persists into the F2 generation, rendering F2 offspring susceptible to metabolic syndrome and related metabolic diseases (Luo et al., 2014). These findings suggest that the intrauterine and early postnatal environment, such as nutritional imbalances and exposure to endocrine disruptors, can impact the DNA methylation and histone acetylation status of specific gene promoters. Such modifications not only affect offspring throughout their lives but can also be stably inherited to the next generation, increasing the risk of abnormal lipid metabolism in offspring (Table 13.2).

Table 13.2 Epigenetic effects of adverse intrauterine environments on lipid metabolic genes of fetal liver

Adverse intrauterine environment	Genes	Epigenetic modification	Site	Species	Main references
Food restriction	*Igf1*	Histone methylation	H3K4me2	Rat	Tosh et al. (2010)
	Lpl	Histone acetylation	H3K9ac H3K14ac	Rat	Zhu et al. (2016)
Caffeine exposure	*Srebf1*	Histone acetylation	H3K14ac H3K27ac	Rat	Hu et al. (2019)
	Fasn	Histone acetylation	H3K14ac H3K27ac	Rat	Hu et al. (2019)
	Igf1	Histone methylation, DNA methylation	H3K4me3, H3K36me3, CpG	Rat	Tan et al. (2012), Fu et al. (2009)
Protein restriction	*Ppara*	DNA methylation	CpG	Rat	Lillycrop et al. (2008)
Maternal diabetes	*Srebf2*	DNA methylation	CpG	Rat	Golic et al. (2018)
Dexamethasone exposure	*Yy1*	miR-122	N/A	Rat	Liu et al. (2021)

4.2 The "Two-Programming" Mechanisms of Fetal-Originated Nonalcoholic Fatty Liver Disease

The basal level of glucocorticoid (GC) plays a crucial role in determining fetal maturation and postnatal fate, and excessive exposure to GC during fetal development can result in developmental abnormalities such as IUGR. The placenta is a vital organ that maintains normal fetal development, and placental 11β-hydroxysteroid dehydrogenase type 2 (11β-HSD2) is responsible for oxidizing and inactivating excess maternal GC. This is a critical regulatory point that protects the fetus from maternal GC interference. Clinical studies have revealed that cord blood cortisol concentrations are elevated in IUGR offspring. Animal experiments conducted by the authors' team have confirmed that exposure to various xenobiotics during pregnancy can inhibit the expression of placental 11β-HSD2, thereby neutralizing the placental GC barrier and exposing the fetus to excessive maternal GC (Chen et al., 2007; Yu et al., 2018). GC can impact hepatic lipid metabolism in numerous ways, including promoting fatty acid uptake and de novo synthesis and inhibiting fat mobilization and fatty acid β-oxidation (Woods et al., 2015).

GC binding to the GR can recruit specific transcription factors and epigenetic modification enzymes to synergistically regulate gene transcription. Research has shown that GC can significantly downregulate the expression of Sirtuin 1 (SIRT1), and the inhibition of SIRT1 expression in the liver can increase lipogenesis and reduce lipolysis in neonatal/adult mice with IUGR (Wolfe et al., 2012). Studies conducted by the authors' team on animal models have also demonstrated that prenatal exposure to nicotine in female fetal mice led to increased blood GC levels and hepatic FASN expression levels that persisted into adulthood (Xu et al., 2013). Similarly, female fetal mice with prenatal caffeine exposure exhibited increased blood GC levels, enhanced hepatic GR and C/EBPα expression, decreased SIRT1 expression in the liver, and GC-dependent enhanced expression of hepatic SREBP1c and FASN. At 28 weeks after birth, these offspring mice displayed a sustained increase in their liver's TG synthetic capability (Hu et al., 2019). These studies suggest that overexposure to maternal-originated GC mediates the "first programming" mechanism of enhanced hepatic lipid synthesis function in offspring following prenatal exposure to xenobiotics.

During the intrauterine period, the liver is the largest metabolic organ in the human body and the primary source of IGF1 in circulation. Approximately 80% of IGF1 is synthesized by the liver. IGF1 is one of the most important growth factors in the body, promoting cell differentiation, proliferation, and metabolism and mediating body lipid metabolism. IGF1 is considered a key molecule in the regulation of IUGR and postnatal catch-up growth. Epidemiological studies have found that IGF1 levels in both IUGR fetal blood and cord blood are lower than those in fetuses with normal weight, suggesting that the reduction in IGF1 may mediate the occurrence of IUGR. Literature suggests that liver IGF1 can promote TG synthesis by inducing SREBP1c expression through the IGF1R and protein kinase B signaling

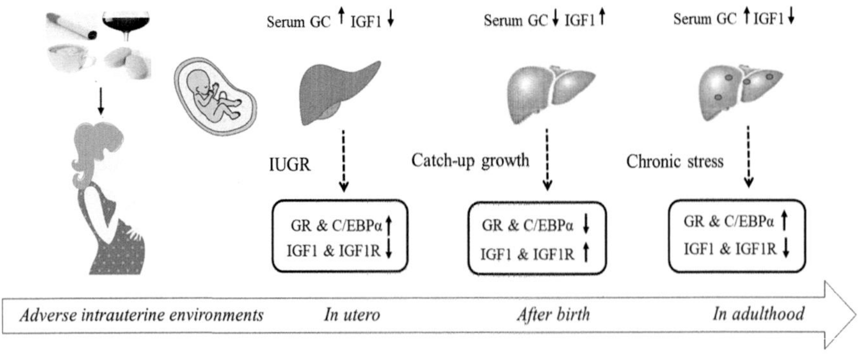

Fig. 13.1 "GC-IGF1" axis programming alterations in fetal-originated nonalcoholic fatty liver disease

pathways (Hagiwara et al., 2012). Studies have also shown that high blood GC levels can inhibit the expression and secretion of IGF1, which may be related to the promotion of C/EBPβ binding by GC, thereby inhibiting transcription of the IGF1 promoter region (Delany et al., 2001). Previous research by the authors' team has confirmed that exposure to caffeine, nicotine, and ethanol during pregnancy can increase blood GC levels and decrease blood IGF1 levels in IUGR fetuses, while adult offspring displayed decreased blood GC levels and increased blood IGF1 levels, accompanied by catch-up growth.

When subjected to chronic stress, these adult offspring exhibit increased blood GC levels and decreased blood IGF1 levels. This negative correlation between blood GC and blood IGF1 before and after birth, along with changes in body weight before and after birth, is referred to as the "GC-IGF1 axis" programming (Xu et al., 2013; Shen et al., 2014; Wang et al., 2014; Hu et al., 2020a; Hu et al., 2020b) (Fig. 13.1). High blood GC levels and suppressed liver IGF1 signaling pathways due to prenatal exposure to xenobiotics can lead to the occurrence of IUGR in offspring. IUGR reflects adaptive mechanisms by which the body changes endocrine hormone levels of GC and IGF1 to adapt to the adverse intrauterine environment, thereby ensuring fetal growth and development. However, the postnatal catch-up growth, lipid metabolism disorders, and susceptibility to metabolic syndrome in IUGR offspring may also be related to intrauterine GC-IGF1 axis programming. The changes in GC-IGF1 axis programming-induced decrease in fetal metabolic rate and postnatal catch-up growth and glucose and lipid metabolism disorders are the "second programming" mechanism leading to the development of fetal-originated NAFLD.

The authors' team has proposed a theory of "two-programming" mechanism for fetal-originated NAFLD. The first programming mechanism involves overexposure to maternal-originated GC induced by prenatal exposure to xenobiotics, which can permanently enhance hepatic lipid synthesis function through epigenetic regulation. The second programming mechanism involves inhibition of the liver IGF1 signaling pathway and weakened hepatic glucose and lipid metabolic function under intrauterine maternal GC overexposure, while the hepatic IGF1 signaling pathway and glucose and lipid metabolic function are compensatory enhanced after birth, resulting in the "GC-IGF1 axis" programming.

4.3 The "Two-Strike" Mechanisms of Fetal-Originated Nonalcoholic Fatty Liver Disease

The factors that induce fetal-originated NAFLD are referred to as intrauterine induction factors. These factors encompass all adverse intrauterine environments and related factors that can cause changes in lipid metabolism in the fetal liver. Intrauterine adverse environments are mainly caused by medication use during pregnancy, exposure to environmental toxins, or nutritional restrictions. Commonly used drugs during pregnancy include GC (e.g., adrenocorticotropic hormone, dexamethasone, and betamethasone), central stimulant drugs (e.g., caffeine and methamphetamine), and environmental toxic substances including nicotine, bisphenol A, and arsenic. As previously mentioned, the adverse environment during pregnancy renders offspring susceptible to NAFLD in adulthood, while postnatal individuals live in a complex environment, often affected by various environmental factors. For instance, multiple chronic stimuli after birth (e.g., high-fat diet and chronic stress) can induce the occurrence of multiple chronic diseases in the offspring, such as metabolic syndrome and related metabolic diseases (e.g., NAFLD).

The authors' team has conducted previous studies that suggest various xenobiotic exposure during pregnancy can inhibit neuroendocrine metabolism changes related to the hypothalamic-pituitary-adrenal (HPA) axis in fetal mice. This inhibition leads to time-related changes in the basic activity of HPA axis in IUGR offspring, resulting in increased activity before weaning, decreased activity in infancy, pursuit in adolescence, and near or even exceeding normal activity in adulthood. These offspring also exhibit a continuous high GC status. Additionally, female offspring with prenatal caffeine exposure show decreased blood corticosterone level and inhibited hepatic GR and FASN expression at 12 weeks after birth, while hepatic SIRT1 mRNA expression is induced. However, after chronic stress, with the increase in blood corticosterone level, the changes in the expression levels of the above genes are reversed. Similarly, offspring with xenobiotic exposure and feeding restriction during pregnancy exhibit increased lipid synthesis, enhanced IGF signaling gene expression, decreased lipid output, and fatty acid oxidation in the liver under postnatal high-fat diet, promoting the development of NAFLD. In summary, the metabolic changes observed in IUGR offspring make them more susceptible to induction by a high-fat diet.

Based on the above findings, we put forth the hypothesis that the "first strike (prenatal xenobiotic exposure)" can regulate hepatic metabolic functions in offspring through the "two-programming" mechanism. As a result, these offspring become susceptible to NAFLD after birth. Subsequently, postnatal environmental factors, such as chronic stress and high-fat diet, serve as the "second strike," which ultimately leads to the development of NAFLD.

In conclusion, we proposed that there are "two-programming and two-strike" mechanisms for the occurrence of fetal-originated NAFLD caused by the adverse environment during pregnancy. Prenatal xenobiotic exposure serves as the "first strike" and induces fetal overexposure to maternal GC, which increases fatty acid de novo synthesis in the fetal liver, manifested as increased expression of the key lipid synthesis enzymes FASN and ACC and their transcriptional factor SREBP1c. These changes can continue after birth and even into adulthood (the first programming). On the other hand, fetal overexposure to maternal GC inhibits the hepatic IGF1 signaling pathway and the lipid metabolic function. While under a nutrition-rich status after birth, hepatic IGF1signaling pathway is activated in the IUGR offspring, which induces compensatory enhancement of hepatic lipid metabolic function, leading to catch-up growth and increased susceptibility to metabolic diseases (the second programming). The body is in the susceptible state of NAFLD after undergoing the "first strike." Postnatal environmental changing factors (such as mental stress and excessive nutrition state) serve as the "second strike," which can aggravate the hepatic lipid metabolism disorder and lead to the aggravation of oxidative stress and inflammatory response, eventually resulting in the development and progression of NAFLD (Fig. 13.2).

To summarize, our research proposes that there are "two-programming and two-strike" mechanisms for the development of fetal-originated NAFLD caused by adverse environments during pregnancy. Prenatal exposure to xenobiotics acts as

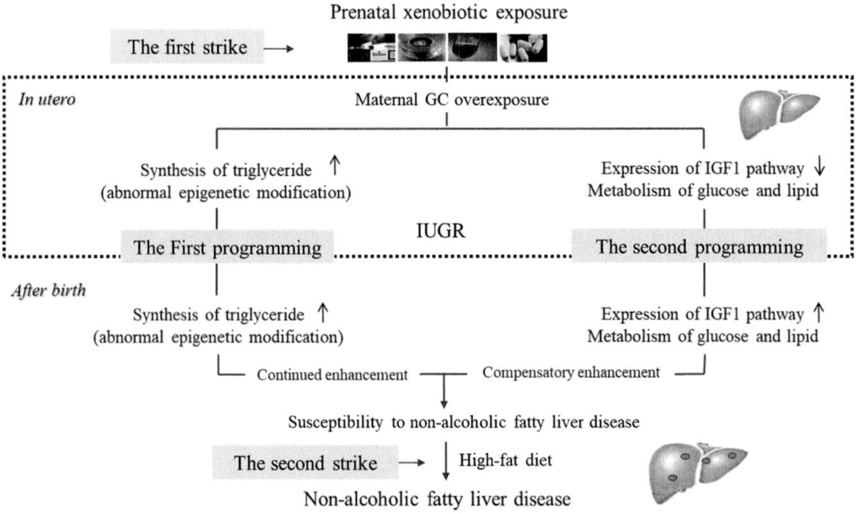

Fig. 13.2 "Two-programming and two-strike" mechanism of fetal-originated nonalcoholic fatty liver disease.

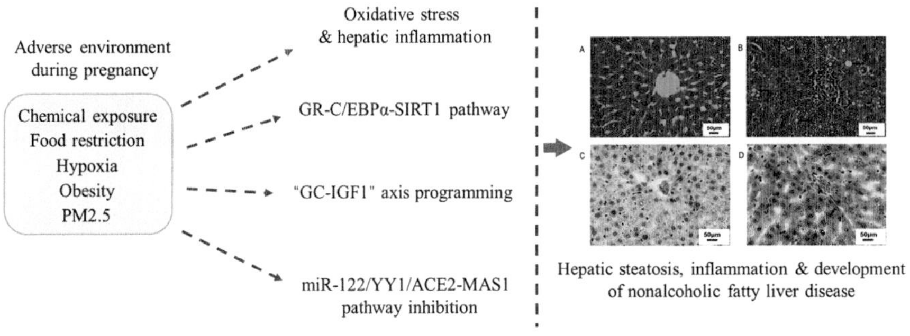

Fig. 13.3 Intrauterine programming mechanisms of fetal-originated nonalcoholic fatty liver disease

the "first strike" and induces fetal overexposure to maternal GC. This overexposure increases de novo fatty acid synthesis in the fetal liver, as evidenced by increased expression of key lipid synthesis enzymes FASN and ACC and their transcriptional factor SREBP1c. These changes can persist after birth and into adulthood (the first programming). Additionally, fetal overexposure to maternal GCs inhibits hepatic IGF1 signaling and lipid metabolic function. However, under a nutrition-rich postnatal environment, hepatic IGF1 signaling is activated in IUGR offspring, inducing compensatory enhancement of hepatic lipid metabolic function, leading to catch-up growth and increased susceptibility to metabolic diseases (the second programming). The "first strike" renders the body susceptible to NAFLD, and postnatal environmental factors, such as mental stress and excessive nutrition, act as the "second strike," exacerbating hepatic lipid metabolism disorder and leading to increased oxidative stress and inflammatory response, ultimately resulting in the development and progression of NAFLD (Fig. 13.2).

5 Summary and Prospect

Epidemiological and experimental evidence supports the notion that NAFLD has an intrauterine origin, and its underlying mechanisms primarily involve epigenetic modification mechanisms and neuroendocrine metabolic programming mechanisms (Fig. 13.3). The "two-programming and two-strike" mechanisms proposed by the authors' team provide potential targets for early warning and intervention of fetal-originated NAFLD. For instance, the detection of epigenetic markers (such as IGF1 epigenetic markers) in peripheral blood mononuclear cells can provide an early warning of fetal-originated NAFLD. Intervention strategies for fetal-originated NAFLD must be implemented in early life, during the critical time window of developmental plasticity. A recent study by the authors' team found that early postnatal intervention with angiotensin-converting enzyme can effectively reverse the development of osteoporosis in adult offspring caused by dexamethasone exposure during pregnancy (Xiao et al., 2020). This successful case provides new research ideas and directions for selecting effective therapeutic targets during the appropriate early life stage and preventing and treating the development and progression of multiple chronic diseases after birth in this special group.

References

Alisi A, Panera N, Agostoni C, et al. 2011. Int J Endocrinol, 2011: 269853.
Bruce K D, Cagampang F R, Argenton M, et al. 2009. Hepatology, 50(6): 1796–1808.
Brumbaugh D E, Tearse P, Cree-Green M, et al. 2013. J Pediatr, 162(5): 930–936 e931.
Chen M, Wang T, Liao Z X, et al. 2007. Exp Toxicol Pathol, 59(3–4): 245–251.
Delany A M, Durant D, Canalis E. 2001. Mol Endocrinol, 15(10): 1781–1789.
Efstathiou SP. 2015. Paediatr Int Child Health, 35(2): 81–2.
Faienza M F, Brunetti G, Ventura A, et al. 2013. Horm Res Paediatr, 79(2): 103–109.
Fu Q, Yu X, Callaway CW, et al. 2009. FASEB J, 23: 2438–2449.
Gale C, Thomas E L, Jeffries S, et al. 2014. Am J Clin Nutr, 99(5): 1034–1040.
Gao X, Lin S H, Ren F, et al. 2016. Nat Commun, 7: 11960.
Golic M, Stojanovska V, Bendix I, et al. 2018. Hypertension, 71: 911–920.
Goodspeed D, Seferovic M D, Holland W, et al. 2015. FASEB J, 29(3): 807–819.
Grun F, Watanabe H, Zamanian Z, et al. 2006. Mol Endocrinol, 20(9): 2141–2155.
Hagiwara A, Cornu M, Cybulski N, et al. 2012. Cell Metab, 15(5): 725–738.
Hu S, Xia L, Luo H, et al. 2019. Toxicology, 417: 23–34.
Hu W, Wang G, He B, et al. 2020a. Toxicology, 432: 152378.
Hu W, Yuan C, Luo H, et al. 2020b. Toxicol Lett, 331: 167–177.
Jiang Y, Xia W, Zhu Y, et al. 2014. Toxicol Lett, 228(2): 85–92.
Lillycrop K A, Phillips E S, Torrens C, et al. 2008. Br J Nutr, 100: 278–282.
Liu H, He B, Hu W, et al. 2021. Biochem Pharmacol, 185: 114420.
Luo H, Deng Z, Liu L, et al. 2014. Toxicol Appl Pharmacol, 274(3): 383–392.
Lv Z, Li G, Li Y, et al. 2013. Environ Toxicol, 28(9): 532–542.
Maranghi F, Lorenzetti S, Tassinari R, et al. 2010. Reprod Toxicol, 29(4): 427–32.
McCurdy C E, Bishop J M, Williams S M, et al. 2009. J Clin Invest, 119(2): 323–335.
Modi N, Murgasova D, Ruager-Martin R, et al. 2011. Pediatr Res, 70(3): 287–291.
Nohara K, Tateishi Y, Suzuki T, et al. 2012. Toxicol Sci, 129(2): 293–304.
Oben J A, Mouralidarane A, Samuelsson A M, et al. 2010. J Hepatol, 52(6): 913–920.
Ortiz L, Nakamura B, Li X, et al. 2013. Toxicol Lett, 223(2): 260–267.
Quist E M, Filgo A J, Cummings C A, et al. 2015. Toxicol Pathol, 43(4): 546–557.
Salamone F, Bugianesi E. 2010. J Hepatol, 53(6): 1146–1147.
Sayiner M, Koenig A, Henry L, et al. 2016. Clin Liver Dis, 20(2): 205–214.
Shen L, Liu Z, Gong J, et al. 2014. Toxicol Appl Pharmacol, 274(2): 263–273.
Tan Y, Liu J, Deng Y, et al. 2012. Toxicol Lett, 214: 279–287.
Tosh D N, Fu Q, Callaway C W, et al. 2010. Am J Physiol Gastrointest Liver Physiol, 299(5): G1023–1029.
Wang L, Shen L, Ping J, et al. 2014. Toxicol Lett, 224(3): 311–318.
Wei J, Sun X, Chen Y, et al. 2014. J Endocrinol, 222(3): 313–325.
Wolfe D, Gong M, Han G, et al. 2012. Am J Obstet Gynecol, 207(4): 308 e301–306.
Woods C P, Hazlehurst J M, Tomlinson J W. 2015. J Steroid Biochem Mol Biol, 154: 94–103.
Xia L, Jiao Z, Pei L, et al. 2020. Reprod Toxicol, 94: 48–54.
Xiao H, Xie X, Wen Y, et al. 2020. Bone, 133: 115245.

Xu D, Xia L P, Shen L, et al. 2013. Acta Pharmacol Sin, 34(12): 1526–1534.
Xu D, Bai J, Zhang L, et al. 2015. Toxicol Res, 4(1): 112–120.
Yu L, Zhou J, Zhang G, et al. 2018. Toxicol Appl Pharmacol, 352: 77–86.
Zhang L, Shen L, Xu D, et al. 2016. Reprod Toxicol, 65: 236–247.
Zhu W F, Zhu J F, Liang L et al. 2016. Mol Med Rep, 13(5): 4487–4493.

Chapter 14
Fetal Origin of Hypercholesterolemia

Ze Chen and Kexin Liu

Abstract Numerous studies have demonstrated the existence of fetal developmental origin of adult hypercholesterolemia. However, a comprehensive theory underlying the pathogenesis of fetal-originated hypercholesterolemia is still lacking. Recent studies have revealed that an adverse intrauterine environment, such as xenobiotic exposure, can induce the fetus to become overexposed to maternal-originated glucocorticoids by opening the placental glucocorticoid barrier. This overexposure can directly affect liver cholesterol metabolic function in utero or indirectly through maternal-originated glucocorticoids, resulting in offspring susceptibility to hypercholesterolemia after birth. Epigenetic modifications also participate in regulating the intrauterine programming mechanism of fetal-originated hypercholesterolemia. This chapter summarizes the research progress of fetal-originated hypercholesterolemia and its intrauterine programming mechanism, providing a theoretical and experimental foundation for clinical early warning and treatment.

1 Introduction

Hypercholesterolemia is a component of the metabolic syndrome and an independent disease that increases the risk of cardiovascular, cerebrovascular, and metabolic diseases, such as atherosclerosis (Shahwan et al. 2019; Song et al. 2016). Based on epidemiological investigations, the British scholar Barker proposed the theory of the "fetal origin" of adult diseases (Hales and Barker 1992; Walker and Walker 1993). This theory posits that an adverse intrauterine environment can cause permanent or programming changes in the structure and function of fetal tissues and organs, significantly increasing the offspring's susceptibility to various chronic

Z. Chen (✉)
Department of Cardiology, Zhongnan Hospital of Wuhan University, Wuhan, Hubei, China
e-mail: chenze19@whu.edu.cn

K. Liu
Department of Pharmacy, Zhongnan Hospital of Wuhan University, Wuhan, Hubei, China

H. Wang et al. (eds.), *Fetal Origin of Diseases*,
https://doi.org/10.1007/978-981-97-5730-5_14

diseases after birth. Intrauterine growth retardation (IUGR) is a perinatal complication characterized by embryonic or fetal growth and development restriction caused by an adverse environment during pregnancy (Valsamakis et al. 2006). Epidemiological investigations have confirmed that an adverse environment during pregnancy can induce various adverse pregnancy outcomes in the offspring, including IUGR. Offspring with IUGR induced by an adverse intrauterine environment are more likely to develop diseases such as hyperlipidemia, metabolic syndrome, and coronary atherosclerosis in adulthood (Sohi et al. 2011; Szostakwęgierek and Szamotulska 2011). Animal studies have demonstrated that offspring with IUGR induced by an adverse intrauterine environment have high blood cholesterol levels after birth and are susceptible to metabolic syndrome and multiple metabolic diseases in adulthood (Zhou et al. 2019; Zinkhan et al. 2016; Chen et al. 2015). These findings suggest that hypercholesterolemia has an intrauterine developmental origin. However, the pathogenic mechanism of fetal-originated hypercholesterolemia has not been systematically elucidated. This chapter summarizes the research progress of fetal-originated hypercholesterolemia and its intrauterine programming mechanism. It provides a theoretical and experimental foundation for exploring early clinical warning, diagnosis, and treatment strategies.

2 Research Status of Hypercholesterolemia

Hypercholesterolemia, as an integral component of metabolic syndrome, is closely associated with the development of various cardiovascular, cerebrovascular, and metabolic diseases. The pathogenic mechanism of hypercholesterolemia mainly involves the enhanced de novo cholesterol synthesis and cholesterol export pathway, as well as the decreased reverse cholesterol transport function of the liver. Due to the ongoing economic development and the fast-paced lifestyle of the twenty-first century, hypercholesterolemia has become increasingly prevalent, and controlling its epidemic trend has become a prominent public health concern.

2.1 Epidemiological Status of Hypercholesterolemia

Hypercholesterolemia, defined as elevated levels of total blood cholesterol (TCH) or low-density lipoprotein cholesterol (LDL-C) (TCH $\geq$ 5.20 mM or LDL-C 3.38 $\geq$ mM), with or without decreased high-density lipoprotein cholesterol (HDL-C) levels (HDL-C $\leq$ 1.04 mM), has become increasingly prevalent globally over the past three decades. According to an epidemiological survey, the prevalence of hypercholesterolemia was 17.3% in the Chinese population aged 45–74 years and 45.6% in the US population of the same age group (Lu et al. 2018). Numerous epidemiological investigations have demonstrated that hypercholesterolemia is closely associated with the development of various cardiovascular, cerebrovascular, and metabolic diseases (Shahwan et al. 2019; Song et al. 2016). Elevated plasma LDL-C level is a significant risk factor for atherosclerosis (Ference et al. 2017), and cholesterol deposition in blood vessels or other tissues contributes to the progression of diseases such as diabetes, xanthoma, osteoarthritis, and tenosynovitis (Shahwan et al. 2019; Farnaghi et al. 2017; Salen and Steiner 2017; Song et al. 2016). Therefore, the development of hypercholesterolemia in a population warrants attention and interventions to prevent and manage the associated health risks.

2.2 *The Major Pathogenic Mechanisms of Hypercholesterolemia*

The liver plays a vital role in maintaining cholesterol homeostasis in the body by regulating cholesterol synthesis, uptake, and efflux. The primary pathogenic mechanisms of hypercholesterolemia include increased hepatic de novo cholesterol synthesis, enhanced hepatic cholesterol output, and decreased hepatic reverse cholesterol transport. Genetic and environmental factors such as diet and physical activity can increase the risk of hypercholesterolemia by affecting hepatic cholesterol metabolism pathways. Recent studies have revealed that hypercholesterolemia has an intrauterine developmental origin. Maternal exposure to adverse environmental factors during pregnancy can affect the hepatic cholesterol metabolic function of the offspring, leading to hypercholesterolemia under the influence of secondary postnatal environmental factors. Epidemiological investigations have confirmed that IUGR offspring are more susceptible to developing hypercholesterolemia, metabolic syndrome, and atherosclerosis (Sohi et al. 2011). Therefore, it is essential to analyze the early intrauterine predisposing factors of fetal-originated hypercholesterolemia and explore the regulatory mechanisms of its intrauterine development and postnatal influencing factors. These studies are of great significance in understanding the pathogenesis of hypercholesterolemia and developing effective interventions to prevent and manage its associated health risks.

3 Evidence of the Intrauterine Origin of Hypercholesterolemia

Numerous epidemiological studies, clinical investigations, and laboratory research have provided evidence that hypercholesterolemia has an intrauterine developmental origin, and exposure to adverse environmental factors during pregnancy is a significant predisposing factor.

3.1 Epidemiological Investigations

Numerous epidemiological investigations have demonstrated that IUGR increases the susceptibility of adult offspring to metabolic syndrome, and they are more likely to experience blood lipid disorders leading to the development of diseases such as hyperlipidemia and atherosclerosis (Sohi et al. 2011). For instance, exposure to famine during pregnancy or early life significantly increases the risk of developing hypercholesterolemia, characterized by elevated TCH and LDL-C levels and decreased HDL-C levels (Ding et al. 2020). Epidemiological evidence suggests that low birth weight increases the risk of elevated blood lipid levels and lipid metabolic disorders in adulthood (Efstathiou et al. 2012). Moreover, poor maternal living habits or disease status during pregnancy may result in low birth weight, the development of metabolic syndrome in adolescence, and elevated blood triglyceride and TCH levels in adulthood (Chen et al. 2017; Efstathiou et al. 2012; Godfrey et al. 1996).

Maternal smoking during pregnancy has been found to cause increased blood lipid and blood pressure levels and increased waist circumference of the offspring in childhood, as well as disordered blood cholesterol phenotype in adulthood (Seal et al. 2013). These findings suggest that an adverse intrauterine environment can result in lipid metabolic disorders in individuals after birth. Individuals who experience malnutrition during development will undergo catch-up growth. This means that the fetus has developmental limitations during the intrauterine period due to exposure to an adverse environment. When the postnatal environment changes from malnutrition to a nutrition-excess environment too quickly, it can overload the normal metabolic function of the offspring, leading to metabolic syndrome and cholesterol metabolic disorders.

3.2 Clinical and Basic Laboratory Studies

Numerous clinical and basic laboratory studies have demonstrated that the incidence of adult hypercholesterolemia is associated with the environment before and after birth, including the intrauterine period (Table 14.1). For example, Ohara discovered in 1993 that arsenic exposure in pregnant mice resulted in hyperglycemia, hypercholesterolemia, and nonalcoholic fatty liver disease in their adult offspring. Recent experiments have shown that an adverse intrauterine environment can lead to an increase in postnatal blood TCH levels in rats with IUGR (Hu et al. 2019; Luo et al. 2019; Zhou et al. 2019; Gardebjer et al. 2018; Xu et al. 2018; Abruzzese et al. 2016; Zinkhan et al. 2016; Liu et al. 2014). These results suggest that the adverse environment during pregnancy is a critical factor leading to increased susceptibility to hypercholesterolemia in adulthood.

The increased susceptibility to hypercholesterolemia in offspring induced by an adverse intrauterine environment is related to changes in their intrauterine

Table 14.1 Adverse intrauterine environments induced lipid metabolic disorders in offspring

Research type	Intrauterine environment	Findings in offspring	Main references
Epidemiological studies	Maternal smoking	Hyperlipidemia, high blood pressure, and increased waist circumference in childhood	Godfrey et al. (1996)
	Maternal smoking, alcohol drinking, diabetes, hypertension, or obesity	Puberty metabolic syndrome (MS), weight gain, high pulmonary blood pressure, elevated blood total cholesterol (TCH), low-density lipoprotein, and triglyceride (TG) levels	Efstathiou et al. (2012)
	Maternal obesity or diabetes	Childhood MS, obesity, insulin resistance, hypertension, and dyslipidemia (elevated TG level or decreased high-density lipoprotein level)	Boney et al. (2005)
Experimental studies	Polystyrene exposure	Disorders in serum lipids and hepatic lipid metabolism in rats at 42 days after birth	Luo et al. (2019)
	Low-protein diet	IUGR in male rats with increased TCH at 40 weeks old	Liu et al. (2014)
	Arsenic exposure	High blood glucose, hypercholesterolemia, and nonalcoholic fatty liver disease in offspring rats	Ohara et al. (1993)
	Hexachlorocyclohexane or malathion exposure	Hypercholesterolemia and hypertriglyceridemia in offspring rats	Prabhakaran and Devi (1993)
	Caffeine exposure	IUGR in offspring rats with increased TCH at 12 weeks old	Xu et al. (2018) Hu et al. (2019)
	Nicotine exposure	IUGR in offspring rats with increased TCH at 24 weeks old	Zhou et al. (2019)
	Ethanol exposure	Serum TCH elevated at postnatal weeks 8 and 24 in offspring rats	Gardebjer et al. (2018) Hu et al. (2019)
	Hyperandrogen	Impaired glucose tolerance, dyslipidemia, and increased risk of MS in pubertal rats	Abruzzese et al. (2016)

cholesterol metabolic function. During the intrauterine period, fetal cholesterol sources include fetal liver cholesterol synthesis, maternal cholesterol transport through the placenta, and placental cholesterol synthesis (Baardman et al. 2013). In the first few weeks of life, the fetus mainly relies on the placental transport of maternal cholesterol as its cholesterol source when most organs, including the liver, are developing. In contrast, in the third trimester, maternal cholesterol accounts for 22%–40% of the fetal cholesterol pool (Larque et al. 2013), and fetal cholesterol synthesis becomes more involved. Therefore, the maternal-placental-fetal pathways work synergistically to regulate fetal blood cholesterol levels during pregnancy.

3.2.1 Changes in Maternal-Placental Cholesterol Metabolism

During pregnancy, maternal lipid profile often exhibits elevated levels of TCH and triglycerides. Despite differences in dietary intake, plasma TCH levels during the third trimester are typically 50% higher than those observed prior to pregnancy. The increase in TCH levels is attributed to the higher concentrations of estrogen, progesterone, and prolactin in maternal plasma during pregnancy. Elevated cholesterol levels are important during pregnancy as they serve as a source of reserves for fetal growth and steroid synthesis in placental tissue. Additionally, the placenta establishes a cholesterol transport pathway between the fetus and mother through both receptor-dependent and receptor-independent mechanisms, such as free diffusion. The regulation of placental cholesterol synthesis primarily involves the activity of HMG-CoA reductase (HMGCR), which secretes cholesterol to the umbilical cord blood for transport to the fetus. On the other hand, maternal-fetal cholesterol transport is mediated by specific placental transporters, including scavenger receptor class B type I (SR-BI), members of the ATP-binding cassette transporter family ABCA1 and ABCG1, and the low-density lipoprotein receptor (LDLR) (Brett et al. 2014).

The expression of SR-BI is predominantly observed in the syncytiotrophoblast cells of the placenta, which is responsible for the maternal cholesterol transport to the placenta. In contrast, ATP-binding cassette transporters ABCA1 and ABCG1 are mainly expressed in the placental vascular endothelial cells, which are responsible for cholesterol transport from the placenta to the fetal circulation. Studies have shown a decrease in human placental ABCA1 expression with gestational age at term, suggesting a shift in the sources of fetal cholesterol from maternal to fetal liver synthesis in late pregnancy (Baumann et al. 2013). In mice, placental ABCA1 knockout resulted in a reduction of approximately 30% in maternal cholesterol transfer to fetal mice, indicating the crucial role of ABCA1 in cholesterol transport across the placenta (Lindegaard et al. 2008). Taken together, these cholesterol transporters in the placenta play a critical role in regulating maternal blood cholesterol transport to fetal blood, thus maintaining blood cholesterol levels during early fetal development (Fig. 14.1).

Clinical studies have demonstrated that adverse environments during pregnancy, such as exposure to exogenous environmental factors and maternal health factors, can directly or indirectly alter maternal blood lipid levels and affect the placental microenvironment, ultimately affecting placental material transport, including lipid transport function, and resulting in abnormal fetal blood cholesterol levels (Nasioudis et al. 2019; Leiva et al. 2015). Adverse exogenous environmental factors that can impact the fetus during pregnancy include exposure to xenobiotics, feeding

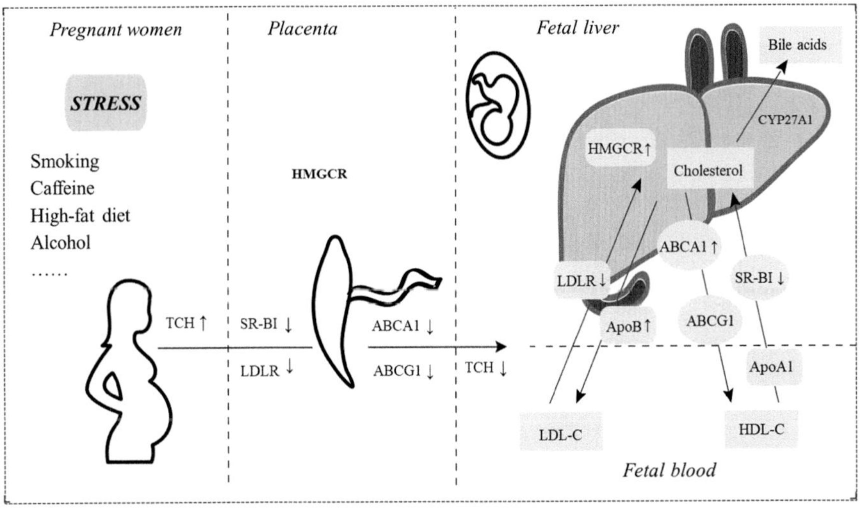

Fig. 14.1 Maternal-placental-fetal regulation of fetal blood cholesterol levels during pregnancy

restriction, and hypoxia. Maternal factors, on the other hand, mainly include nutritional status (e.g., severe maternal malnutrition) and maternal diseases (e.g., maternal eclampsia, depression, thyroid diseases, diabetes, hypercholesterolemia, microbial infection, and stress status). Studies have shown that increased maternal blood TCH and LDL-C levels are associated with disordered placental endothelial cell function and suppressed LDLR expression (Ethier-Chiasson et al. 2007; Leiva et al. 2015). Additionally, maternal preeclampsia has been shown to decrease placental ABCA1 expression, resulting in decreased cholesterol transport (Mistry et al. 2017). Animal experiments have also demonstrated that maternal blood cholesterol levels increase under prenatal nicotine exposure, resulting in decreased expression of placental LDLR, SR-BI, ABCA1, and ABCG1 in female offspring, inhibition of cholesterol transport from maternal blood to fetal blood, and decreased fetal blood cholesterol levels (Zhang et al. 2018). These findings suggest that adverse environments during pregnancy can impair placental cholesterol transport function, leading to decreased blood cholesterol levels in offspring (Fig. 14.1).

3.2.2 Changes in Fetal Cholesterol Metabolic Function

Under physiological conditions, hepatic cholesterol synthesis is primarily regulated by the rate-limiting enzyme HMGCR. Apolipoprotein B (ApoB) and transporters ABCA1 and ABCG1 are responsible for the output of hepatic cholesterol to peripheral tissues. Sterol regulatory element-binding factor 2 (SREBF2), which codes for protein SREBP2, promotes the expression of HMGCR and ApoB (Xu et al. 2018). Since fetal development requires a significant amount of cholesterol, the rate of sterol synthesis in the fetal liver is much higher than that in adults, and the rate of cholesterol synthesis is not as strictly controlled as in adults. Therefore, de novo synthesis of fetal hepatic cholesterol plays a crucial role in the development of fetal-originated hypercholesterolemia. Additionally, the fetus can transport cholesterol from peripheral tissues to the liver through the reverse transport of cholesterol. LDLR mediates LDL-C uptake, while SR-BI primarily transports HDL-C, which binds to apolipoprotein A1 (apoA-1) (Larque et al. 2013). Cholesterol-27 alpha-hydroxylase (CYP27A1) degrades cholesterol to bile acids. However, bile acid production in the fetal liver is significantly lower than in adults, resulting in very little cholesterol loss from the fetus in the form of bile acids (Hardy et al. 1980). Through the regulation of the above processes, the body ensures sufficient cholesterol intake while avoiding excessive accumulation, thus maintaining the dynamic balance of cholesterol in the liver (Fig. 14.1c).

Studies have shown that dietary restriction during pregnancy in rats increased the serum TCH level of adult offspring, as well as liver HMGCR expression and cholesterol level (Cong et al. 2012; Liu et al. 2014). Exposure to caffeine during pregnancy upregulated the expressions of SREBP2 and HMGCR in fetal liver (Hu et al. 2019). In female fetal liver, caffeine exposure also upregulated the expressions of HMGCR, ApoB, and ABCA1, leading to fetal hepatic cholesterol accumulation (Xu et al. 2018). These findings suggest that the development of fetal-originated hypercholesterolemia may be due to the enhanced de novo cholesterol synthesis and cholesterol output function of the offspring's liver. Additionally, studies have demonstrated that proanthocyanidin intake during pregnancy can impair the reverse transport of cholesterol in the offspring in utero, increasing the risk of atherosclerosis in adulthood (Del Bas et al. 2015). The authors' team also found that dexamethasone exposure during pregnancy inhibited hepatic LDLR expression in fetal rats (Li et al. 2020), indicating that one of the causes of fetal-originated hypercholesterolemia is the inhibiting effect of an adverse environment during pregnancy on the reverse cholesterol transport function of the fetal liver. However, due to the inhibition of placental cholesterol transport caused by various adverse environments during pregnancy, fetuses still exhibit low blood TCH levels despite the enhancement of fetal cholesterol synthesis and output function or the inhibited reverse cholesterol transport function (Spencer et al. 1997). Importantly, changes in cholesterol metabolism observed in these studies can persist after birth, leading to hypercholesterolemia in adult offspring.

4 Intrauterine Programming Mechanism of Fetal-Originated Hypercholesterolemia

Intrauterine programming refers to the permanent alterations in tissue morphology and function caused by early developmental damage. These changes in the expression patterns of functional genes in tissues and organs are typically maintained from puberty to adulthood and throughout life. The adverse environment during pregnancy can lead to the early onset of lipid metabolic disorders and susceptibility to hypercholesterolemia in adult offspring through the "two-programming and two-strike" mechanism.

4.1 Epigenetic Mechanism and Multigenerational Genetic Effects of Cholesterol Metabolic Programming

Epigenetic modification refers to the regulation of genomic expression through DNA methylation, histone modification, and noncoding RNAs. Epigenetic regulation helps maintain the stability of the internal environment and ensures normal physiological function. However, epigenetic regulation is susceptible to adverse intrauterine environments. Studies have shown that exposure to toxic substances during pregnancy can cause abnormal epigenetic modification of genes related to fetal growth and development, resulting in permanent phenotype changes after birth (Jirtle and Skinner 2007). Epigenetic modification abnormalities play a crucial role in the intrauterine programming of many adult diseases. Additionally, epigenetic changes are involved in programming the effects of environmental changes during pregnancy on postnatal function and may persist after birth, leading to fetal-originated hypercholesterolemia. Studies have confirmed that epigenetic modification abnormalities are closely related to fetal hepatic cholesterol metabolism (Table 14.2). For example, maternal gestational diabetes mellitus and a low-protein diet can cause changes in the DNA methylation levels of liver cholesterol metabolic enzymes in fetal and juvenile rats (Golic et al. 2018; Lillycrop et al. 2008) and histone methylation modification changes in HMGCR and CYP27A1 in adult offspring liver (Cong et al. 2012; Sohi et al. 2011).

Previous studies investigate the role of histone deacetylase SIRT1 and acetylase p300 in mediating the acetylation modification level and expression changes in functional genes of cholesterol metabolism in intrauterine and postnatal rats exposed to xenobiotics during pregnancy (Hu et al. 2019; Zhou et al. 2019; Xu et al. 2018). Moreover, the study also explores the involvement of various miRNAs in the disorder of offspring cholesterol metabolism under adverse pregnancy environments, such as maternal hypercholesterolemia, proanthocyanidins, and dexamethasone exposure (Li et al. 2020; Dumolt et al. 2019; Del Bas et al. 2015). The consumption of betaine during pregnancy has been found to reduce the methylation of CpG in the promoter region of Cyp27a1 in the liver of chicks, thereby improving liver cholesterol accumulation (Hu et al. 2017). Furthermore, the authors' team has recently discovered that the H3K27ac level of LDLR promoter in peripheral blood mononuclear cells (PBMCs) of male neonates was decreased by predelivery use of dexamethasone in the clinic, suggesting the potential of epigenetic modification as a biomarker and an early warning and intervention target for the susceptibility to hypercholesterolemia in offspring (to be published).

Table 14.2 Adverse intrauterine environment induced epigenetic modifications of genes related to hepatic cholesterol metabolism in the fetus

Intrauterine environment	Genes	Epigenetic modifications	Sites	Main references
Maternal low-protein diet	*HMGCR, CYP27A1*	Histone methylation	H3K9me1 H3K27me3	Sohi et al. (2011) Cong et al. (2012)
	PPARα	DNA methylation	CpG	Lillycrop et al. (2008)
	IGF1	Histone methylation	H3K4me2	Tosh et al. (2010)
Maternal diabetes	*SREBP2*	DNA methylation	CpG	Golic et al. (2018)
Betaine exposure	*CYP27A1*	DNA methylation	CpG	Hu et al. (2017)
Nicotine exposure	*SR-B1, LDLR*	Histone acetylation	H3K27ac	Zhou et al. (2019)
Caffeine exposure	*HMGCR*	Histone acetylation	H3K9ac H3K14ac H3K27ac	Xu et al. (2018) Hu et al. (2019)
	SREBP2, HMGCS	Histone acetylation	H3K9ac H3K14ac	Hu et al. (2019)
	IGF1	Histone methylation	H3K4me3 H3K36me3	Fu et al. (2009) Tan et al. (2012)
Grape proanthocyanidin exposure	*NPC1, ABCG1, ABCA1, NR1H3, SEBP2, LCAT, SCARB1, PLTP*	miR-33a	N/A	Del Bas et al. (2015)
Dexamethasone exposure	*LDLR*	miR-148a	N/A	Li et al. (2020)
Maternal hypercholesterolemia	N/A	miR-27a, miR-200c	N/A	Dumolt et al. (2019)

It is worth noting that adult hypercholesterolemia with fetal origin may be inherited through multiple generations. Exposure to various xenobiotics or chronic stress during pregnancy can result in adult hypercholesterolemia in the F1 generation, which can then continue to the F2, F3, and F4 generations (Anway et al. 2005, 2006; Buchwald et al. 2012; Luo et al. 2014). For instance, exposure to vastazoline or methoxyDDT during pregnancy led to adult hypercholesterolemia in F1 to F4 male offspring (Anway et al. 2005). In a monkey model of pregnancy stress or dexamethasone exposure, the serum TCH level increased in the F1 generation and persisted into the F2 and F3 generations (Buchwald et al. 2012; Anway et al. 2006). In F1 generation IUGR rats, caffeine exposure during pregnancy induced phenotypic changes in lipid metabolism, which increased TCH synthesis in F2 generation male rats after birth (Luo et al. 2014). Previous studies by the authors' team have shown that caffeine exposure during pregnancy induces glucolipid metabolic phenotype changes and increases susceptibility to metabolic syndrome in IUGR rats of the F1 generation, which may be related to changes in neuroendocrine metabolic programming related to the hypothalamic-pituitary-adrenal (HPA) axis. The metabolism of the F2 generation rats showed the characteristics of metabolic syndrome and some related diseases (Kou et al. 2014; Luo et al. 2014). An increasing body of evidence suggests that abnormal epigenetic marks transmitted through germ cells may be a major cause of the transgenerational inheritance of adverse environmental effects. DNA methylation, histone modification, and noncoding RNA may all play vital roles in this process. Pregnancy and the early postnatal environment can lead to epigenetic changes in offspring's specific genes that can persist into adulthood and even into the next generation. These findings indicate that fetal-originated hypercholesterolemia may be inherited through multiple generations and may be related to epigenetic modification, although the specific mechanism requires further investigation.

4.2 The "Two-Programming and Two-Strike" Mechanisms of Fetal-Originated Hypercholesterolemia

The regulation of fetal tissue morphology and functional maturity is highly dependent on the level of intrauterine glucocorticoid (GC). However, a high concentration of intrauterine GC may lead to fetal developmental abnormalities. Adverse environmental conditions during pregnancy, such as maternal nutritional restriction, prenatal distress, depression and anxiety, exposure to LPS, and eclampsia, can inhibit the expression of placental 11β-hydroxysteroid dehydrogenase type 2 (11β-HSD2), leading to maternal-originated GC overexposure in fetuses. This overexposure may mediate changes in fetal developmental programming caused by adverse events during pregnancy (Moisiadis and Matthews 2014). Animal studies conducted in our laboratory have shown that caffeine and nicotine exposure during pregnancy can cause intrauterine maternal-originated GC overexposure and activate the fetal liver glucocorticoid receptor (GR). Under the regulation of epigenetic enzymes, fetal liver cholesterol synthesis function is enhanced, while its reverse cholesterol transport function is suppressed, leading to fetal hepatic cholesterol accumulation (Zhou et al. 2019; Xu et al. 2018). Abnormal epigenetic modifications that enhance liver cholesterol synthesis in utero may persist after birth and eventually lead to continuous enhancement of cholesterol synthesis in adult offspring, making them susceptible to diseases such as hypercholesterolemia and metabolic syndrome. This mechanism is referred to as the "first programming" mechanism for the development of fetal-originated hypercholesterolemia.

Furthermore, GC can also inhibit the expression and secretion of insulin-like growth factor 1 (IGF1), a core factor of the endocrine regulatory system (Kou et al. 2014; Wang et al. 2014; Xu et al. 2012). Studies have shown that an increased GC level and decreased IGF1 level were observed in the blood circulation of fetuses with IUGR, while decreased blood circulation GC level and increased liver IGF1 level were shown in adults with catch-up growth (Gicquel and Le Bouc 2006; Kenyon 2001; Kamei et al. 2011; Tosh et al. 2010). The GC-IGF1 axis exhibits adaptive changes. A high level of IGF1 can affect hepatic cholesterol metabolism through the PI3K/Akt signaling pathway, leading to susceptibility to hypercholesterolemia in the body (Shen et al. 2014; Hagiwara et al. 2012). This programming effect is the programming change of the liver "GC-IGF1 axis" under adverse environmental exposure during pregnancy (such as xenobiotics), which causes low intrauterine IGF1 levels in offspring and high IGF1 levels after birth, leading to intrauterine IUGR and catch-up growth after birth. It reflects the overall adaptation and compensatory change of IUGR individuals in the face of different living environments. The "second programming" mechanism of fetal-originated hypercholesterolemia is the reduction of fetal metabolic rate and the catch-up growth and cholesterol metabolic disorders after birth mediated by the programming changes of the "GC-IGF1 axis."

Offspring with IUGR are highly sensitive to the HPA axis under chronic stress after birth (He et al. 2015; Li et al. 2015). Once exposed to chronic stimulation, such as a high-fat diet after birth, IUGR offspring experience enhanced catch-up growth, increased liver IGF1 level, and cholesterol synthesis (Xu et al. 2018; Zhang et al. 2016; Li et al. 2013). The authors' team has analyzed and found that the occurrence of hypercholesterolemia in offspring rats induced by caffeine or ethanol exposure during pregnancy is related to the interaction between xenobiotics during pregnancy, high-fat diet after birth, and different genders. This is mainly due to the enhanced hepatic cholesterol synthesis function and weakened reverse transport function (Guo et al. 2018; Qi et al. 2017). Based on these findings, the authors' team proposes a "two-programming and two-strike" mechanism for fetal-originated hypercholesterolemia (Fig. 14.2). The "first programming" refers to the changes in fetal hepatic cholesterol metabolic function caused by intrauterine maternal-originated GC overexposure during pregnancy, which involves epigenetic modification and stable expression changes that can continue into adulthood. The "second programming" refers to compensatory programming changes in the liver GC-IGF1 axis that mediate intrauterine inhibitory development, postnatal catch-up growth, and enhanced cholesterol metabolism in the offspring. The adverse environment during pregnancy acts as the "first strike," increasing the susceptibility of offspring to hypercholesterolemia after birth. Meanwhile, environmental changes after birth (such as high-fat diet and chronic stress) act as the "second strike," aggravating the disorder of cholesterol metabolism. The "two-strike" interact and eventually lead to the occurrence of adult hypercholesterolemia.

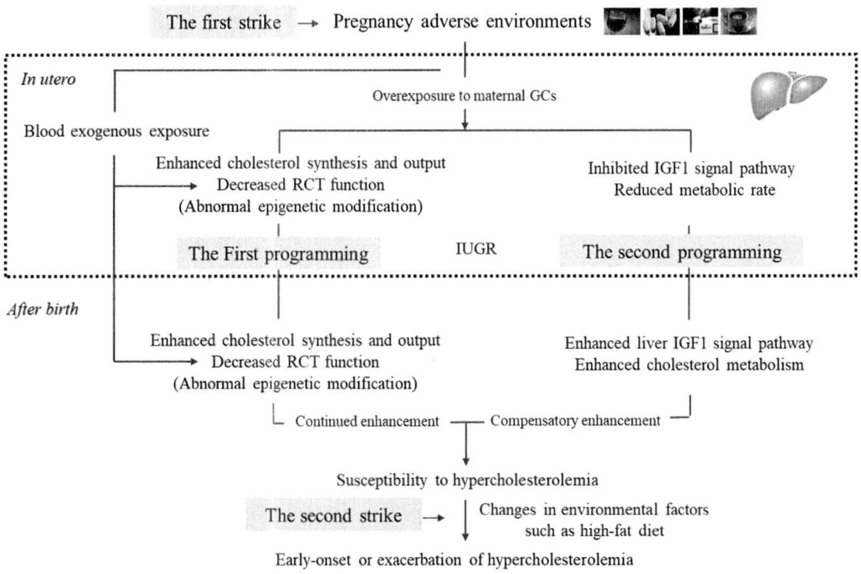

Fig. 14.2 The "two-programming and two-strike" mechanism of fetal-originated hypercholesterolemia

5 Summary and Prospect

This chapter provides a comprehensive review of fetal-originated hypercholesterolemia, focusing on its intrauterine origin, the impact of adverse prenatal environment on fetal cholesterol metabolism, and proposing a "two-programming and two-strike" mechanism (Fig. 14.3). The first programming is characterized by changes in fetal hepatic cholesterol metabolic function due to fetal overexposure to maternal-originated GC, leading to epigenetic modifications that can persist into adulthood. The second programming is related to changes in the "GC-IGF1 axis," resulting in reduced fetal metabolic rate, postnatal catch-up growth, and enhanced hepatic cholesterol metabolism. The "first strike" during pregnancy can regulate liver cholesterol metabolism through these two kinds of programming. In contrast, the "second strike" caused by environmental changes after birth can aggravate cholesterol accumulation and lead to early hypercholesterolemia. The identification of related biomarkers, early evaluation system of developmental toxicity, and harmful factors during pregnancy can help in the early warning and intervention technology of fetal-originated hypercholesterolemia. The authors' team recently found that the expression of HMGCR in clinical PBMC samples was increased in female neonates exposed to dexamethasone during pregnancy (Li et al. 2022), while the LDLR promoter H3K27ac of PBMCs in male neonates was decreased (to be published). The detection of key PBMC gene expression and epigenetic markers can help in the

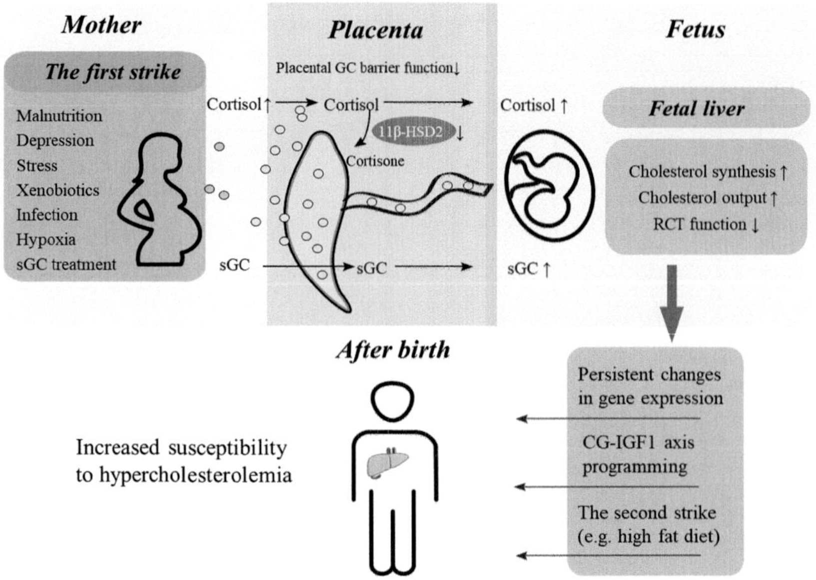

Fig. 14.3 Pathogenetic mechanisms of fetal-originated hypercholesterolemia

early warning of fetal-originated hypercholesterolemia. For the early prevention and treatment of the disease, epigenetic modification intervention in the early postnatal period may be an effective means to prevent long-term effects.

References

Abruzzese G A, Heber M F, Ferreira S R, et al. 2016. J Endocrinol, 230(1): 67–79.
Anway M D, Cupp A S, Uzumcu M, et al. 2005. Science, 308(5727): 1466–1469.
Anway M D, Memon M A, Uzumcu M, et al. 2006. J Androl, 27(6): 868–879.
Baardman M E, Kerstjens-Frederikse W S, Berger R M, et al. 2013. Biol Reprod, 88(1): 24.
Baumann M, Korner M, Huang X, et al. 2013. Placenta, 34(11): 1079–1086.
Boney C M, Verma A, Tucker R, et al. 2005. Pediatrics, 115: e290–e296.
Brett K E, Ferraro Z M, Yockell-Lelievre J, et al. 2014. Int J Mol Sci, 15(9): 16153–16185.
Buchwald U, Teupser D, Kuehnel F, et al. 2012. J Med Primatol, 41(4): 231–240.
Chen L H, Chen S S, Liang L, et al. 2017. J Chin Med Assoc, 80(1): 44–49.
Chen L H, Liang L, Zhu W F, et al. 2015. Chin J Contemp Pediatr, 17(10): 1124–1130.
Cong R, Jia Y, Li R, et al. 2012. J Nutr Biochem, 23(12): 1647–1654.
Del Bas J M, Crescenti A, Arola-Arnal A, et al. 2015. J Nutr Biochem, 26(12): 1670–1677.
Ding X Y, Yang Z Y, Zhao L Y, et al. 2020. Nutrients, 12(8): 2266.
Dumolt J H, Ma M, Mathew J, et al. 2019. Am J Physiol-Endoc M, 317(5): e831–e838.
Efstathiou S P, Skeva, II, Zorbala E, et al. 2012. Circulation, 125(7): 902–910.
Ethier-Chiasson M, Duchesne A, Forest J C, et al. 2007. Biochem Bioph Res Co, 359(1): 8–14.

Farnaghi S, Crawford R, Xiao Y, et al. 2017. Int J Rheum Dis, 20(2): 131–140.
Ference B A, Ginsberg H N, Graham I, et al. 2017. Eur Heart J, 38(32): 2459–2472.
Fu Q, Yu X, Callaway C W, et al. 2009. FASEB J, 23(8): 2438–2449.
Gardebjer E M, Cuffe J S M, Ward L C, et al. 2018. Am J Physiol-Endoc M, 315(4): e694–e704.
Gicquel C, Le Bouc Y. 2006. Horm Res, 65 Suppl 3: 28–33.
Godfrey K, Robinson S, Barker D J, et al. 1996. BMJ, 312(7028): 410–414.
Golic M, Stojanovska V, Bendix I, et al. 2018. Hypertension, 71(5): 911–920.
Guo Y, Luo H, Wu Y, et al. 2018. Reprod Toxicol, 79: 47–56.
Hagiwara A, Cornu M, Cybulski N, et al. 2012. Cell Metab, 15(5): 725–738.
Hales C N, Barker D J. 1992. Diabetologia, 35: 595–601.
Hardy K J, Hoffman N E, Mihaly G, et al. 1980. J Physiol, 309: 1–11.
He Z, Li J, Luo H, et al. 2015. Sci Rep, 5: 17679.
Hu Y, Sun Q, Zong Y, et al. 2017. Gen Comp Endocrinol, 246: 241–248.
Hu S, Liu K, Luo H, et al. 2019. Toxicology, 418: 11–21.
Jirtle R L, Skinner M K. 2007. Nat Rev Genet, 12(7): 620.
Kamei H, Ding Y, Kajimura S, et al. 2011. Development, 138(4): 777–786.
Kenyon C. 2001. Cell, 105(2): 165–168.
Kou H, Liu Y, Liang G, et al. 2014. Toxicol Appl Pharm, 275(2): 79–87.
Larque E, Ruiz-Palacios M, Koletzko B. 2013. Curr Opin Clin Nutr, 16(3): 292–297.
Leiva A, Salsoso R, Saez T, et al. 2015. Placenta, 36(8): 895–902.
Li J, Luo H, Wu Y, et al. 2015. Toxicol Appl Pharm, 284(3): 345–353.
Li L, Hu W, Liu K, et al. 2020. Toxicol Appl Pharm, 395: 114979.
Li X, Hu W, Li L, et al. 2022. Biochem Pharmacol, 206:115306.
Li Z, Xu D, Zhang B, et al. 2013. Arch Med Res, 44(5): 335–345.
Lillycrop K A, Phillips E S, Torrens C, et al. 2008. Brit J Nutr, 100(2): 278–282.
Lindegaard M L, Wassif C A, Vaisman B, et al. 2008. Hum Mol Genet, 17(23): 3806–3813.
Liu X, Qi Y, Tian B, et al. 2014. J Clin Biochem Nutr, 55(1): 40–47.
Lu Y, Wang P, Zhou T, et al. 2018. J Am Heart Assoc, 7(3): e007462.
Luo H, Deng Z, Liu L, et al. 2014. Toxicol Appl Pharm, 274(3): 383–392.
Luo T, Zhang Y, Wang C, et al. 2019. Environ Pollut, 255(Pt 1): 113122.
Mistry H D, Kurlak L O, Mansour Y T, et al. 2017. J Lipid Res, 58(6): 1186–1195.
Moisiadis V G, Matthews S G. 2014. Nat Rev Endocrinol, 10(7): 391–402.
Nasioudis D, Doulaveris G, Kanninen T T. 2019. Minerva Ginecol, 71(2): 155–162.
Ohara Y, Peterson T E, Harrison D G. 1993. J Clin Invest, 91(6): 2546–2551.
Prabhakaran S, Devi K S. 1993. Indian J Biochem Biophys, 30: 234–238.
Qi Y, Luo H, Hu S, et al. 2017. Cell Physiol Biochem, 44(2): 657–670.
Salen G, Steiner R D. 2017. J Inherit Metab Dis, 40(6): 771–781.
Seal N, Krakower G, Seal J. 2013. Jnp-J Nurse Pract, 9(10): 695–705.
Shahwan M J, Jairoun A A, Farajallah A, et al. 2019. Diabetol Metab Syndr, 13(4): 2387–2392.
Shen L, Liu Z, Gong J, et al. 2014. Toxicol Appl Pharm, 274(2): 263–273.
Sohi G, Marchand K, Revesz A, et al. 2011. Mol Endocrinol, 25(5): 785–798.
Song Y, Liu X, Zhu X, et al. 2016. Sci Rep, 6: 36093.
Spencer J A, Chang T C, Crook D, et al. 1997. Arch Dis Child-Fetal, 76(1): 21–25.
Szostakwęgierek D, Szamotulska K. 2011. Medycyna Wieku Rozwojowego, 15(3): 203–215.
Tan Y, Liu J, Deng Y, et al. 2012. Toxicol Lett, 214(3): 279–287.
Tosh D N, Fu Q, Callaway C W, et al. 2010. Am J Physiol-Gastr L, 299(5): 1023–1029.
Valsamakis G, Kanaka-Gantenbein C, Malamitsi-Puchner A, et al. 2006. Ann Ny Acad Sci, 1092: 138–147.
Walker A R, Walker B F. 1993. Lancet, 341(8857): 1421–1421.
Wang L, Shen L, Ping J, et al. 2014. Toxicol Lett, 224(3): 311–318.
Xu D, Luo H W, Hu W, et al. 2018. Faseb J, 32(10): 5563–5576.
Xu D, Wu Y, Liu F, et al. 2012. Toxicol Appl Pharm, 264(3): 395–403.

Zhang G, Zhou J, Huang W, et al. 2018. Toxicol Lett, 296: 31–38.
Zhang L, Shen L, Xu D, et al. 2016. Reprod Toxicol, 65: 236–247.
Zhou J, Zhu C, Luo H, et al. 2019. FASEB J, 33(1): 1110–1123.
Zinkhan E K, Zalla J M, Carpenter J R, et al. 2016. Physiol Rep, 4(13): e12862.

Chapter 15
Fetal Origin of Osteoporosis

Liaobin Chen, Hao Xiao, and Hangyuan He

Abstract Osteoporosis is a systemic bone disease characterized by decreased bone mass and destruction of bone microstructure. Studies have identified the fetal origin of osteoporosis, where the bone status in the early developmental period determines the onset and progression of osteoporosis in adulthood and old age. Intrauterine programming is the process of damage to fetal organs during intrauterine development, which ultimately leads to permanent changes in tissue structure and function. Previous studies by the authors' team have identified the involvement of the bone-localized renin-angiotensin system and the glucocorticoid-insulin-like growth factor 1 axis in the intrauterine programming of fetal-originated osteoporosis. This chapter summarizes the phenomenon and the intrauterine programming mechanisms of fetal-originated osteoporosis to provide the theoretical and experimental basis for the early prevention and treatment of osteoporosis.

1 Introduction

Osteoporosis is a systemic bone disease characterized by decreased bone mass and destruction of bone microstructure, leading to an increased risk of fractures. As a result, osteoporosis and fractures can significantly reduce the quality of life for patients and impose a heavy economic burden on society. In China, authoritative statistics from 2001 indicate that the incidence of osteoporosis in elderly women aged 60 to 69 years was as high as 50% to 70%. The US Department of Health and Human Services estimated that more than ten million women were affected by osteoporosis by 2020 (Durnell et al. 2011). With the development of the etiology for osteoporosis and the theory of the developmental origin of the disease, research has

L. Chen (✉) · H. Xiao · H. He
Division of Joint Surgery and Sports Medicine, Department of Orthopedic Surgery, Zhongnan Hospital of Wuhan University, Wuhan, Hubei, China
e-mail: lbchen@whu.edu.cn

H. Wang et al. (eds.), *Fetal Origin of Diseases*,
https://doi.org/10.1007/978-981-97-5730-5_15

confirmed the relationship between osteoporosis and early life development, which has attracted wide attention.

The theory of the developmental origin of health and disease posits that the intrauterine environment has a lasting and decisive influence on the developmental process of offspring. Nutritional imbalances during critical time windows in early life may permanently influence and "program" the development of fetal-originated disease in later life. Adverse environmental factors during pregnancy have become a predisposing factor for bone-related diseases in children and adults. Numerous studies suggest that the status of bone in early development affects the occurrence and development of osteoporosis in adulthood and old age. Intrauterine programming is a process in which fetal organs are damaged during the intrauterine period, and their organizational structure and function are permanently programmed to change. These changes in tissue and organ function or gene expression are usually maintained from intrauterine to adulthood or even throughout life, leading to a series of changes in adulthood (Wolski et al. 2014). Despite many studies showing that osteoporosis has a fetal origin, the specific mechanism remains unclear. This chapter summarizes the occurrence and intrauterine programming mechanisms of fetal-originated osteoporosis, which is expected to provide the theoretical and experimental basis for the early prevention and treatment of osteoporosis.

2 Pathogenesis of Adult Osteoporosis

Osteoporosis is characterized by decreased bone mass and qualitative changes in bone tissue, leading to increased bone fragility and fracture risk (Minisola et al. 2017). Due to its high incidence worldwide, osteoporosis has become a significant health and economic burden on society. In the United Kingdom, approximately 50% of women and 20% of men over the age of 50 are at risk of osteoporotic fractures (van Staa et al. 2001). In China, about 20.7% of women and 14.4% of men over the age of 50 suffer from osteoporosis (Wang et al. 2009). The systemic effects of osteoporosis on the bone are well documented, with low bone density increasing the risk of almost all types of fractures, especially at the hip, vertebra, and distal forearm (Svedbom et al. 2013). Hip and vertebral fractures are among the most devastating consequences of osteoporosis, causing severe disability and even death, resulting in an increased socioeconomic burden (Svedbom et al. 2013). It is estimated that the cost of treating osteoporotic fractures in the United States will reach $20 billion by 2050, while in the European Union, the figure will reach $30 billion (Wright et al. 2014). The incidence of osteoporotic fractures has been increasing in Western Europe, North America, and Asia in recent years (Lau et al. 2009). Therefore, adult osteoporosis has a high incidence and causes a significant economic burden worldwide.

Osteoporosis is classified into primary and secondary categories based on its pathogenesis. Primary osteoporosis includes postmenopausal osteoporosis (type I), senile osteoporosis (type II), and idiopathic osteoporosis (including juvenile type).

Secondary osteoporosis refers to osteoporosis caused by other diseases, drugs, and factors affecting bone metabolism. The causes and mechanisms of adult osteoporosis are numerous, but the understanding of its pathogenesis is controversial. The most important factor in adult osteoporosis is an endocrine system disorder. Various endocrine diseases can lead to the development of osteoporosis, where the lack or excess of a certain hormone in the human body directly or indirectly affects and regulates the change of bone tissue and bone mass, leading to the occurrence of osteoporosis (Weitzmann 2014). Autoimmune disorders are another significant cause of adult osteoporosis. The link between rheumatoid arthritis, systemic lupus erythematosus, ankylosing spondylitis, and osteoporosis has been well-established. In 2000, the concept of osteoimmunology was introduced, suggesting that the skeletal and immune systems are cross-linked, and immune disorders could lead to abnormal bone metabolism (Arron et al. 2000). T lymphocytes and B lymphocytes could interact with osteoblasts and osteoclasts to jointly regulate the formation and absorption of bone, ultimately leading to the occurrence of osteoporosis. Genetic factors also play a role in the occurrence of adult osteoporosis. Many genes are closely related to the development of osteoporosis in adults, and approximately 80% of the variation in peak bone mass in the population can be explained by genetics. Gene polymorphism could lead to the failure to reach optimal peak bone mass during the growth and development of individuals, ultimately leading to the occurrence of osteoporosis (Mckenna et al. 2004).

3 Evidence for Intrauterine Developmental Origins of Osteoporosis

With the ongoing research into fetal-originated disease, more attention has been given to the influence of early growth and development on adult osteoporosis. During fetal bone development, bone accumulation occurs at a faster rate than during adulthood, providing the possibility of interaction between genetic information and the early environment (Berendsen and Olsen 2015). In recent years, several epidemiological and laboratory studies have demonstrated that adult osteoporosis has an intrauterine origin.

3.1 *Epidemiological Evidence*

The correlation between osteoporosis and early life growth and development has been widely acknowledged in literature. The initial proposition of the link between early life environment and susceptibility to osteoporosis was made in a 1995 epidemiological study by Cooper et al. on 21-year-old women born in Bath, England, between 1968 and 1969 (Cooper et al. 1995). The study showed that there was a statistically significant correlation between 1-year-old female body weight and adult femoral neck bone mineral content, independent of adult body weight and body mass index. Subsequently, Louey et al. conducted an epidemiological survey in 2005 on a cohort study of older adults aged 65–75 years in Hertfordshire, UK (Louey et al. 2005). The study found a positive correlation between birth weight and population bone mineral content and a more significant positive correlation with male bone mineral density. This association persisted even after controlling for adult lifestyle influences such as physical activity, calcium intake, smoking, and alcohol consumption. This finding was later confirmed by epidemiological surveys in the United States, Australia, Sweden, and New Zealand as reported by Harlay in 2007. Other epidemiological studies from mother-offspring have shown that neonatal bone mass was significantly positively correlated with birth weight, birth length, and placental weight after adjusting for postnatal sex and gestational age. Antoniades et al. conducted a cohort study of 445 identical and 966 fraternal female white twins with an average age of 47.5 years and found that bone mineral content and mineral density of adult femoral neck and lumbar vertebra were positively correlated with birth weight (Antoniades et al. 2003). This suggested that different birth weights can affect adult bone mass even with a high degree of genetic uniformity. These epidemiological investigations all indicate that adult osteoporosis has an intrauterine origin.

3.2 Clinical and Basic Laboratory Research

Numerous clinical studies have provided evidence supporting the intrauterine origin of osteoporosis. In a study analyzing the bone mineral content of newborns and their mothers' habits, it was observed that women who smoked during pregnancy had 11% less bone mineral content in their newborns than those who did not. Further research indicated a negative association between maternal smoking during pregnancy and child growth and bone mass, suggesting that exposure at different developmental times was more critical to bone mass than exposure dose or duration (Harvey et al. 2010). Low levels of maternal active vitamin D in late pregnancy led to a decrease in body and lumbar bone mineral content in offspring at 9 years of age (Harvey et al. 2010). Moreover, several laboratory studies demonstrated that adult osteoporosis has an intrauterine origin. For instance, alcohol exposure during pregnancy affected the bone development of fetal rats, especially the ossification process of offspring bones, and the growth plates showed significant changes in histomorphology (Snow et al. 2007). After protein intake restriction of pregnant mice, the bone mineral content of their offspring decreased in adulthood (Mehta et al. 2002). Adult offspring of pregnant rats fed a diet rich in saturated fat had structural abnormalities in the distal femoral trabecular bone and showed characteristics of osteoporosis (Liang et al. 2009). Studies have also shown that changes in maternal estrogen levels during pregnancy can affect the early stages of fetal bone tissue development, leading to permanent bone changes (Migliaccio et al. 1996). In conclusion, the early

Table 15.1 Intrauterine origin of osteoporosis

Study	Risk factors	Results	Main references
Clinical research	Low weight at 1 year old	Decreased mineral content of adult lumbar vertebrae	Cooper et al. (1995)
	Low birth weight	Decreased mineral content of adult lumbar vertebrae	Louey et al. (2005)
	Prenatal alcohol exposure	Fetal maxillofacial bone dysplasia	Muggli et al. (2017)
	Low birth weight	Decreased mineral content of fatal lumbar vertebrae	Harlay (1978)
	Prenatal ethanol exposure	Dysplasia of rat growth plate	Harvey et al. (2010)
	Prenatal low-protein diet	Decreased mineral content of adult lumbar vertebrae	Snow et al. (2007)
	Prenatal caffeine exposure	Low bone mass at different stages	Mehta et al. (2002)
Animal study	Prenatal fumonisin exposure	Low bone mass of rat	Ewa et al. (2021)
	Prenatal bisphenol A exposure	Low bone mass of rat	Wang et al. (2021)
	Prenatal electromagnetic field exposure	Decreased area of primary ossification center in fetal rats	Adem et al. (2020)
	Prenatal cocaine exposure	Low bone mass of rat	Seifert et al. (2019)
	Prenatal alcohol exposure	Dysplasia of the mandible	Isabel et al. (2016)
	Prenatal teratogenic agent exposure	Decreased peak bone mass of rat	Arkady et al. (2016)

life environment has a long-term influence on the later bone development of individuals, and osteoporosis has an intrauterine origin (Table 15.1).

4 Intrauterine Programming Mechanism of Fetal-Originated Osteoporosis

Glucocorticoid (GC) have a crucial role in regulating fetal tissue morphology and functional maturation. However, excessive concentrations of GC can result in abnormal fetal development (Fowden et al. 1998). The authors' team previous studies have indicated that exposure to xenobiotics (such as caffeine, nicotine, and ethanol) during pregnancy can lead to fetal exposure to high levels of maternal GC (Xu et al. 2012; Zhou et al. 2018; Yu et al. 2018). High levels of maternal GC exposure have been associated with susceptibility to fetal-originated disease, including osteoporosis.

4.1 Epigenetic Mechanism and Multigenerational Effect of Bone Development

Epigenetic modification, which involves DNA methylation, histone modification, and noncoding RNAs, plays a crucial role in regulating genome expression and is heritable. Intrauterine programming is significantly impacted by epigenetic modifications. Epigenetic modifications are highly sensitive to adverse intrauterine environments. Exposure to toxic substances during pregnancy can lead to permanent changes in the epigenetic modifications of genes associated with fetal development. However, the mechanism underlying epigenetic modifications induced by adverse intrauterine environments remains unknown. Histones contain numerous amino acids that can regulate various DNA transcription processes through reversible covalent bonds. Adverse intrauterine environments can lead to epigenetic changes in genes related to fetal bone development, which can affect gene expression and bone development in adult offspring. For instance, histone methylation in the promoter region of the insulin-like growth factor 1 (IGF1) gene has been observed in animals with intrauterine growth retardation (IUGR). These epigenetic modification changes are well-correlated with the intrauterine programming of IGF1 and the occurrence of adult metabolic diseases. Low IGF1 caused by adverse intrauterine environments is closely related to the occurrence of adult osteoporosis. Noncoding RNAs, including miRNAs and lncRNAs, play a regulatory role in gene expression at the transcriptional and posttranscriptional levels. Noncoding RNA plays an essential role in the differentiation of osteoblasts. MiR-20a, a member of the miR-17-92 group, is critical for tissue and organ development and is upregulated during the induction of human bone marrow-derived mesenchymal stem cells (BMSCs) into bone cells. High expression of miR-20a significantly promotes the osteogenic differentiation of BMSCs, and the mechanism is related to the Runx2 signaling pathway (Santos 2016).

Multigenerational inheritance refers to the transmission of epigenetic marks and phenotypes to offspring without direct environmental exposure (Skinner 2016). The adverse environment during pregnancy not only leads to fetal-originated osteoporosis but also to the multigenerational inheritance of various fetal-originated disease, including osteoporosis. Studies have shown that a low-protein diet during pregnancy can cause abnormalities in pancreatic development and glucose metabolism in offspring, and similar abnormalities persist in their offspring and grandchildren (Frantz et al. 2011). Zinc deficiency during pregnancy can lead to immune suppression in offspring, passed on to the third generation. Similarly, fetal-originated osteoporosis is inherited in multiple generations (Beach et al. 1982). The authors' team observed long bone dysplasia and decreased peak bone mass in the second and third generations exposed to dexamethasone during pregnancy (Xiao et al. 2018; Han et al. 2022). Epigenetic modification plays a crucial role in the transgenerational inheritance of fetal-originated disease. Environmental abnormalities during pregnancy and early after birth can lead to epigenetic changes in offspring genes that persist into adulthood and even into the next generation (Rando 2016). The multigenerational inheritance of metabolic syndrome phenotypes in the offspring and grandchildren of diabetic pregnant women was also related to the epigenetic modification of oocytes (Ge et al. 2014). The authors' team found a predisposition to osteoporosis in the offspring and great-grandchildren of dexamethasone exposure during pregnancy. Furthermore, an increased expression of miR-98-3p was observed in the oocytes of the offspring and grandoffspring exposed to dexamethasone prenatal. Animal and in vitro experiments confirmed that miR-98-3p mediated transgenerational inheritance of fetal-originated osteoporosis by targeting inhibition of JAG1/Notch1 signaling pathway expression (Han et al. 2022). These results suggest that fetal-originated osteoporosis has multigenerational inheritance and is associated with abnormal epigenetic modification. However, there are limited reports on the multigenerational inheritance of fetal-derived osteoporosis, and the specific cause and mechanism have not been explained in detail, requiring further study.

4.2 The "Two-Programming and Two-Strike" Mechanisms of Fetal-Originated Osteoporosis

The renin-angiotensin system (RAS) plays a crucial role in humoral regulation in the human body, affecting cardiovascular activities and regulating the blood pressure system mainly through paracrine and autocrine modes. Local RAS in bone tissue also regulates bone development and metabolism (Zhang et al. 2014a). Recent studies have shown that adverse factors during pregnancy, such as nutritional restriction, GC exposure, caffeine, and nicotine exposure, can lead to changes in RAS, increasing the susceptibility of offspring to certain diseases after birth. Overactivated local bone RAS has been found to lead to decreased bone formation and mass, which is a significant mechanism of osteoporosis (Zhang et al. 2014b). The authors' team has suggested that RAS is involved in the "first programming" mechanism of fetal-originated osteoporosis. Prenatal exposure to betamethasone has been shown to upregulate angiotensin-converting enzyme (ACE) in fetal sheep's cerebrospinal fluid, suggesting that high-level glucocorticoid induced by prenatal xenobiotic exposure regulates local RAS activity in fetal tissue (Wilson et al. 2014). Studies have also shown that prenatal exposure to xenobiotics, such as caffeine, dexamethasone, ethanol, and nicotine, can cause long bone dysplasia in fetal rats and significantly reduce the peak bone mass of offspring rats after birth, thus increasing their susceptibility to osteoporosis in adulthood (Wu et al. 2020; Xiao et al. 2019; Wen et al. 2019; Shangguan et al. 2018; Xiao et al. 2018; Shangguan et al. 2017). Increased GC levels in fetal blood and activated local bone RAS have been observed in these animal models, mainly represented by high ACE expression, increased Ang II content, and decreased AT2R expression. These observations suggest that adverse environmental factors during pregnancy lead to sustained inhibition of osteogenic differentiation through activation of local bone RAS, ultimately decreasing peak bone mass in offspring rats and increasing their susceptibility to osteoporosis in adulthood. In vitro experiments have confirmed that prenatal exposure to exogenous substances activates the glucocorticoid receptor and its co-transcription factor C/EBP-α, which induces increased histone acetylation in the ACE promoter region of offspring bone tissue through the recruitment of epigenetic enzyme CBP/p300, leading to continuous activation of RAS.

The IGF1 signaling pathway is a fundamental component of the endocrine regulatory system of the body, involved in regulating cell proliferation, differentiation, and metabolism in various tissues (Magner et al. 2013). During the intrauterine period, the level of IGF1 directly determines fetal birth weight, organ structure, and functional development (Magner et al. 2013). Studies have confirmed that the levels of IGF1 in the blood and liver tissues of IUGR fetuses were decreased, while the catch-up growth after birth was often accompanied by an increase in IGF1 levels in the blood and liver tissues (Kamei et al. 2011). Furthermore, high levels of GC have been observed to inhibit the expression of IGF1 in various tissues and cells. Prenatal exposure to xenobiotics such as caffeine, nicotine, and ethanol has been found to increase the level of fetal GC while decreasing the corresponding blood and liver IGF1 levels (Shangguan et al. 2018; Xu et al. 2015; Huang et al. 2015). It has been observed that under a normal diet, the expression of IGF1 in the liver was increased, while the level of GC was decreased after birth. However, the decrease in GC level and the increase of IGF1 expression in the liver were more apparent when the offspring was given a high-fat diet. This negative correlation between GC level and IGF1 expression in liver tissue suggests the existence of an axial relationship between GC level and IGF1 in multiple tissues, mainly the liver, known as the "GC-IGF1 axis." The GC-IGF1 axis is a compensatory regulatory axis mediating the adaptive reduction of fetal metabolic rate under adverse conditions during pregnancy and the catch-up effect after birth. In bone tissue, intrauterine exposure to high GC due to xenobiotics during pregnancy can lead to the inhibition of the local IGF1 signaling pathway in bone, thus affecting osteogenic differentiation. After birth, the level of GC is decreased, and the IGF1 signaling pathway is highly expressed, promoting the early differentiation of osteoblasts and causing the "catch-up growth" of bone length. However, because bone mass is in a relatively underdeveloped state, the catch-up growth of bone eventually leads to an increased susceptibility to osteoporosis in adult offspring (Shangguan et al. 2018).

To summarize, recent studies have led us to propose the "two-programming" mechanism for offspring susceptibility to osteoporosis caused by prenatal exposure to xenobiotics. These mechanisms contribute to an increased susceptibility to osteoporosis in adult offspring and constitute the "first strike" to the development of long bone, providing the potential basis for osteoporosis susceptibility. Subsequently, after birth, the "second strike" treatment may induce and exacerbate the occurrence of osteoporosis.

5 Summary and Prospect

This chapter presents an innovative perspective on fetal-originated osteoporosis induced by prenatal exposure to xenobiotics. The proposed "two-programming and two-strike" mechanism suggests that the "first programming" involves the continuous activation and inhibition of osteogenesis in local RAS in bone caused by maternal glucocorticoid. The "second programming" is the postnatal catch-up growth resulting from intrauterine bone development inhibition caused by long bone GC-IGF1 axis programming. However, this programming cannot fully compensate for the bone dysplasia caused by the "thrifty phenotype" after birth. These "two-programming" constitute the "first strike" against the long bone, which underlies the predisposition of IUGR individuals to osteoporosis after birth. The "second strike" in adulthood, such as the castration-induced aging model, induces and accelerates the onset of osteoporosis (Fig. 15.1).

With the development of genetic technology, the relationship between diseases and genetic inheritance has become clearer. Many chronic diseases that occur in adulthood are believed to have a fetal origin and are closely linked to early-life environmental factors. Traditionally, osteoporosis has been attributed to poor calcium and vitamin D absorption and changes in hormone levels, and treatment strategies have focused on calcium supplementation and hormone replacement therapy in the elderly. However, these treatments are often initiated at a later stage of disease progression and have limited efficacy. Through the study of intrauterine

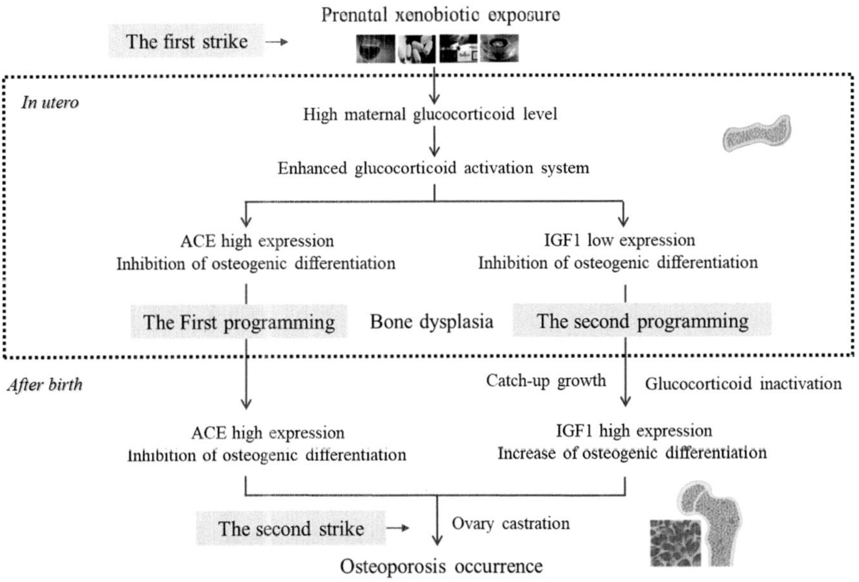

Fig. 15.1 Intrauterine programming mechanisms of fetal-originated osteoporosis in offspring induced by prenatal xenobiotic exposure

programming mechanisms, it has become apparent that the high sensitivity of osteoporosis in the elderly is determined by the intrauterine environment during embryonic development. Furthermore, its mechanism is related to changes in organ development programming and functional homeostasis caused by changes in the fetal endocrine axis and related cell signaling pathways during intrauterine development. Identifying common programming strategies from animal models has become a method for translating basic research results on fetal-originated disease into clinical applications. By assessing adverse environmental factors during pregnancy, primary and secondary prevention of fetal-originated osteoporosis can be carried out. Moreover, by elucidating the intrauterine programming mechanism of fetal-originated disease, biomarkers can be sought, and corresponding measures can be taken at the early stage of the disease to prevent its occurrence. This revolutionary approach to the prevention and treatment of osteoporosis has profound medical significance and great economic value.

References

Adem B, Vieira P F, Melo S A. 2020. Trends Cancer, 6(1): 20–30.
Antoniades L, MacGregor A J, Andrew T, et al. 2003. Rheumatology (Oxford), 42(6): 791–796.
Arkady R J, Schwartz H N, Scully S, et al. 2016. Prog Neuropsychopharmacol Biol Psychiatry, 70: 170–218.
Arron J R, Choi Y. 2000. Nature, 408: 535–536.
Beach R S, Gershwin M E, Hurley L S, et al. 1982. Science, 218(4571): 469–471.
Berendsen AD, Olsen BR. 2015. Bone, 80: 14–18.
Cooper C, Cawley M, Bhalla A, et al. 1995. J Bone Miner Res, 10(6): 940–947.
Durnell S K, Kristi R, Rachel N. 2011. J Midwi Women's Heal, 56(6): 615–627.
Ewa. 2021. Biol Trace Elem Res, 199(12): 4458–4465.
Fowden A L, Li J, Forhead A J.1998. Proc Nutr Soc, 57: 113.
Frantz E D, Aguila M B, Pinheiro-Mulder A R, et al. 2011. Mech Ageing Dev, 132(3): 110–116.
Ge Z J, Zhang C L, Schatten H, et al. 2014. Biol Repord, 90(6): 139.
Han H, Xiao H, Wu Z, et al. 2022. Exp Mol Med, 54(3): 298–308.
Harlay A. 1978. Linfirmiere Francaise, 61: 17.
Harvey N C, Javaid M K, Arden N K, et al. 2010. J Devel Ori Heal Disea, 1(1): 35–41.
Huang H G, He Z, Zhu C Y, et al. 2015. Toxicol Appl Pharmacol, 28(8): 84–94.
Isabel C, Calvet D, Mas J L. 2016. Presse Med, 45(12 Pt 2): e457–e471.
Kamei H, Ding Y, Kajimura S, et al. 2011. Development, 138(4): 777–786.
Lau C S, Mak A. 2009. Nat Rev Rheumatol, 5(7): 400–404.
Liang C, Oest M E, Prater M R, et al. 2009. Birth Defects Res B Dev Reprod Toxicol, 86(5): 377–384.
Louey S, Cock M L, Harding R. 2005. J Reprod Dev, 51(1): 59–68.
Magner N L, Jung Y, Wu J, et al. 2013. Stem cells, 31(10): 295–310.
Mckenna M J, Nguyen-Huynh A T, Kristiansen A G. 2004. Otol & Neurotol, 25: 447–450.
Mehta G, Roach H I, Langley-Evans S, et al. 2002. Calcif Tissue Int, 71(6): 493–498.
Migliaccio S R, Newbold B C, Bullock W J, et al. 1996. Endocrinology, 137(5): 2118–2125.
Minisola S, Cipriani C, Occhiuto M, et al. 2017. Inter Emer Med, (1–2): 1–7.
Muggli E, Matthews H, Penington A, et al. 2017. JAMA Pediatr, 171(8): 771–780.
Rando O J. 2016. Cold Spring Harbor Perspect Med, 6(5): 222–233.
Santos M C M M. 2016. Development, 22(3): 223–245.

Seifert O, Baerwald C. 2019. Best Pract Res Clin Rheumatol, 33(3): 101435.
Shangguan Y, Jiang H, Pan Z, et al. 2017. Cell Death Dis, 8(10): 157–168.
Shangguan Y F, Wen Y X, Tan Y, et al. 2018. Am J Pathol, 9(12): 158–188.
Skinner M K. 2016. Nat Rev Endocril, 12(2):68–70.
Snow M E, Keiver K. 2007. Bone, 41(2): 181–187.
Svedbom A, Hernlund E, Ivergard M, et al. 2013. Arch Osteop, 4(2): 8–18.
van Staa T P, Leufkens H G, Cooper C. 2001. J Bone Miner Res, 16(3): 581–588.
Wang Y, Tao Y, Hyman ME, et al. 2009. Osteoporos Int, 20(10): 1651–62.
Wang T, Xia X, Yeung W S B. 2021. Fertil Steril, 115(4): 907–908.
Weitzmann M N. 2014. Curr Opin Endocrinol, 21(6): 461–467.
Wen Y X, Shangguan Y F, Pan Z Q, et al. 2019. Toxicol Appl Pharmacol, 363: 1–10.
Wilson B, Marshall A, Pirro N, et al. 2014. FASEB J 28(8): 422–438.
Wolski H, Barlik M, Pieńkowski W, et al. 2014. Archi Perinatal Med, 20(4): 224–228.
Wright N C, Looker A C, Saag K G, et al. 2014. J Bone Miner Metab, 29(11): 520–526.
Wu Z X, Pan Z Q, Wen Y X, et al. 2020. Food Chem Toxicol, 136: 111083.
Xiao H, Wen Y X, Pan Z Q, et al. 2019. FASEB J, 33(11): 12972–12982.
Xiao H, Wen Y, Pan Z, et al. 2018. Cell Death Dis, 9(6): 638–641.
Xu D, Bai J, Zhang L, et al. 2015. Toxicol Res, 4(1): 112–120.
Xu D, Zhang B J, Liang G, et al. 2012. PloS One, 7(9): e44497.
Yu L T, Zhou J, Zhang G H, et al. 2018. Toxicol Appl Pharmacol, 352: 77–86.
Zhang Y, Wang K, Song Q, et al. 2014a. Molecul Med Rep, 9(9): 128–134.
Zhang Y, Wang K, Zheng J, et al. 2014b. Endocrine, 47(2): 598–611.
Zhou J, Liu F, Yu L, et al. 2018. Toxicol Appl Pharmacol, 344: 1–12.

Chapter 16
Fetal Origin of Osteoarthritis

Liaobin Chen, Yang Tan, and Qingxian Li

Abstract Traditionally, osteoarthritis has been viewed as an age-related degenerative disease primarily affecting the elderly population. However, recent studies have suggested that osteoarthritis has a fetal origin and is associated with metabolic syndrome. Low birth weight infants who experienced fetal chondrodysplasia during the intrauterine period due to maternal exposure to adverse environmental conditions showed an increased incidence of osteoarthritis in adulthood, which was further exacerbated by external stimulation after birth. The adverse environment during pregnancy may cause programming changes in cartilage matrix metabolism, glycolipid metabolic function, and subchondral bone osteogenesis hypofunction through high intrauterine maternal glucocorticoids, leading to persistent reduction in articular cartilage quality and susceptibility to adult osteoarthritis even over multiple generations. These programming changes involve abnormal epigenetic modifications. Thus, the proposed fetal developmental origin of osteoarthritis and its intrauterine programming mechanism provide a theoretical and experimental basis for the early comprehensive prevention and treatment of fetal-originated osteoarthritis. This highlights the importance of addressing the intrauterine environment to prevent the onset of osteoarthritis in later life.

1 Introduction

Osteoarthritis (OA) is a chronic joint disease that causes degenerative lesions of articular cartilage, leading to pain and disability. In China, the prevalence of OA is around 10% and increases with age, with up to 50% of individuals over the age of 60 affected. In the USA, there are over 40 million patients with OA, and nearly 500,000 people undergo joint replacement surgery annually due to this disease,

L. Chen (✉) · Y. Tan · Q. Li
Division of Joint Surgery and Sports Medicine, Department of Orthopedic Surgery, Zhongnan Hospital of Wuhan University, Wuhan, Hubei, China
e-mail: lbchen@whu.edu.cn

H. Wang et al. (eds.), *Fetal Origin of Diseases*,
https://doi.org/10.1007/978-981-97-5730-5_16

which poses a considerable socioeconomic burden and health risk for middle-aged and elderly individuals (Krasnokutsky et al. 2008). While OA has traditionally been considered a degenerative disease of the elderly, recent academic opinion suggests that it also belongs to the category of metabolic syndrome (MS). Barker (2003) proposed the "Developmental Origins of Health and Disease (DOHaD)" theory, suggesting that nutritional imbalances during critical early life windows may permanently influence development and disease onset later in life (Goldenberg et al. 2008). Epidemiological surveys and retrospective clinical analyses have demonstrated a strong association between OA and low birth weight (Hussain et al. 2017), indicating that OA may also have an intrauterine developmental origin. In the last decade, scholars have increasingly focused on the etiology and intrauterine programming mechanisms of fetal-originated OA through epidemiological and laboratory studies. This chapter reviews the progress of research on fetal-originated OA, taking into account international retrospective studies and the series of experimental studies on developmental origin diseases conducted by the authors' team in recent years. The analysis of causative factors and pathogenesis of fetal-originated OA can provide a theoretical basis for exploring early diagnosis and corrective treatment strategies for fetal-originated OA.

2 Current Research Status of OA

The impact of OA on public health has become increasingly significant with socioeconomic development and population aging. Numerous large-scale epidemiological surveys and laboratory studies have been conducted to investigate OA, revealing a variety of risk factors that contribute to its progression.

2.1 Current Status of the Development of OA

OA is a prevalent chronic joint disease that commonly affects middle-aged and elderly individuals, characterized by degenerative changes in chondrocytes and cartilage matrix, reactive hyperplasia of joint margins and subchondral bone, and eventual structural damage and dysfunction of joint cartilage, leading to joint pain, stiffness, swelling, activity limitation, and deformity. Globally, approximately 250 million people suffer from OA, making it the 12th leading cause of years lived with disability (YLD), with a significant increase in global YLD due to OA by 31.5% from 2006 to 2016 (Liu et al. 2018). In 2018, 249,000 total knee replacement surgeries were performed in China, costing about 12.948 billion yuan, highlighting the significant burden of OA on patients, their families, and society. The prevalence of knee OA varies based on classification, age, geography, and gender, with numerous epidemiological studies demonstrating an increase in prevalence with age (Fu et al.

2022). In Johnston, USA, a survey demonstrated a higher prevalence of imaging knee OA than symptomatic knee OA, with geographic differences observed (Jordan et al. 2007). Similarly, a population-based survey conducted in a community in Beijing, China, revealed that the prevalence of imaging lateral knee OA was more than twice as high as that studied in the Framingham, USA, area for OA (Felson et al. 2002). Furthermore, significant gender differences in the prevalence of knee OA have been reported in various population-based surveys conducted in Beijing, Inner Mongolia, Heilongjiang, and Korea, where the prevalence of imaging knee OA was found to be 42.8% in women and 21.5% in men (Lee and Kim 2017). A survey of OA in 1030 individuals aged 50 years or older in eight villages in the Wuchuan region of Inner Mongolia also found a higher prevalence of knee OA in women than in men. A study on the prevalence of knee OA in China from 2010 to 2017 reported a higher annual standardized prevalence in women than in men, with the prevalence of knee OA increasing with age, particularly after the age of 55 years, including 9.4% in men and 18.0% in women. The study also revealed a significant increase in prevalence among younger populations under 35 years of age, indicating a trend toward a younger age of OA (Chen et al. 2021).

2.2 *Main Mechanisms of OA*

Articular cartilage is primarily formed during the embryonic period and serves a crucial role in facilitating painless joint movement and restoring joint flexibility. Following birth, articular cartilage growth is primarily sustained through matrix synthesis, which is generally permanent and undergoes minimal cell proliferation and apoptosis. Chondrocytes are responsible for producing matrix components, including type II collagen, aggregated proteoglycans, and other proteoglycans, which provide tensile strength and cushioning capacity to the cartilage matrix. Adult chondrocytes are metabolically active and carry out functions such as glucose transport and cholesterol efflux, which are linked not only to normal cartilage self-stability but also to the pathophysiology of OA. Overall, pathological changes in biochemical molecules and cytokines in the three parts of the joint joints, including the cartilage, subchondral bone, and synovium, are significant contributors to the development of OA.

The pathogenesis of adult OA is multifactorial and not yet fully understood. The mechanisms underlying its development primarily include ①cartilage lesions: on the one hand, cytokine cascades and the production of inflammatory mediators disrupt the balance between intra- and extracellular matrix synthesis and degradation in cartilage. For instance, increased production of inflammatory cytokines such as IL-1β and TNFα can lead to decreased collagen anabolism and increased catabolism (including matrix metalloproteinases (MMPs)) as well as elevated levels of other inflammatory mediators (e.g., IL-8, IL-6, PGE2, and NO), thereby increasing cartilage susceptibility to oxidative damage. On the other hand, reactive oxygen species

(ROS) directly promote chondrocyte apoptosis, catabolism, and matrix degradation. ②Subchondral bone remodeling: OA patients with subchondral bone clearly exhibit osteosclerosis, characterized by graded and uneven increases in subchondral bone stiffness, subchondral bone remodeling, increased bone conversion, osteophytes, and even bone redundancy. ③Synovial lesions: active synovial inflammation releases proteases and inflammatory cytokines, which may promote adjacent cartilage destruction and accelerate the progression of OA.

3 Evidence for Intrauterine Developmental Origins of OA

Intrauterine growth retardation (IUGR) refers to the limitation of embryonic or fetal growth caused by adverse environmental conditions during pregnancy, resulting in multiple organ dysfunction, growth retardation, and low birth weight. To date, numerous epidemiological investigations and clinical and basic laboratory studies have suggested an intrauterine developmental origin of OA.

3.1 Epidemiological Evidence

Several retrospective cohort studies have reported a positive association between low birth weight and the development of osteochondrosis in different body parts such as hip and lateral knee (Clynes et al. 2014), lumbar spine (Jordan et al. 2005), and clinical hand OA (Sayer et al. 2003). Furthermore, a cohort study by Hussain et al. (2015) revealed that low birth weight was independently associated with a higher incidence of adult hip OA requiring arthroplasty, where individuals with low birth weight had a 2.14 times higher likelihood of undergoing hip replacement for OA in adulthood than those with normal birth weight. Additionally, a cross-sectional survey conducted during the Chinese famine in the mid-twentieth century showed that the prevalence of arthritis was significantly higher in individuals exposed to famine during childhood and fetal life (Xu et al. 2017). These findings suggest that low birth weight and prematurity may increase the risk of developing OA, thus highlighting the developmental origin of OA.

3.2 Clinical and Basic Laboratory Research

Currently, there is limited clinical research exploring the relationship between adverse environmental exposures during pregnancy and the development of OA in offspring. One clinical study by Ghassemi Jahani et al. (2014) suggests that thalidomide use during pregnancy may increase the risk of hip and knee OA in offspring. However, numerous basic experimental studies have demonstrated that cartilage dysplasia may result in abnormal articular mechanical characteristics and articular cartilage matrix composition in adulthood, which is a potential cause of OA. There is a significant association between poor environmental exposure during pregnancy and cartilage dysplasia, which may increase the susceptibility of offspring to OA.

Hong et al. (2017) conducted an animal study that revealed that oral administration of nano-titanium dioxide to pregnant mice caused placental lesion and cartilage dysplasia in fetal mice due to the transfer of the nanoparticle across the placental barrier, which led to a significant reduction of Ca^{2+} and Zn^{2+} in the mother, placenta, and fetus, resulting in developmental delays. Similarly, exposure to chloroacetonitrile in fetal rats caused skeletal dysplasia, low birth weight, delayed ossification of cartilage, and destruction of calcified areas (Ahmed et al. 2008). These findings demonstrate that prenatal adverse factors can lead to cartilage dysplasia in offspring with IUGR. Multiple animal studies conducted by the authors' team indicate that prenatal dietary restriction or exposure to foreign substances, such as caffeine, nicotine, and ethanol, during pregnancy can result in decreased extracellular matrix synthesis in IUGR offspring, poor articular cartilage quality, and increased susceptibility to OA in adulthood (Tan et al. 2016; Tie et al. 2016a; Tie et al. 2016b; Luo et al. 2015; Ni et al. 2015; Liu et al. 2012). Tan et al. (2012) found that prenatal caffeine exposure can reduce the synthesis of extracellular matrix in fetal cartilage, delay growth plate growth, and differentiate growth plate chondrocytes into calcified cells, leading to shortened length of fetal bone. Similarly, prenatal nicotine exposure can directly induce low differentiation of bone mesenchymal stem cells (BMSCs) into chondrocytes (Deng et al. 2013) and reduce cartilage matrix synthesis in fetal growth plates of male offspring, leading to cartilage dysplasia (Deng et al. 2021). Moreover, prenatal ethanol exposure can directly affect the differentiation of BMSCs into articular cartilage in offspring, resulting in reduced tolerance of articular cartilage to external stimulation in adulthood due to the inhibitory effect that can continue into adulthood (Ni et al. 2018).

In vitro studies have demonstrated that nicotine can directly reduce extracellular matrix synthesis by acting on chondrocytes of the finger growth plates of human infants and young children through neuronal nicotinic acetylcholine receptors (Kawakita et al. 2008). Additionally, our team has shown that nicotine can dose-dependently reduce chondrogenic differentiation and the expression of aggrecan (Deng et al. 2012). Moreover, by constructing a three-dimensional culture model of BMSC differentiation into chondrocytes in vitro, it has been observed that ethanol can directly inhibit the differentiation of BMSCs into articular cartilage (Ni et al. 2018). These studies suggest that exposure to foreign substances during pregnancy can lead to intrauterine cartilage dysplasia and have long-lasting effects on articular cartilage function.

4 Intrauterine Programming Mechanism of Fetal-Originated OA

The phenomenon of prenatal adverse factors leading to cartilage dysplasia that persists after birth is similar to the concept of "programming" disease development (Moisiadis and Matthews 2014). Adverse environmental exposures during pregnancy can "program" developmental cartilage through various mechanisms, including epigenetic modifications. This intrauterine programming can have lifelong effects on offspring and can be stably inherited for several generations, known as transgenerational genetic effects (Zhao et al. 2020; Xie et al. 2018a). Therefore, understanding the mechanisms underlying intrauterine programming of cartilage dysplasia due to adverse environmental exposures during pregnancy is crucial for developing early prevention and treatment strategies for fetal-originated OA.

4.1 Epigenetic Mechanism and Multigenerational Effect of Cartilage Development

Epigenetic modification is a common gene regulation phenomenon that plays a critical role in embryonic development and the maintenance of normal mammalian life activities. DNA methylation, histone modification, and noncoding RNA are the primary modes of epigenetic modifications, which work together to regulate gene expression and are susceptible to the influence of various environmental factors. Abnormal epigenetic modifications during embryonic development can result in abnormal embryonic development and multiple diseases in adulthood. It is generally believed that the embryonic origin of adult diseases is mainly related to DNA methylation and histone modification. Research has shown that abnormal DNA methylation and histone acetylation modifications in the promoter region of the insulin-like growth factor 1 (IGF1) gene are present in IUGR animals, and these epigenetic modifications are closely related to the intrauterine programming of IGF1 and the susceptibility to metabolic diseases in adult offspring (Regan et al. 2010). Intrauterine and early postnatal environments can have an impact on DNA methylation and histone acetylation of particular gene promoters, which can persist throughout the offspring's life and be transmitted to the next generation (Heard and Martienssen 2014). Previous research has demonstrated that fetal exposure to caffeine can lead to abnormal histone modification and reduced expression of the IGF1 gene in the liver, as well as delayed functional development of growth plate cartilage due to decreased expression of IGF1 downstream signaling pathways (Tan et al. 2012). Similarly, prenatal nicotine exposure has been found to reduce cartilage matrix synthesis in male offspring's fetal growth plates through the Snail family transcriptional repressor 1/histone deacetylase 1/2 (Snail/HDAC1/2) pathway, which results in decreased H3K9ac and H3K14ac levels in aggrecan (ACAN) and collagen type II alpha 1 chain (Col2A1) promoter regions of cartilage (Deng et al. 2021). Furthermore, nicotine exposure during pregnancy can lead to a decrease in the H3K9ac level in the promoter region of the TGFβ signaling pathway in the articular cartilage of both F1 and F2 female adult rats, resulting in poor cartilage quality (Xie et al. 2018). Despite the transient nature of epigenetic modifications, a few individuals may harbor stable epigenetic markers across generations. However, most intrauterine programming-related diseases do not persist beyond the F2 generation (Byrnes et al. 2013). Our recent investigation has demonstrated that prenatal caffeine exposure can lead to articular cartilage pathological lesions in male offspring rats of F1–F3 generations. This condition is associated with reduced expressions of TGFβR1, SOX9, and COL2A1 genes in offspring cartilage of multiple generations, which is caused by the decreased H3K9ac level in the promoter regions of these genes (Zhao et al. 2020). Similarly, prenatal dexamethasone exposure can inhibit the expression of miRNA-92a-3p in oocytes of F1 and F2 offspring and cartilage of F1–F3 offspring. This inhibition results in increased expression of HDAC2 in offspring cartilage of multiple generations and downregulated H3K9ac levels in the promoter region of the transforming growth factor beta 1 (TGFβ) signaling pathway and its expression, leading to poor cartilage quality in multiple generations (Tie et al. 2022). Based on the aforementioned findings, we postulate that osteoarthritis has intrauterine origins, and its etiology may be linked to anomalous epigenetic modification and low expression programming of IGF1 and TGFβ signaling pathways caused by prenatal exposure to harmful exogenous substances.

4.2 The "Two-Programming and Two-Strike" Mechanisms of Fetal-Originated OA

Fetal-originated OA is a localized manifestation of the impact of adverse prenatal factors on the overall health of the offspring. Although the inducements of fetal-originated OA occur prenatally, the pathological symptoms only emerge postnatally. As such, the onset of fetal-originated OA is linked to both the "local" and "whole body" effects, as well as the "prenatal" and "postnatal" factors. In light of this "space-time relationship," we propose a "two-programming and two-strike" mechanism that underlies the onset of fetal-originated OA (Fig. 16.1).

The "two-programming" mechanism, which is mediated by intrauterine overexposure to maternal-derived glucocorticoids (GC), has been found to be involved in the programming alterations of multiple organs in offspring, including the susceptibility to fetal-originated OA. The first programming is the inhibitory development programming, which occurs due to high maternal-derived GC levels in the intrauterine environment caused by prenatal adverse factors. This inhibits the expression of

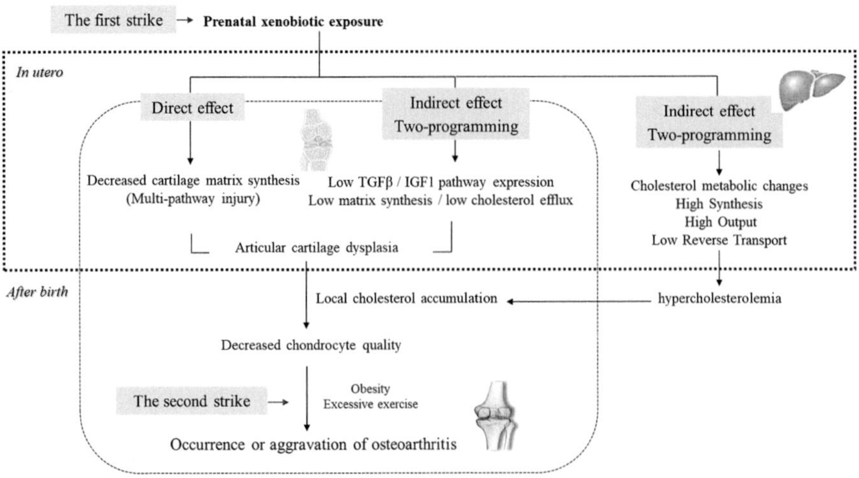

Fig. 16.1 Intrauterine programming mechanisms of osteoarthritis in offspring induced by prenatal exogenous substance exposure

the TGFβ signaling pathway in cartilage and the differentiation of subchondral osteoblasts in offspring, leading to dysplasia of cartilage and subchondral bone (Li et al. 2020; Xie et al. 2018; Xie et al. 2020). The second programming is the hepatic and cartilaginous "GC-insulin-like growth factor 1 (GC-IGF1) axis" programming. In utero, overexposure to maternal-derived GC induced by prenatal adverse factors can inhibit the expression of IGF1 and glucose and lipid metabolism genes in the fetal liver. After birth, the expression of the IGF1 signaling pathway and its downstream gluconeogenesis and lipid synthesis genes increases in the liver of offspring separated from the intrauterine high GC environment, resulting in their gradual increase in blood glucose and lipid levels (Wang et al. 2014; Shen et al. 2014; Hu et al. 2019). Moreover, intrauterine overexposure to maternal-derived GC can induce "GC-IGF1 axis" programming in cartilage, leading to increased glucose transporter 1 (Glut1)-dependent glucose intake and liver X receptor (LXR)-related cholesterol efflux disorder in offspring cartilage (Ni et al. 2015a, b; Li et al. 2020; Tie et al. 2016a). The abnormal glucose and lipid metabolism in the liver and cartilage caused by the "GC-IGF1 axis" programming can result in cartilaginous advanced glycation end products (AGEs) and cholesterol accumulation in offspring. In summary, cartilage dysplasia caused by "the first programming" and cartilage glucose and lipid accumulation induced by "the second programming" can jointly lead to poor cartilage quality in offspring. The "two-programming" mechanism's interaction between the liver, blood circulation, and cartilage reflects the relationship between the "local" and the "whole body" in the onset of fetal-originated OA.

The "two-strike" mechanism reflects the connection between "prenatal" and "postnatal" factors in the onset of fetal-originated OA. The "first strike" refers to prenatal exposure to various adverse factors, which is the fundamental cause of inducing fetal-originated OA. The "second strike" refers to various OA risk factors that offspring are exposed to after birth, which act as the "fuse" for the onset of fetal-originated OA. High-fat diets, long-distance running, chronic stress, ovariectomy, and intra-articular papain injection simulate hypernutrition, excessive exercise, long-term high stress, aging, and chemical stimulation in real society, respectively. Numerous studies have demonstrated that postnatal exposure to these adverse factors can further aggravate poor cartilage quality in offspring exposed to adverse environments during pregnancy, leading to the onset of OA. However, when not exposed to the above risk factors after birth, offspring exposed to adverse factors during pregnancy will not exhibit a typical OA phenotype (Luo et al. 2015; Ni et al. 2015a; Xiao et al. 2020a; Li et al. 2020; Tan et al. 2016; Ni et al. 2015b). In summary, the second strike can transform the fetal-originated OA susceptibility state created by the first strike into disease onset.

In summary, prenatal adverse factors, as the first strike, can induce "two-programming" related to maternal-derived GC, leading to poor cartilage quality in offspring through multiple targets, and postnatal adverse factors, as the second strike, promote the onset of OA. Additionally, prenatal adverse factors may directly inhibit cartilage development (Tie et al. 2016b; Ni et al. 2018) or downregulate local 11β-hydroxysteroid dehydrogenase 2 (11β-HSD2), opening the cartilage and subchondral osteogenic glucocorticoid barrier and exacerbating the effect of maternal-derived GC (Shi et al. 2022; Xiao et al. 2020b). Therefore, the mechanism of fetal-originated OA is characterized by multiple pathways and targets.

5 Summary and Prospect

In conclusion, adverse prenatal environments have a significant impact on fetal cartilage development, leading to cartilage matrix synthesis disorder, chondrocyte dysplasia, and increased susceptibility to OA after birth. This is mainly related to the "two-programming and two-strike" mechanism mediated by intrauterine overexposure to maternal-derived GC. Although some progress has been made in OA treatment, therapeutic methods to delay its occurrence and development are limited. Therefore, in-depth research on the fetal origins of OA and its epigenetic programming mechanism provides the possibility of intervening in the progression of fetal-originated OA from the molecular programming perspective. Improving living habits to reduce or avoid the "two-strike" after birth, such as avoiding high-fat diets and excessive exercise, can also prevent, delay, or even treat fetal-originated OA. Moreover, early warning technology development is essential for early intervention in fetal-originated OA. Our team has recently established an in vitro cell culture model based on the "two-strike" theory, suggesting that decreased H3K9ac levels of TGFβR1 in human umbilical cord mesenchymal stem cells may be a potential predictive biomarker for neonatal cartilage dysplasia and OA susceptibility (Qi et al. 2021). To establish an early-warning marker screening system that is stable, reliable, and simple, it is necessary to further promote relevant research in this area.

References

Ahmed A E, El-Mazar H M, Nagy A A, et al. 2008. Toxicol Ind Health, 24(8): 511–518.
Barker D J. 2003. Eur J Epidemiol, 18(8): 733–736.
Byrnes J J, Johnson N L, Carini L M, et al. 2013. Psychopharmacology (Berl), 227(2): 263–272.
Clynes MA, Parsons C, Edwards MH, et al. 2014. J Dev Orig Health Dis, 5: 453–458.
Chen H B, Wu J H, Wang Z J, et al. 2021. Int J Environ Res Public Health, 23;18(16): 8864.
Deng Y, Cao H, Cu F, et al. 2013. Toxicol Appl Pharmacol, 269(1): 25–33.
Deng Y, Gao H, Hui Wang H, Liaobin Chen L B. 2021. Toxicology, 459: 152847.
Deng Y, Li T Q, Yan Y E, et al. 2012. Biomed Mater Eng, 22(1–3): 81–87.
Felson D T, Nevitt M C, Zhang Y, et al. 2002. Arthritis Rheum, 46(5): 1217–1222.
Fu M, Zhou H M, Li Y S, et al., 2022. Arthritis Res Ther, 24(1): 8.
Goldenberg R L, Culhane J F, Iams J D, et al. 2008. Lancet, 371(9606): 75–84.
Ghassemi Jahani SA, Danielson B, Karlsson J, et al. 2014. J Child Orthop, 8: 423–433.
Heard E, and Martienssen R A. 2014. Cell, 157(1): 95–109.
Hong F, Zhou Y, Zhao X, et al. 2017. Int J Nanomedicine, 12: 6197–6204.
Hussain S M, Wang Y, Wluka A E, et al. 2015. Arthritis Care Res (Hoboken), 67(4): 502–508.
Hussain S., Ackerman I., Wang Y., Zomer E., Cicuttini F.M. 2017. Osteoarthritis Cartilage 25, S202–203.
Hu S W, Qin J, Zhou J, et al. 2019. Toxicol Appl Pharmacol, 375: 46–56.
Jordan J M, Helmick C G, Renner J B, et al. 2007. J Rheumatol, 34(1): 172–180.
Jordan K M, Syddall H, Dennison E M, et al. 2005. J Rheumatol, 32(4): 678–683.
Kawakita A, Sato K, Makino H, et al. 2008. PLoS One, 3(12): e3945.
Krasnokutsky S, Attur M, Palmer G, et al. 2008. Osteoarthritis Cartilage, 16 Suppl 3: S1–3.
Lee S, and Kim S J. 2017. Int J Rheum Dis, 20(7): 809–817.
Li J, Xiao H, Luo H W, et al. 2020b. Food Chem Toxicol, 140: 111279.
Li Q X, Wang L L, Wang Y Y, et al. 2020a. Pharmacol Res, 151, 104555.
Liu L, Liu F, Kou H, et al. 2012. Toxicol Lett, 214(3): 307–313.
Liu Q, Wang S, Lin J, et al. 2018. Osteoarthritis Cartilage, 26(12): 1636–1642.
Luo H, Li J, Cao H, et al. 2015. Sci Rep, 5: 17746.
Moisiadis V G, and Matthews S G. 2014. Nat Rev Endocrinol, 10(7): 391–402.
Ni Q B, Wang L L, Wu Y P, et al. 2015b. Toxicol Lett, 238(2): 117–125.
Ni Q B, Tan Y, Zhang X R, et al. 2015a. Sci Rep, 5: 14711.
Ni Q B, Lu K H, Li J, et al. 2018. Toxicol Sci, 164: 179–190.
Regan F M, Williams R M, McDonald A, et al. 2010. J Clin Endocrinol Metab, 95(5): 2113–2122.
Sayer A A, Poole J, Cox V, et al. 2003. Arthritis Rheum, 48(4): 1030–1033.
Shen L, Liu Z F, Gong J, et al. 2014. Toxicol Appl Pharmacol, 274: 263–273.
Shi H S, Li B, Gao H, et al. 2022. Ecotoxicol Environ Saf, 239: 113624.
Tan Y, Liu J, Deng Y, et al. 2012. Toxicol Lett, 214(3): 279–287.
Tan Y, Wu Y, Ni Q, et al. 2016. Br J Nutr, 116(8): 1346–1355.
Tie K, Tan Y, Deng Y, et al. 2016a. Reprod Toxicol, 60: 11–20.
Tie K, Zhang X, Tan Y, et al. 2016b. FASEB J, 30(2): 785–797.
Tie K, Zhao Z, Wu Z X, et al. 2022. Biochem Pharmacol, 203: 115196.
Wang L L, Shen L, Ping J, et al. 2014. Toxicol Lett, 224: 311–318.
Xie Z, Zhao Z, Yang X, et al. 2018. Toxicol Appl Pharmacol, 352: 107–118.
Xie X K, Tan Y, Xiao H, et al. 2020. Toxicol Lett, 321: 122–130.
Xu X, Liu L, Xie W, et al. 2017. Medicine (Baltimore), 96(13): e6496.

Xiao H, Xie X K, Wen Y X, et al. 2020a. *Bone*, 133: 115245.
Xiao H, Wu Z X, Li B, et al. 2020b. *Br J Pharmacol*, 177: 4683–4700.
Qi Y J, Li B, Wen Y X, et al. 2021. Stem Cell Res Ther, 12: 163.
Zhao Z, Qin J, Pei LG, et al. 2020. Toxicology, 442: 152533.

Chapter 17
Fetal Origin of Glomerulosclerosis

Ying Ao and Hangyuan He

Abstract The kidney is a highly vulnerable organ during development, and exposure to adverse environmental conditions during pregnancy can result in intrauterine growth retardation, leading to fetal kidney damage and increased susceptibility to adult glomerulosclerosis. This chapter provides a comprehensive summary of the pathophysiology, mechanism, and intrauterine origin of glomerulosclerosis. Furthermore, it reviews the kidney's developmental toxicity and long-term damage caused by adverse environmental conditions during pregnancy, considering both kidney morphology and function development. The chapter also examines the intrauterine epigenetic programming mechanism of fetal-originated glomerulosclerosis. As such, this chapter lays a solid theoretical foundation for the early prevention and treatment of fetal-originated glomerulosclerosis.

1 Introduction

In 1986, Barker proposed the theory of "fetal origin of adult diseases" which suggests that unfavorable environmental conditions during fetal development can lead to permanent changes in body structure and function, increasing the susceptibility to chronic diseases in adulthood (Barker and Osmond 1986). The intrauterine environment can significantly affect the development of the kidney, which is the primary excretory organ of the body. Brenner proposed that adverse environmental conditions during the embryonic period can result in impaired kidney development, decreased nephron number, and increased susceptibility to kidney disease in

Y. Ao (✉)
Department of Pharmacology, School of Basic Medical Science, Wuhan University, Wuhan, Hubei, China
e-mail: yingao@whu.edu.cn

H. He
Division of Joint Surgery and Sports Medicine, Department of Orthopedic Surgery, Zhongnan Hospital of Wuhan University, Wuhan, Hubei, China

H. Wang et al. (eds.), *Fetal Origin of Diseases*,
https://doi.org/10.1007/978-981-97-5730-5_17

individuals with intrauterine growth retardation (IUGR) (Dorey et al. 2014). Glomerulosclerosis is the most common pathological manifestation of glomerular disease. Recent studies have indicated that the occurrence of glomerulosclerosis is closely associated with adverse intrauterine environment and low birth weight of the fetus. This chapter provides a review of the intrauterine origin and programming mechanism of glomerulosclerosis.

2 Current Research Status of Glomerulosclerosis

Glomerulosclerosis is a clinicopathologic syndrome that is characterized by the proliferation of glomerular mesangial cells and the excessive deposition of the extracellular matrix. The primary clinical manifestations of glomerulosclerosis are massive albuminuria and progressive deterioration of renal function. It is a common pathologic pathway and a necessary stage for many chronic kidney diseases to progress to the terminal stage. In China, glomerular sclerosis accounts for approximately 5%–14% of glomerular diseases (Feng et al. 2015) and is one of the primary causes of end-stage kidney disease in Chinese children and young adults (Chen et al. 2015b). Depending on the cause, there are two types of glomerulosclerosis: primary and secondary. Primary glomerulosclerosis accounts for approximately 80% of cases, and the etiology is unknown. Secondary glomerulosclerosis accounts for approximately 20% of cases and is mostly caused by hypertension, diabetes, viral infection, drug toxicity, and gene mutation.

2.1 Pathophysiology of Glomerulosclerosis

Glomerulosclerosis is a disease that involves all cellular components of glomeruli, including endothelial cells, podocytes, mesangial cells, and parietal epithelial cells.

In physiological conditions, a polysaccharide-protein complex covers the surface of glomerular capillary endothelial cells, which plays a protective role for endothelial cells. However, under pathological conditions such as hyperglycemia, this complex is destroyed, leading to endothelial cell damage such as swelling, mitochondrial damage, oxidative stress, and the disappearance of window-like connections between cells (Peng et al. 2016). The damaged endothelial cells release inflammatory cytokines, which stimulate an inflammatory response in the glomeruli (Zheng et al. 2017). Additionally, endothelial cells interact with podocytes through the vascular endothelial growth factor A - endothelial nitric oxide synthase/nitric oxide - endothelin-1 (VEGFA-eNOS/NO-ET-1) axis, which induces and exacerbates podocyte injury (Sol et al. 2020). Since endothelial cells are the first cellular barrier in direct contact with all circulating factors, it has been suggested that endothelial cell injury is the initial step in the development of glomerulosclerosis (Sol et al. 2020).

The primary target of glomerulosclerosis is podocyte injury. Podocytes are terminally differentiated cells with limited proliferation capacity. Injury, loss, and reduction of podocytes caused by various pathogenic factors lead to decreased glomerular filtration barrier function and result in renal impairment, such as albuminuria. Furthermore, leaked urine protein and other components act as pathogenic factors that stimulate the epithelial-mesenchymal transition of podocytes and promote the shedding and reduction of podocytes (Sun et al. 2021).

Under normal conditions, mesangial cells provide structural support to the glomerular capillary network and regulate blood flow due to their high content of skeletal proteins. However, under the influence of high glucose and other pathogenic factors, mesangial cells become activated and transform into fibroblasts, which secrete α-smooth muscle actin and collagen, resulting in the proliferation of extracellular matrix and thickening of the glomerular basement membrane (Patel et al. 2003). Additionally, activated mesangial cells secrete inflammatory cytokines such as interleukin-1 beta, interleukin 6 and fibrogenic factors such as transforming growth factor-β (TGF-β), fibronectin, which mediate oxidative stress and promote the development of glomerulosclerosis (Zhang et al. 2012).

The loss and reduction of podocytes result in the dilation of the exposed glomerular capillary network, which is partially devoid of podocyte support. This leads to the interaction between endothelial cells and the glomerular parietal epithelial cells (PECs), which in turn stimulates the activation of the PECs. Activated PECs are characterized by increased migration, proliferation, and matrix deposition. PECs can migrate along Bowman's capsule to the vascular stalk, invade segments of the glomerular tuft, and promote the deposition of the extracellular matrix (Smeets et al. 2011). However, some studies have indicated that PECs may also express podocyte markers after glomerular injury and potentially contribute to podocyte regeneration (Andeen et al. 2015; Sakamoto et al. 2014).

In summary, under the influence of injury factors, the interactions among podocytes, capillary endothelial cells, mesangial cells, and parietal epithelial cells collectively promote the development of glomerulosclerosis.

2.2 *Main Mechanisms of Glomerulosclerosis*

The occurrence and development of glomerulosclerosis are regulated by multiple signaling pathways.

2.2.1 The TGF-β Pathway

One major signaling pathway that regulates glomerulosclerosis is the TGF-β pathway (Meng et al. 2016). TGF-β binds to transforming growth factor-β receptor 2 (TGFR2) to recruit and activate TGFR1, which induces the phosphorylation of Smad2 and Smad3 proteins and their entry into the nucleus, where they activate the expression of collagen and other target genes. Additionally, the TGF-β/TGFR pathway also promotes target gene synthesis via transforming growth factor-β-activated kinase 1 (TAK1), phosphoinositide 3-kinases/protein kinase B (PI3K/AKT), mitogen-activated protein kinase (MAPK), and other pathways. Studies have shown that both TGF-β and MAPK are activated in renal fibrosis (Hewitson 2009), and blocking the TGF-β, p38, and c-Jun N-terminal kinase pathways inhibits fibroblast generation and collagen deposition, thereby ameliorating the development of glomerulosclerosis (Mathew et al. 2011).

2.2.2 The Wnt/β-Catenin Pathway

The role of the wingless (Wnt)/β-catenin pathway in glomerulosclerosis is still a topic of debate. In adult kidneys, the Wnt/β-catenin pathway is inactive. However, under the influence of injury factors, the Wnt/β-catenin pathway can be activated to participate in tissue repair. Studies have shown that the expression of β-catenin is upregulated in mesangial cells stimulated by tumor necrosis factor (TNF)-α, and knockdown of β-catenin can inhibit TNF-α-induced apoptosis and the production of fibrosis factors (Lin et al. 2017). However, it has also been found that high glucose can promote β-catenin degradation and mediate mesangial cell apoptosis. The activation of the Wnt/β-catenin pathway can alleviate TGF-β-induced glomerulosclerosis (Ho et al. 2016). Therefore, the role of the Wnt/β-catenin pathway in glomerulosclerosis requires further investigation.

2.2.3 The EGFR Pathway

The epidermal growth factor receptor (EGFR) belongs to the tyrosine-protein kinase superfamily, and its ligands include EGF and TGF-α. Upon activation by ligands, EGFR undergoes self-phosphorylation of tyrosine residues, which activates downstream signaling pathways such as transcription and activation of extracellular regulated protein kinases (ERK). Injury factors such as high fat can induce EGFR pathway activation and stimulate mesangial cell proliferation (Mishra et al. 2002). Inhibition of the EGFR pathway can block aldosterone-induced glomerular extracellular matrix expression, cell proliferation, and migration (Sheng et al. 2016). Erlotinib, an EGFR inhibitor, significantly reduces the glomerulosclerosis index and renal impairment in a model of renal damage induced by subtotal nephrectomy (Yamamoto et al. 2018). Therefore, the EGFR pathway plays a role in mediating the occurrence and development of glomerulosclerosis.

2.2.4 The VEGFA/eNOS Pathway

VEGFA is a protein produced and secreted by podocytes. Its receptors, VEGFR1 and VEGFR2, are expressed in glomerular endothelial cells. In physiological conditions, VEGFA can induce the expression of eNOS in endothelial cells, resulting in increased production of NO and inhibition of ET-1 expression. However, under the influence of injury factors, the expression of ET-1 is upregulated, leading to podocyte injury, and activation and proliferation of mesangial cells. Studies have shown that knockdown or overexpression of VEGFA in podocytes can cause endothelial cell swelling, necrosis, and capillary network obstruction (Eremina et al. 2008) and that eNOS knockout can lead to injury of podocytes and endothelial cells, resulting in albuminuria and glomerulosclerosis (Yuen et al. 2012). Therefore, the VEGFA/eNOS pathway plays a crucial role in the interaction between podocytes and endothelial cells, and an imbalance in its expression may contribute to glomerular injury.

2.2.5 Others

Glomerulosclerosis is a multifaceted disease where the expression of multiple signaling pathways is implicated in its pathogenesis. The nuclear factor kappa-B pathway activation has been found to promote the production of inflammatory cytokines, mesangial cell activation and proliferation, and fibroblast transformation (Ji et al. 2016). On the other hand, the nuclear factor erythroid 2 related factor 2/antioxidant response element pathway, which plays a critical role in regulating oxidative stress, has been shown to be downregulated in glomerulosclerosis. This downregulation leads to the accumulation of reactive oxygen species and the expression of inflammatory cytokines, ultimately promoting the formation of glomerulosclerosis (Gong et al. 2017).

3 Evidence of the Intrauterine Origin for Glomerulosclerosis

Various adverse environmental factors during pregnancy, such as maternal nutritional deficiency, placental insufficiency, hypoxia, caffeine intake, alcohol consumption, and glucocorticoid and cigarette smoke exposure, have been found to negatively impact intrauterine fetal development, with IUGR being a common developmental toxic outcome. The Brenner hypothesis proposes that IUGR affects fetal kidney development, resulting in a decrease in the number of nephrons and a reduction in kidney volume. Although IUGR offspring experience catch-up growth that restores kidney volume, they do not regenerate nephrons, which leads to glomerular ultrafiltration issues. The permanent loss of nephrons eventually causes the decompensation of glomerular filtration function, which significantly increases the incidence of adult glomerulosclerosis, end-stage renal disease, and albuminuria. These findings suggest that glomerulosclerosis has an intrauterine origin.

3.1 Epidemiological Investigations

Epidemiological studies have demonstrated that adverse environments during pregnancy can significantly increase the risk of developing glomerulosclerosis and renal dysfunction in adulthood. For example, offspring of mothers who experienced the Dutch famine during pregnancy have been found to have a significantly higher risk of developing albuminuria in adulthood (Painter et al. 2005). Maternal deficiency of folic acid, vitamin A, or energy during pregnancy has been associated with abnormal renal structure, proteinuria, and decreased creatinine clearance in offspring (Lee et al. 2018). Maternal alcohol consumption during pregnancy has also been found to increase the risk of kidney damage in adult offspring (Das et al. 2019). Additionally, meta-analyses have shown that low birth weight individuals have a 70% increased risk of renal impairment and chronic kidney disease in adulthood compared to normal birth weight individuals, with this risk being more pronounced in Asian and Australian populations (Silverwood et al. 2013). A Japanese epidemiological study further indicated that low birth weight is a definite cause of glomerulosclerosis and severe albuminuria since the incidence of glomerulosclerosis was significantly higher in low birth weight individuals (37.5%) than in normal birth weight individuals (9.7%) (Ikezumi et al. 2013). These findings suggest that adverse environments during pregnancy and low birth weight are closely associated with glomerulosclerosis and renal dysfunction in adulthood.

3.2 Clinical and Basic Laboratory Research

Numerous clinical and basic studies have conducted comprehensive and in-depth investigations on the effects of various adverse prenatal environments on kidney development, glomerular morphology, and function.

3.2.1 Changes of Glomerular Structure Before Birth

Nephrons are formed during intrauterine development, and adverse environmental factors during pregnancy can negatively impact fetal kidney development, leading to structural damage. Clinical studies have reported that maternal smoking during pregnancy is associated with reduced fetal kidney volume, and smoking volume is negatively correlated with fetal kidney volume (Taal 2011). Several animal studies have also shown that adverse prenatal environments, such as a low-protein diet during pregnancy (Langley-Evans 1999), smoke exposure (Stangenberg et al. 2015), glucocorticoid exposure (Moritz et al. 2011), and uterine artery ligation (Wlodek et al. 2008), result in a decrease in the number of embryonic nephrons. Moreover, neonatal rats exposed to melamine in utero have been found to exhibit abnormal pathological changes such as glomerular congestion, basement membrane thickening, and precocious glomerular development (Chu et al. 2018). In fetal zebrafish, doxorubicin treatment for 48 hours after 9 hours of fertilization has been shown to reduce and irregularize embryonic podocytes, and the proteins of podocyte marker genes nephrin and wilms tumor gene 1 were reduced, indicating injury to podocyte development (Zennaro et al. 2014). Recent studies conducted by the authors' team have shown that exposure to exogenous substances, such as caffeine, nicotine, ethanol, and dexamethasone, during the second and third trimesters of pregnancy can result in the thinning of the renal cortex, a decrease in the number of nephrons, and the emptiness of glomerular balloons in fetal rats. Electron microscopy has revealed the fusion of foot processes, the thickening of the glomerular basement membrane, and the formation of autophagosomes (Ao et al. 2015; Sun et al. 2015; Zhu et al. 2018; Li et al. 2019; He et al. 2019). These findings confirm that the development of glomeruli is sensitive to the intrauterine environment, and adverse environmental factors during pregnancy can cause histological and ultrastructural damage to embryonic glomeruli. Such damage may serve as the developmental origin of glomerulosclerosis and abnormal renal function in adults.

3.2.2 Changes of Glomerular Structure and Function After Birth

The adverse maternal environment during pregnancy can also affect the structure and function of glomeruli after birth. Clinical data has shown that the incidence of glomerulosclerosis and chronic kidney disease in low birth weight adults is significantly higher than in normal weight individuals (Al Salmi and Hannawi 2020). Several animal studies have confirmed similar findings. For instance, offspring rats exposed to insecticide during pregnancy exhibited pathological changes such as renal cortical area congestion, glomerular hyperplasia, and interstitial fibrosis 60 days after birth (Alves et al. 2021). Adult rats exposed to lipopolysaccharide (LPS) in utero showed abnormal renal morphology and function, such as renal interstitial fibrosis and albuminuria (Farias et al. 2020). The kidneys of offspring rats exposed to a low-protein diet in utero showed foot process fusion under electron microscopy, and the mRNA expression of desmin, an early marker of podocyte injury, was significantly increased (Gross et al. 2006). Adult offspring rats exposed to di(2-ethylhexyl)phthalate during the perinatal period showed swelling or disappearance of the foot process, glomerular sclerosis, renal interstitial fibrosis, and other pathological changes (Wei et al. 2012). Our team confirmed that exposure to exogenous substances (such as caffeine, nicotine, ethanol, and dexamethasone) during the second and third trimester of pregnancy resulted in a decreased number of podocytes, foot process fusion, significantly increased glomerular sclerosis index, and elevated albuminuria, creatinine, and urea nitrogen kidney damage in adult offspring (Ao et al. 2015; Sun et al. 2015; Zhu et al. 2018; Li et al. 2019; He et al. 2019; Zhao et al. 2022). Table 17.1 summarized the literature in the past 10 years related to the effects of the adverse prenatal environment on the structure and function of adult glomeruli.

Table 17.1 Intrauterine origin of glomerulosclerosis

Prenatal adverse environment	Species	Phenotype	Reference
Prenatal caffeine exposure	Rat	Renal dysplasia and adult glomerulosclerosis in the F1 female offspring and transgenerational toxicity on kidney to F2 generation	Zhao et al. (2022)
Biological insecticide exposure during pregnancy and lactation	Rat	Presence of cortical congestion and reduction of the subcapsular space, increase of serum urea and creatinine level	Alves et al. (2021)
Gestational diabetes	Rat	Apparent glomerulosclerosis with podocyte loss in adult rats	Liao et al. (2021)
Maternal obesity	Rat	Glomerulosclerosis, renal fibrosis, and oxidative stress in adult rats	Larkin et al. (2021)
Prenatal lipopolysaccharide (LPS) exposure	Rat	Maternal LPS administration increased proteinuria, systolic arterial pressure, and renal collagen deposition in adult offspring	Farias et al. (2020)
Perinatal iron deficiency	Rat	Male offspring exhibited glomerular basement membrane thickening, increased collagen deposition, and glomerular hypertrophy with the second strike of high-salt diet	Woodman et al. (2020)
Maternal hypoxia in late gestation	Mice	Low nephron number in male offspring is associated with low podocyte endowment per glomerulus of the male offspring at postnatal day 21	Gonçalves et al. (2020)
Subchronic exposure to arsenite and fluoride from gestation to puberty	Rat	Disrupted kidney ultrastructure, inhibited the activity of antioxidant enzymes, and increased the generation of malondialdehyde in the adult offspring kidney	Tian et al. (2019)
Maternal high-fat diet	Mice	Increased kidney weight, higher levels of renal triglycerides, and increased expression of oxidative stress, inflammatory, and fibrotic markers, as well as increased albuminuria	Nguyen et al. (2019)
Prenatal ethanol exposure	Rat	Glomerular hyperplasia and glomerulosclerosis of the female adult offspring	He et al. (2019)
Prenatal dexamethasone exposure	Rat	Glomerulosclerosis, elevated levels of serum creatinine and urine protein, ultrastructural damage of podocytes, the reduced expression levels of podocyte marker genes	Li et al. (2019)
Prenatal ethanol exposure	Rat	Glomerulosclerosis as well as interstitial fibrosis	Zhu et al. (2018)
Sleep restriction during pregnancy	Rat	Reduced creatinine clearance and glomerular area, increased macrophage infiltration, collagen deposit	Argeri et al. (2016)

(continued)

Table 17.1 (continued)

Prenatal adverse environment	Species	Phenotype	Reference
Maternal diabetes	Mice	Activation of renal transforming growth factor-β1 (TGFβ1) and collagen type IV expression, increased oxidative stress, and enhanced renal lipid deposition of the offspring with the second strike of high-salt diet	Aliou et al. (2016)
Maternal protein restriction	Rat	Increased glomerular sclerosis index, interstitial fibrosis score, and ratio of ED1-positive cell of the offspring 20 weeks after birth	Yuasa et al. (2016)
Pregnant rats underwent bilateral uterine artery ligation	Rat	Long-term kidney dysfunction, more severe glomerulosclerosis, and greater glomerular diameters in the intrauterine growth retardation offspring at postnatal week 32	Murano et al. (2015)
Prenatal testosterone administration	Rat	Increased proteinuria in adult male offspring rats	Bábíčková et al. (2015)
Maternal nicotine exposure during gestation and lactation	Rat	Induced kidney fibrosis and increased connective tissue growth factor expression of the offspring on postnatal days 7 and 21	Chen et al. (2015)
Maternal smoking exposure	Rat	Pronounced glomerular TGFβ1 and fibrotic marker expression, promotion of a glomerular epithelial-mesenchymal transition-activated process	Block et al. (2015)
Prenatal caffeine exposure	Rat	Glomerulosclerosis as well as interstitial fibrosis of the male adult offspring, accompanied by elevated levels of serum creatinine and urine protein	Ao et al. (2015)
Prenatal nicotine exposure	Rat	Glomerular sclerosis of the adult offspring	Sun et al. (2015)
Maternal vitamin A deficiency		Decreased renal function and number of nephrons	Bhat et al. (2008)
High-salt diet during pregnancy	Ewe	The offspring had 20% fewer glomeruli but larger glomerular radii	Tay et al. (2012)
Perinatal di(2-ethylhexyl) phthalate exposure	Rat	Swelling and disappearance of the podocytic process, glomerular sclerosis, renal interstitial fibrosis	Wei et al. (2012)
Prenatal high-sucrose exposure	Rat	The adult offspring had thicker glomerular basal membrane	Wu et al. (2011)
Prenatally subjected to sodium overload	Rat	Increased 24-h urinary protein, reduced glomerular filtration rate of the adult offspring	Cardoso et al. (2010)
Prenatal dexamethasone exposure on days 15 and 16 of gestation	Rat	Reduction in glomerular number, glomerulosclerosis, and hypertension	Ortiz et al. (2003)
Maternal protein restriction	Rat	Swelling of the podocytic process, podocytes damage	Gross et al. (2006)

4 Intrauterine Programming Mechanism of Fetal-Originated Glomerulosclerosis

Intrauterine programming refers to the phenomenon of persistent alterations in the morphology and physiology of tissues due to disruptions occurring during the early stages of development. These changes are characterized by modifications in the transcriptional activity of genes responsible for tissue and organ function, which tend to endure from embryonic stages to the later stages of life.

4.1 Epigenetic Mechanism of Glomerulosclerosis

Epigenetics pertains to heritable alterations in gene expression that do not entail modifications in the underlying DNA sequence. These modifications play a vital role in the regular development of complex multicellular organisms and represent the principal mechanism underlying intrauterine programming changes. Epigenetic modifications are diverse and include DNA methylation, histone modification, and noncoding RNAs, such as miRNAs.

4.1.1 DNA Methylation

DNA methylation is a chemical process that involves the covalent bonding of a methyl group to specific bases on DNA sequences, catalyzed by DNA methyltransferase (DNMT). This epigenetic modification is a crucial mechanism for regulating gene expression in mammals, particularly during embryonic development. Additionally, an imbalance in DNA methylation regulation has been implicated in the onset of fetal-originated disease. Research has demonstrated that in neonatal mice exposed to in utero smoke, the kidneys show increased levels of DNA methylation, reduced expression of developmental genes such as glial cell line-derived neurotrophic factor (GDNF) and recombinant fibroblast growth factor 3, and stunted kidney development (Stangenberg et al. 2019). Similarly, in neonatal rats subjected to uterine artery ligation during pregnancy, the DNA methylation levels of the p53 gene in the kidney decreased, and p53 expression increased, leading to increased apoptosis (Pham et al. 2003). According to Larkin et al. (2021), adult offspring models with glomerular sclerosis induced by prenatal high-fat diet exhibited increased renal methylation levels and expressions of DNA methyltransferases DNMT1 and DNMT3a. In addition, Wang et al. (2017) found that adult offspring rats exposed to LPS in utero had increased levels of inflammatory factors in their blood, upregulated expressions of DNMT1 and DNMT3B in the kidney, and increased levels of DNA methylation in the renal cortex. Taken together, these studies suggest that DNA methylation plays a crucial role in regulating the occurrence of offspring renal dysplasia, glomerulosclerosis, and kidney injury caused by adverse environmental conditions during pregnancy.

4.1.2 Histone Acetylation

Histone acetylation pertains to the acetylation of histone lysine residues and the subsequent activation of gene transcription under the influence of histone acetyltransferase (HAT). On the other hand, histone deacetylase (HDAC) deacetylates histones, thereby inhibiting gene transcription. The literature has demonstrated that histone acetylation is closely associated with fetal kidney development, and HDAC plays a significant role in regulating the expression of multiple renal development signaling pathways, such as GDNF and Wnt (Rosenberg et al. 2011). Recent studies have also revealed that blockers of HDAC1, 2, 3, 6, and 8 can impede fetal renal proliferation and differentiation (Fuhrmann et al. 2021). However, there are only a few reports on the effect of histone acetylation on fetal-originated renal impairment. One study found that the histone acetylation level/gene expression in the promoter region of angiotensin-converting enzyme (ACE) 1 in adult renal offspring was upregulated by prenatal LPS exposure, and the mechanism may involve the activation and inflammatory response of the renal renin-angiotensin system (RAS) (Wang et al. 2016a). The authors' team conducted a series of studies that found that prenatal exposure to alcohol and dexamethasone resulted in reduced histone acetylation levels and expression in the promoter region of angiotensin II receptor type 2 (AT2R) in the kidneys of offspring from pre-birth to post-birth (Zhu et al. 2018; Li et al. 2018). Additionally, caffeine exposure during pregnancy inhibited the expression of podocyte phenotypic genes by reducing histone acetylation and expression of Kruppel-like factor 4 (KLF4) in offspring kidneys, leading to podocyte developmental toxicity (Zhu et al. 2019). These studies provide evidence that histone acetylation may also be one of the epigenetic mechanisms underlying fetal-originated glomerulosclerosis.

4.1.3 MicroRNAs

MicroRNAs (miRNAs) are a class of endogenous small noncoding RNAs consisting of 19–25 nucleotides. They negatively regulate gene expression at the posttranscriptional level by binding to complementary target sequences in target mRNA. MiRNAs play a crucial role in various biological processes, including organ development. Recent studies have revealed that some miRNAs are highly expressed in the kidney and are closely associated with kidney development and disease, indicating that miRNAs are important regulators in renal physiology and pathology (Zhao et al. 2020). It has been reported that the expression of miR-192 and miR-200 families is upregulated in the kidneys of adult offspring in the model of prenatal low-protein diet, accompanied by epithelial-interstitial trans-differentiation, increased collagen production, glomerulosclerosis, and interstitial fibrosis (Sene et al. 2018). This suggests that miRNAs may be involved in mediating glomerulosclerosis caused by adverse environmental conditions during pregnancy. Recent studies by our team have also found that dexamethasone exposure during pregnancy can upregulate the expression of miR-135a in the kidneys of male offspring rats and inhibit the expression of KLF4, thereby inhibiting the expression of podocyte phenotypic genes and leading to podocyte developmental toxicity.

4.2 Intrauterine Programming Mechanism of RAS of Fetal-Originated Glomerulosclerosis

The synthesis of all components of the RAS within the kidney is crucial for the maintenance of normal fetal kidney development and renal structure and function. Disruption of RAS activity during embryonic development can result in renal dysplasia, including reduced glomerular volume and nephron number, as well as glomerular sclerosis and albuminuria in adulthood (Almeida and Coimbra 2019). Clinical studies have demonstrated that fetal kidneys affected by intrauterine growth retardation exhibit decreased RAS expression, increased apoptosis, and reduced nephron numbers (Wang et al. 2016b). However, exposure to prenatal adverse environments, such as a low-protein diet during pregnancy (Watanabe et al. 2018), anoxia (Walton et al. 2017), or uterine artery ligation (Murano et al. 2015), can lead to increased renal RAS activity in adult offspring, which can be alleviated by treatment with losartan, an AT1R receptor blocker (Watanabe et al. 2018). In the study conducted by the authors' team, it was observed that prenatal exposure to xenobiotics, including caffeine, nicotine, ethanol, and dexamethasone, led to a reduction in the expression of AT1aR and AT2R in the fetal kidney (Ao et al. 2015; Sun et al. 2015; Zhu et al. 2018; Li et al. 2018). Furthermore, it was observed that, after birth, the expression of AT2R continued to decline persistently until adulthood, while the expression of AT1aR and the AT1aR/AT2R ratio were both increased. Since AT2R plays a crucial role in promoting fetal kidney development and safeguarding adult renal function, it is postulated that the low-function programming of AT2R may be involved in mediating the development of fetal kidney dysplasia and adult glomerulosclerosis caused by prenatal exposure to xenobiotics.

In recent years, the study of RAS has revealed the existence of nonclassical RAS pathways in the body. ACE2, an ACE cognate, converts Angi to Ang (1–9) and AngII to Ang (1–7), which then functions through Mas receptors. ACE2 deficiency during pregnancy has been linked to maternal weight gain and fetal growth restriction (Bharadwaj et al. 2011), indicating that the nonclassical RAS pathway may play a role in embryonic development similar to that of the classical RAS pathway. Our team's research found that the mRNA expressions of ACE2 and Mas receptors in fetal kidney tissue were significantly decreased in the group prenatally exposed to alcohol (Zhu et al. 2018), suggesting that the nonclassical RAS pathway may also contribute to mediating the kidney damage in offspring caused by prenatal exposure to adverse environments.

4.3 Intrauterine Programming Mechanism of the GC-IGF1 Axis of Fetal-Originated Glomerulosclerosis

The insulin-like growth factor 1 (IGF1) signaling pathway is a core endocrine regulation system that plays a crucial role in the proliferation, differentiation, development, and metabolism of multi-tissue organs before and after birth. Upon binding to its receptor, IGF1R, IGF1 activates the MAPK/ERK pathway, which regulates cell proliferation and anti-apoptosis. Additionally, IGF1 activates the PI3K/AKT pathway, which regulates cell differentiation and metabolic functions. IGF1 expression has been detected in nearly all embryonic tissues since implantation. Studies have demonstrated that IGF1 plays a critical role in regulating kidney development in rodents, and a decrease in blood IGF1 levels can lead to glomerulosclerosis in adulthood. Conversely, treating IGF1-deficient mice with IGF1 can increase the number of nephrons (Rogers et al. 1999). Mutations in the IGF1 gene have been linked to low birth weight in some individuals, and fetuses or fetal mice with intrauterine growth retardation exhibit reduced levels of IGF1 in cord blood. However, the regulatory mechanism of IGF1 in intrauterine fetal kidney tissue remains unclear.

Glucocorticoids are a crucial factor in regulating the morphological and functional maturation of fetal tissues and are commonly used to promote lung maturation in premature infants (Marciniak et al. 2011). Additionally, glucocorticoids can inhibit the expression of IGF1 (Hu et al. 2020). Previous research by the authors' team found that maternal glucocorticoid overexposure and low IGF1 expression were present in fetal kidney tissue of IUGR fetuses caused by various adverse environmental conditions during pregnancy, which was accompanied by glomerulosclerosis and other kidney damage (He et al. 2019; Chen et al. 2020). IGF1 plays a crucial role in the structural differentiation, multiple organogenesis, and functional development of tissues during embryonic development. Therefore, an intrauterine low IGF1 status may lead to kidney dysplasia (insufficient nephron reserve) in IUGR fetuses. However, the IGF1 signaling pathway expression in the adult kidney was found to be over-activated, leading to catch-up growth and hypertrophy of IUGR offspring after birth under good nutritional conditions, which further exacerbates the structural and functional abnormalities of the kidney (Bach and Hale 2015). The authors' team also observed that the level of glucocorticoids in IUGR offspring decreased continuously after birth, while the expression of the IGF1 signaling pathway in the kidneys was enhanced. Furthermore, the kidneys showed glomerular hypertrophy and increased susceptibility to glomerular sclerosis (He et al. 2019; Muller et al. 2000). Therefore, the influence of intrauterine adverse environments on progeny's kidney may be regulated by "the programming of the glucocorticoid-insulin-like growth factor 1 (GC-IGF1) axis." Specifically, intrauterine maternal exposure to high levels of glucocorticoids before birth led to inhibition of the fetal kidney IGF1 signaling pathway and fetal kidney dysplasia, while postnatal low glucocorticoid levels led to enhanced renal IGF1 signaling and associated susceptibility to fetal kidney disease (He et al. 2019). Thus, changes in "GC-IGF1 axis" programming may be one of the intrauterine programming mechanisms in renal disease.

4.4 The "Two-Programming and Two-Strike" Mechanisms of Fetal-Originated Glomerulosclerosis

Barker and Brenner's hypothesis of fetal-originated metabolic disease suggests that adverse intrauterine environments or low birth weight leads to fetal renal dysplasia, resulting in the ultrafiltration of individual nephrons. After birth, the offspring is susceptible to kidney disease and high blood pressure after a second blow from a poor lifestyle, such as a high-salt, high-fat diet (Jebasingh and Thomas 2022). Previous research conducted by the authors' team has demonstrated that the local IGF1 level in the kidneys of IUGR offspring exposed to alcohol during pregnancy increased, and mesangial cells proliferated significantly after being given a high-fat diet, resulting in a significant decrease in renal filtration function and the occurrence of glomerulosclerosis (He et al. 2019). These findings indicate that when offspring are subjected to a "second shock" after birth, such as a high-fat or high-protein diet, these factors can further affect the structure and function of the kidney, induce and accelerate the occurrence of glomerulosclerosis, and eventually lead to the development of fetal-originated glomerulosclerosis (Chen et al. 2020). Based on the literature and a series of studies conducted by the authors' team, it is proposed that there are "two-programming and two-strike" mechanisms involved in the development of adult glomerulosclerosis in rats born after exposure to exogenous substances during pregnancy (Fig. 17.1). The "first programming" involves the low expression programming of AT2R in renal tissue and inhibition of podocyte differentiation caused by intrauterine exposure to high levels of glucocorticoids. The "second programming" involves the catch-up growth of intrauterine mesangial cells after birth, caused by the development inhibition of glomerular GC-IGF1 axis programming. The catch-up growth of mesangial cells further increases the susceptibility to glomerular sclerosis. These "two-programming" mechanisms constitute the "first strike" to the kidneys, which underlie the postnatal vulnerability of IUGR individuals. The "second strike" occurs in adulthood, such as a high-fat diet model, which induces and accelerates the development of glomerulosclerosis.

5 Summary and Prospect

An increasing body of evidence suggests that the adverse prenatal environment is a crucial factor in the development of IUGR and can lead to kidney dysplasia and adult glomerulosclerosis. Despite this, the intrauterine mechanisms behind fetal-originated glomerulosclerosis remain largely unexplored. Epigenetic mechanisms

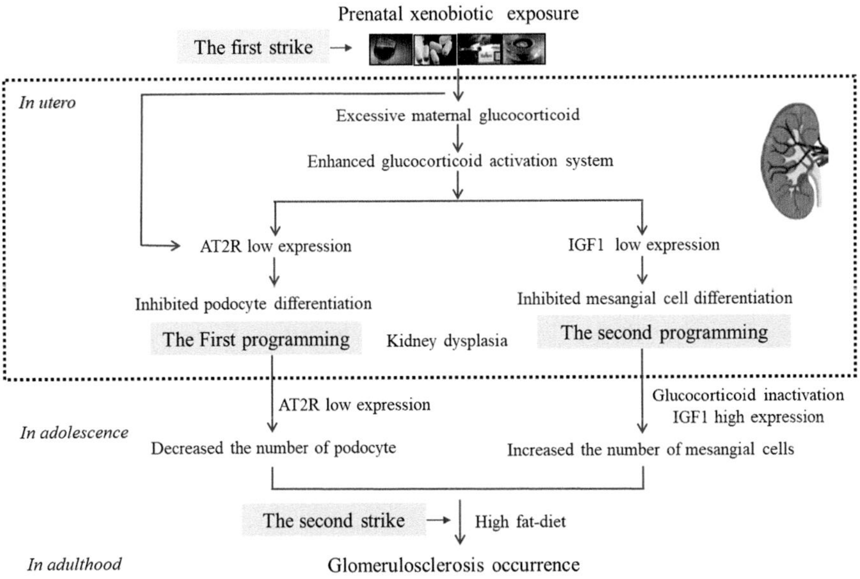

Fig. 17.1 Intrauterine programming mechanism of glomerulosclerosis in offspring induced by prenatal xenobiotic exposure

have been shown to play a crucial role in the occurrence and development of fetal-originated glomerulosclerosis, but the key genes affected and the epigenetic forms regulating kidney development in response to adverse prenatal environments remain unknown. Additionally, it is unclear when the toxic effects of the adverse prenatal environment on kidney development in offspring occur and whether early warning systems or treatment therapies can be established to prevent kidney damage in early life. Addressing these questions will aid in a full understanding of the intrauterine mechanisms of fetal kidney injury and will have far-reaching implications for the guidance of early prevention and treatment of fetal-originated kidney diseases.

References

Aliou Y, Liao M C, Zhao X P, et al. 2016. Pediatr Res, 79(3): 416–24.
Almeida L F, Coimbra T M. 2019. Front Pediatr, 17; 7: 296.
Al Salmi I, Hannawi S. 2020. Saudi J Kidney Dis Transpl, 31(4): 717–726.
Alves RD C, Ferreira C G M, Ferreira de Melo IM, et al. 2021. Acta Histochem, 123(8). 151799.
Andeen N K, Nguyen T Q, Steegh F, et al. 2015. Kidney Int, 88(5): 1099–1107.
Ao Y, Sun Z X, Hu S S, et al. 2015. Toxicol Appl Pharmacol, 287(2): 128–138.
Argeri R, Nishi E E, Volpini R A, et al. 2016. Physiol Rep, 4(16): e12888.
Bábíčková J, Borbélyová V, Tóthová L, et al. 2015. J Urol, 193(5): 1700–1708.
Bach L A, Hale L J. 2015. Am J Kidney Dis, 65(2): 327–336.
Barker D J, Osmond C. 1986. Lancet, 1(8489): 1077–1081.
Bharadwaj M S, Strawn W B, Groban L, et al. 2011. Hypertension, 58(5): 852–858.

Bhat R Y, Sen S. 2008. Saudi Med J, 29(9): 1337–1339.
Block D B, Mesquita F F, Lima I, et al. 2015. Nephron, 129(4): 283–92.
Cardoso H D, Cabral E V, Vieira-Filho L D, et al. 2010. Hypertens Res, 33(1): 76–82.
Chen H, Zhu Y, Zhao X, et al. 2020. Toxicol Lett; 321(5): 44–53.
Chen C M, Chou H C, Huang L T. 2015a. Pediatr Res, 77(1–1): 56–63.
Chen Z, Liu X, Fan J, et al. 2015b. Beijing Med, 37(6): 524–526.
Cheng H W, Chan K Y, Lau H T, et al. 2017. Am J Hosp Palliat Care, 34(4): 380–384.
Chu C, Fung K P, Wang C C. 2018. Environ Toxicol, 33(3): 370–380.
Das S K, McIntyre H D, Alati R, et al. 2019. Nephrology (Carlton), 24(1): 21–27.
Dorey E S, Pantaleon M, Weir K A, et al. 2014. Reproduction, 147(6): R189–198.
Eremina V, Jefferson J A, Kowalewska J, et al. 2008. N Engl J Med, 358 (11): 1129–1136.
Farias J S, Santos K M, Lima N K S, et al. 2020. Arch Biochem Biophys, 684: 108306.
Feng J, Zhang Y, Li Y, et al. 2015. J Shanxi Med Univ, 46(9): 910–913.
Fuhrmann L, Lindner S, Hauser A T, et al. 2021. Int J Mol Sci, 16; 22(8): 4157.
Gonçalves G D, Walton S L, Gazzard S E, et al. 2020. Anat Rec (Hoboken), 303(10): 2668–2678.
Gong W, Li J, Chen Z, et al. 2017. Free Radic Biol Med, 106: 393–405.
Gross M L, Koch A, Muhlbauer B, et al. 2006. Lab Invest, 86(3): 262–274.
He H, Xiong Y, Li B, et al. 2019. Toxicol Lett, 311: 17–26.
Hewitson T D. 2009. Am J Physiol Renal Physiol, 296: F1239–F1244.
Ho C, Hsu Y C, Lei C C, et al. 2016. Am J Med Sci, 351: 286–295.
Hu W, Yuan C, Luo H, et al. 2020. Toxicol Lett, 331.
Ikezumi Y, Suzuki T, Karasawa T, et al. 2013. Am J Nephrol, 38(2): 149–157.
Jebasingh F, Thomas N. 2022. Front Public Health, 21(9): 767545.
Ji X, Li C, Ou Y, et al. 2016. Mol Cell Endocrinol, 437: 268–279.
Langley-Evans SC. 1999. Life Sci, 64(11): 965–974.
Larkin B P, Nguyen L T, Hou M, et al. 2021a. Front Cell Dev Biol, 9: 705263.
Larkin B P, Saad S, Glastras S J, et al. 2021b. PLoS One, 16(3): e0248854.
Lee Y Q, Collins C E, Gordon A, et al. 2018. Nutrients, 10(2): 241.
Li Z, Wang Z. 2018. Adv Exp Med Biol, 1086: 169–187.
Li B, Zhu Y, Chen H, et al. 2019. Toxicology, 411(1): 32–42.
Liao M C, Pang Y C, Chang S Y, et al. 2021. Diabetologia, 64(9): 2108–2121.
Lin X, Zha Y, Zeng X Z, et al. 2017. Mol Med Rep, 15: 3575–3582.
Marciniak B, Patro-Malysza J, Poniedzialek-Czajkowska E, et al. 2011. Curr Pharm Biotechnol, 12(5): 750–757.
Mathew A, Cunard R, Sharma K. 2011. Contrib Nephrol, 170: 217–227.
Meng X M, Nikolic-Paterson D J, Lan H Y. 2016. Nat Rev. Nephrol, 12: 325–338.
Mishra R, Leahy P, Simonson M S. 2002. Am J Physiol Renal Physiol, 283: F1151-F1159.
Moritz K M, De Matteo R, Dodic M, et al. 2011. Am J Physiol Regul Integr Comp Physiol, 301(2): R500-R509.
Muller C, Reddert A, Wassmann S, et al. 2000. J Renin Angiotensin Aldosterone Syst, 1(3): 273–277.
Murano Y, Nishizaki N, Endo A, et al. 2015. Pediatr Res, 78(6): 678–682.
Nguyen L T, Mak C H, Chen H, et al. 2019. Nutrients, 11; 11(1): 146.
Ortiz L A, Quan A, Zarzar F, et al. 2003. Hypertension, 41(2): 328–334.
Painter R C, Roseboom T J, van Montfrans G A, et al. 2005. J Am Soc Nephrol, 16(1): 189–194.
Patel K, Harding P, Haney L B, et al. 2003. J Cell Physiol, 195: 435–445.
Peng H, Li Y, Wang C, et al. 2016. Sci Rep, 6: 20304.
Pham TD, MacLennan NK, Chiu CT, et al. 2003. Am J Physiol Regul Integr Comp Physiol, 285(5): R962-R970.
Rogers S A, Powell-Braxton L, Hammerman M R. 1999. Dev Genet, 24(3–4): 293–298.
Rosenberg S L, Chen S, McLaughlin N, et al. 2011. Pediatr Nephrol, 26(9): 1445–1452.
Sakamoto K, Ueno T, Kobayashi N, et al. 2014. Am J Physiol Renal Physiol, 306(1): F98–104.
Sene L B, Rizzi V H G, Gontijo J A R, et al. 2018. J Exp Biol, 221(Pt 10): jeb171694.

Sheng L, Yang M, Ding W, et al. 2016. Exp Cell Res, 346: 99–110.
Silverwood R J, Pierce M, Hardy R, et al. 2013. Kidney Int, 84: 1262–1270.
Sol M, Kamps J A A M, van den Born J, et al. 2020. Front Pharmacol, 11: 573557.
Stangenberg S, Nguyen L T, Chen H, et al. 2015. Int J Biochem Cell Biol, 64: 81–90.
Stangenberg S, Nguyen L T, Chan Y L, et al. 2019. Clin Exp Pharmacol Physiol, 46(2): 183–193.
Smeets B, Kuppe C, Sicking E-M, et al. 2011. J Am Soc Nephrol, 22(7): 1262–1274.
Sun K, Xie Q, Hao C M. 2021. Kidney Dis (Basel), 7(5): 350–358.
Sun Z, Hu S, Zuo N, et al. 2015. Toxicol Res, 4(4): 1045–1058.
Taal H R. 2011. Pediatr Nephrol, 26(8): 1275–1283.
Tay S H, Blache D, Gregg K, et al. 2012. Animal, 6(11): 1803–1810.
Tian X, Feng J, Dong N, et al. 2019. Sci Total Environ, 686: 1229–1237.
Wang J, Yin N, Deng Y, et al. 2016a. Sci Rep, 6: 39469.
Wang J, Cui J, Chen R, et al. 2017. PLoS One, 12(1): e0169206.
Wang Y P, Chen X, Zhang Z K, et al. 2016b. J Renin Angiotensin Aldosterone Syst, 17(3):1470320316654810.
Walton S L, Bielefeldt-Ohmann H, Singh RR, et al. 2017. Sci Rep, 7(1): 8241.
Watanabe I K M, Jara Z P, Volpini R A, et al. 2018. J Dev Orig Health Dis, 9(5): 530–535.
Wei Z Z, Song L Q, Wei J, et al. 2012. Toxicol Lett, 212(2): 212–221.
Wlodek M E, Westcott K, Siebel A L, et al. 2008. Kidney In, 74(2): 187–195.
Woodman A G, Mah R, Keddie D L, et al. 2020. Cardiovasc Res, 116(1): 183–192.
Wu L, Mao C, Liu Y, et al. 2011. Peptides, 32(1): 104–111.
Yamamoto Y, Iyoda M, Tachibana S, et al. 2018. Nephrol Dial Transplant, 33: 598–606.
Yuasa K, Kondo T, Nagai H, et al. 2016. Congenit Anom (Kyoto), 56(2): 79–85.
Yuen D A, Stead B E, Zhang Y, et al. 2012. J Am Soc Nephrol, 23(11): 1810–1823.
Zhang L, Pang S, Deng B, et al. 2012. Int J Biochem Cell Biol, 44: 629–638.
Zhao Q, Yan T, Chopp M, et al. 2020. J Cereb Blood Flow Metab, 40(2): 246–262.
Zhao X, Li B, Xiong Y, et al. 2022. Food Chem Toxicol, 165: 113082.
Zheng X, Soroush F, Long J, et al. 2017. PloS One, 12(9): e0185250.
Zennaro C, Mariotti M, Carraro M, et al. 2014. PLoS One, 9(5): e98131.
Zhu Y, Zuo N, Li B, et al. 2018. Toxicology, 400–401: 9–19.
Zhu Y, Chen H, Zhao X, et al. 2019. Toxicol Lett, 314(7): 63–74.

Chapter 18
Fetal Origin of Epilepsy

Dan Xu and Tao Jiang

Abstract Epilepsy is a neurological syndrome characterized by abnormally excessive synchronized neuronal electrical activity in the brain, which arises from multiple etiologies and complex pathogenesis. The incidence of epilepsy has increased in recent years, affecting people of all ages and leading to a serious decline in the quality of life for patients. It has been increasingly recognized that epilepsy has an intrauterine developmental origin, with multiple adverse factors during pregnancy leading to developmental impairment of the fetal nervous system and abnormal changes in neuronal ion channels. Furthermore, high-stress sensitivity of the hypothalamic-pituitary-adrenal axis after birth may also contribute to susceptibility to adverse external stresses that can trigger epilepsy. This chapter provides a comprehensive overview of the current pathogenesis and mechanisms of epilepsy, with a particular focus on the evidence of intrauterine origin of epilepsy and the mechanism of intrauterine programming of fetal-originated epilepsy mediated by altered epigenetic modifications. The aim is to explore and summarize early warning markers of offspring susceptibility to epilepsy caused by adverse environmental factors during pregnancy at the molecular level, which provides an important theoretical basis for further early prevention and treatment studies related to fetal-originated epilepsy.

D. Xu (✉)
Department of Pharmacology, School of Pharmaceutical Sciences, Wuhan University, Wuhan, Hubei, China
e-mail: xuyidan70188@whu.edu.cn

T. Jiang
Division of Joint Surgery and Sports Medicine, Department of Orthopedic Surgery, Zhongnan Hospital of Wuhan University, Wuhan, Hubei, China

H. Wang et al. (eds.), *Fetal Origin of Diseases*,
https://doi.org/10.1007/978-981-97-5730-5_18

1 Introduction

Epilepsy is a neurological syndrome characterized by diverse clinical manifestations and multiple etiologies, leading to prolonged course and recurrent seizures that significantly impact patients' quality of life. Epilepsy is highly prevalent in preadolescents, particularly in young children. The age of onset of epilepsy is closely linked to the type of seizures, with infantile spasm signs occurring up to 1 year of age, motor seizures up to 6 years of age, and disoriented seizures commonly observed between 1 and 8 years of age. These findings suggest that altered homeostasis during critical periods of brain development is closely associated with the occurrence and development of epilepsy. The quality of the living environment early in life is a crucial determinant of lifelong health, including the health of offspring. Research has shown that an adverse environment during pregnancy is strongly associated with preterm birth, low birth weight, and intrauterine growth retardation, which in turn can increase susceptibility to neuropsychiatric disorders, metabolic syndrome, and other diseases later in life (Struber et al. 2014). Animal studies have shown that maternal stress or steroid drugs exposure alters the developmental programming of the hypothalamic-pituitary-adrenal (HPA) axis, neuroimmune system function, and neurobehavior in the offspring (Struber et al. 2014; Brunton and Russell 2010). The cellular composition, loop connections, and functional formation of the fetal brain in the developing stage are very different from those of the adult brain. In comparison, the fetal brain in the developing stage is susceptible to maternal nutritional status, acute and chronic stress, and adverse environmental exposures that contribute to postnatal epilepsy susceptibility. This chapter focuses on the effects of adverse environments during pregnancy on offspring epileptic seizures and explores the intrauterine developmental origins of epileptic seizures and associated programming mechanisms.

2 Current Research Status of Epilepsy

Epilepsy is a neurological syndrome characterized by complex clinical manifestations resulting from abnormal cortical neuronal discharges. The incidence of epilepsy is steadily increasing, making it a significant public health concern due to its diverse etiology and complex mechanisms of occurrence.

2.1 Current Status of Epilepsy Morbidity

Seizures are characterized by generalized or focal paroxysmal episodes resulting from abnormally excessive synchronization of neuronal discharges in the brain. The occurrence of seizures is a complex pathophysiological process involving multiple neurobiological factors, and any factor affecting this process may lead to epilepsy. According to statistics, the overall prevalence of epilepsy is 61.44 per 100,000, with a higher incidence and lifetime prevalence in low- and middle-income countries (Fiest et al. 2017). Of these cases, 30–40% of patients with epilepsy are not completely cured, up to 70% have partial nonspecific epilepsy, and approximately 10% of patients with epilepsy are diagnosed with drug-resistant epilepsy. Population survey data in China shows that approximately ten million people have epilepsy, but only about one-third of them receive appropriate or adequate treatment. The annual incidence of epilepsy in urban and rural China is 35 per 100,000 and 25 per 100,000, respectively, with 60% of epilepsy patients originating in early childhood. The incidence of childhood epilepsy is 151 per 100,000, and the prevalence is 345 per 100,000 (Ding et al. 2021; Lin et al. 2021). Currently, epilepsy has become a common chronic disease affecting people of all ages, and its incidence is increasing over time, seriously impacting individuals' physical and mental health and normal daily life.

2.2 *The Main Mechanisms of Epilepsy*

Contemporary research suggests that epilepsy is primarily caused by abnormal neuronal discharges, which are closely associated with cortical developmental disorders, abnormal neuronal excitation/inhibition signals, genetic variations, and immune dysregulation (Table 18.1).

2.2.1 Cortical Developmental Disorders

Cortical morphological and functional developmental disorders are a type of brain developmental defect that includes neuronal ectopia, anencephalic gyrus, megalencephalic gyrus, cerebral penetrance malformation, and hypoplasia or absence of the corpus callosum. These disorders often result in neuropsychiatric conditions such as epilepsy, mental retardation, local neurological dysfunction, and mental retardation. During fetal development, harmful factors in the internal and external environment can cause cortical developmental disorders ranging from local functional developmental abnormalities to whole-brain morphological developmental abnormalities. Cepeda et al. (2007) found that the etiology of patients with intractable epilepsy is mainly due to cortical developmental disorders that can affect the growth, migration, and metabolism of neurons or glial cells. Focal cortical dysplasia is considered the most common cause of cortical developmental malformations and intractable epilepsy in children, with the mechanism related to the PI3K/Akt/mTOR signaling pathway regulating neuronal growth and migration (Philip and Peter 2017). Due to the refractory nature of pharmacological treatment for intractable epilepsy, alternative treatments such as the ketogenic diet, surgical resection, or neuromodulation are often considered clinically (Lily et al. 2018).

Table 18.1 Various etiologies and their influence on diverse types of epilepsy

Etiologies	Pathology/ pathophysiological changes	Phenotypes	Main references
Cortical development disorder	Anencephaly	Neonatal epilepsy	Marie-Coralie et al. (2018)
	Periventricular nodular ectopia	Focal epilepsy	Stephen et al. (2016)
	Corpus callosum dysplasia Mild cortical dysplasia Focal cortical dysplasia Multinodular and vacuolated neuronal tumors	Epilepsy Childhood epilepsy Pediatric intractable epilepsy Late-onset epilepsy	Maria et al. (2020) Roland et al. (2017) Hye et al. (2020) Chao et al. (2020) Maria et al. (2018)
Cranial trauma	Transorbital penetration Severe craniocerebral trauma Acute brain injury Temporal lobe trauma	Tonic epilepsy Intractable epilepsy Epilepsy Epilepsy	Keun and Bo (2020) Keret et al. (2018) Terrone et al. (2020) Tubi et al. (2019)
Abnormal neuronal excitation/inhibition signals	Glutamate decarboxylase abnormalities Glutamate receptor downregulation	Temporal lobe epilepsy Temporal lobe epilepsy	Daif et al. (2018) Tizazu et al. (2020) Pfisterer et al. (2020)
Genetic variation	Abnormal glutamate metabolism SCN1A SCN3A SCN8A GATOR1 DDX3X KMT2E GATOR1 KCNT1	Epilepsy Multiple epilepsy syndromes Childhood epilepsy Epilepsy Epilepsy Epilepsy Autism with epilepsy Focal epilepsy Sleep-related epilepsy	Boison et al. (2018) Ingrid and Rima (2019) Zaman et al. (2020) Denis et al. (2019) Sara et al. (2019) Lennox et al. (2020) Donnell-Luria et al. (2019) Barbora et al. (2021) Rubboli et al. (2018)

2.2.2 Abnormal Neuronal Excitation/Inhibition Signaling

Glutamate and γ-aminobutyric acid (GABA) are the most important excitatory and inhibitory neurotransmitters, respectively, in the central nervous system. Abnormalities in these neurotransmitters and their receptors can cause abnormal neuronal discharge, leading to the disruption of neural circuits and ultimately inducing epilepsy (Cho 2013). Abnormal glutamate metabolic receptors have been suggested to play an important role in epileptic seizures (Guo et al. 2010). Furthermore, an increasing number of studies have focused on the crucial role played by glial cells in the pathogenesis of epilepsy. Under pathological conditions, glial cells continuously activate and secrete active substances that lead to increased synthesis and release of excitatory amino acids, which excite neighboring neurons and activate signaling pathways (Patel et al. 2019; de Lanerolle et al. 2010). In addition, it has been found that the amount of GABA transporter protein in hippocampal tissue is reduced in patients with epilepsy, resulting in the release of significantly less GABA than in normal hippocampal tissue. This means that the GABA released in patients with epilepsy is not sufficient to suppress seizures (During et al. 1995). Nickels (2019) has shown that during epilepsy formation, abnormal synaptic connections are formed between neurons in the brain, establishing pathological neural loops that lead to abnormally increased excitability in the brain, thus repeatedly triggering epilepsy.

2.2.3 Genetic Variants and Immune Dysregulation

Epilepsy is a genetically linked disorder that encompasses both primary and secondary epilepsy. The development of primary epilepsy is significantly influenced by genetic factors, with nuclear gene mutations, mitochondrial gene mutations, and chromosomal abnormalities being the primary modes of inheritance. Furthermore, epilepsy susceptibility is also attributed to inflammation and abnormal immune responses during early development. Walker et al. (1969) were the first to propose that immune system dysfunction may contribute to epileptic seizures, indicating that the pathogenesis of epilepsy is closely linked to the dysfunction of the neuro-immune-endocrine network. Elenkov et al. (2005) found that patients with epilepsy often exhibit immune system dysfunction, suggesting that abnormal immune system function may be a major cause in the development of specific types of epilepsy. Additionally, several studies have indicated a close association between epilepsy and humoral and cellular immunity in adaptive immunity (Mao et al. 2013; Omran et al. 2012).

3 Evidence for the Intrauterine Origin of Epilepsy

In light of the growing concern for the detrimental effects of environmental factors on fetal development, researchers have directed their attention toward exploring the impact of various factors during this stage on the susceptibility to epilepsy after birth. Recent epidemiological investigations, as well as clinical and animal studies, have indicated that epilepsy may have an origin in intrauterine development.

3.1 Epidemiological Investigations

Epidemiological evidence indicates that epilepsy predominantly occurs during early childhood or childhood onset, with its pathogenesis yet to be fully understood. Stressful states during fetal development have been closely linked to the onset of epilepsy, and the formation of the epileptic process may be negatively influenced by stress-related disorders themselves, suggesting that stress might be a crucial trigger for seizures in children (Bosnjak et al. 2002). In patients with epileptic tendencies or low convulsive thresholds, a positive feedback loop of "stress-seizure-stress-related disorders" is established, whereby stress increases seizure frequency, exacerbating stress-related disorders, ultimately leading to seizures through stress again. Research has shown that exposure to adverse life events or altered stress sensitivity during fetal or early postnatal development is significantly associated with seizures in offspring. For instance, prenatal maternal exposure to chronic stress or adverse life events during childhood could increase the risk of seizure onset by affecting HPA axis hyperactivity in offspring (van Campen et al. 2014). A population-based study revealed that children who experienced war early in life had a higher frequency of epilepsy, earlier onset of first seizures, and more severe seizures than children not affected by war (Bosnjak et al. 2002). Additionally, it has been found that maternal emotional stress is associated with the risk of infantile spasms (Shang et al. 2010) and that experiencing stressful stimuli during pregnancy can increase the risk of epilepsy in offspring (Li et al. 2008). These findings suggest that the occurrence of epilepsy is associated with several adverse environmental exposures, such as adverse life events during fetal development and stress during pregnancy, indicating that seizures have an intrauterine developmental origin.

3.2 Clinical and Basic Laboratory Studies

The occurrence of epilepsy is significantly influenced by adverse environmental exposures during fetal development. Current research suggests that the mechanisms underlying this occurrence are centered around abnormalities in pre- and postnatal neuronal ion channels, as well as postnatal changes in the high-stress sensitivity of the HPA axis.

3.2.1 The Fetal Origin of Epilepsy Susceptibility

Animal experiments have demonstrated a strong association between stress during fetal development and postnatal seizures, with the hippocampus playing a crucial role in the initiation and progression of epilepsy (Pitkanen and Lukasiuk 2011). Gestational fasciculation stress was found to increase hairy pilocarpine-induced seizures in offspring by decreasing the expression of the GABA receptor α1 subunit in fetal rats, leading to an imbalance of excitatory and inhibitory functions in the central nervous system (Nejatbakhsh et al. 2018). Rats subjected to daily restraint stress during late gestation had offspring with more tonic-clonic seizures in the stress group (Frye and Bayon 1999). Pregnant rats subjected to bright light restraint stress in early gestation (5–12 days of gestation), or mid-to-late gestation (12–20 days of gestation) had offspring with reduced seizure thresholds in the brain at 10 days postnatal, and electrical ignition rates were significantly higher (Edwards et al. 2002). These findings suggest that pregnancy stress can increase the incidence of epilepsy in littermates. Furthermore, studies have shown that exposure to lithium chloride-pilocarpine during pregnancy can increase susceptibility to postnatal convulsions in littermates, along with increased hippocampal neuronal damage (Cossa et al. 2016). Injection of β2-phenylalanine into rats on days 10–12 of gestation was found to increase the frequency of auditory convulsions in offspring 23 days after birth (Beck and Gavin 1976). Betamethasone exposure during pregnancy was found to increase susceptibility to trifluoroethanol-induced clonic convulsion, while it had no significant effect on seizures induced by mariposa alginate (Velisek 2011). Treatment with betamethasone or dexamethasone on days 15–18 of gestation was found to increase seizure thresholds in both seizure models, with dexamethasone increasing ignition thresholds but not altering seizure thresholds (Young et al. 2006). Moreover, maternal dexamethasone exposure was found to result in temporal lobe epilepsy susceptibility in male offspring, which can be transmitted to grandchildren (Hu et al. 2020).

3.2.2 Mechanisms of Neuronal Ion Channel Abnormalities in Epilepsy

External factors during development can alter the expression of ion channels that regulate neuronal cell excitability, causing an imbalance in the electrical activity of the central nervous system and eventually leading to seizures. Potassium, sodium, and calcium channels are the primary ion channels associated with epilepsy, with the normal distribution of potassium ions playing a crucial role in maintaining resting potentials and developing action potentials in neurons. Abnormal potassium ion distribution is often accompanied by paroxysmal depolarization of brain electrophysiology, resulting in abnormal electrophysiological activity of neurons. External factors, such as neurotransmitters, hormones, and extracellular ions, can modulate voltage-gated potassium channels (KCNQ) through different pathways, generating M-currents and affecting neuronal excitability regulation, neurotransmitter release, and synaptic transmission, leading to decreased potassium channel function, increased neuronal excitability, and seizures. For instance, altered expression of voltage-gated potassium channel genes KCNQ2 and KCNQ3 due to pregnancy stress induced benign familial neonatal convulsions and was closely associated with familial neonatal epilepsy (Biervert et al. 1998).

Jacobs et al. (2000) conducted a review on traumatic brain injury during fetal development or the early postnatal period leading to refractory seizures. The study involved two animal models and a series of experiments to investigate the mechanisms underlying the induction of hyperexcitability in the brain after injury. The findings revealed that the injury can induce hyperexcitability in the brain through three distinct mechanisms. Firstly, direct injury to cortical pyramidal neurons leads to changes in membrane ion channels that increase the sensitivity of these cells to excitatory inputs. Secondly, injury alters the connectivity of cortical circuits, and enhanced excitatory connectivity may increase recurrent excitatory loops within the epileptogenic zone. Finally, when the lesion affects the neocortex, which is at its peak of development, extensive and sustained changes in GABA and glutamate receptors are induced due to reorganization of the thalamocortical, corpus callosum, and intracortical circuits and over-innervation due to immature connections, which eventually induce epilepsy. Zaman et al. (2020) conducted an experimental study demonstrating that severe abnormalities in sodium currents are significantly associated with the development of neonatal epilepsy. The study showed that current density significantly decreased, and abnormal "sustained" current levels were observed in neonatal rats with epilepsy. Additionally, Ahmadzadeh et al. (2011) found that the increased release of inflammatory cytokines and ion channel proteins in the limbic lobe of adult offspring littermates treated with lithium chloride-pilocarpine during pregnancy promotes hippocampal neuronal degeneration, leading to a significant increase in epilepsy susceptibility. Furthermore, fetal rats exposed to external gamma radiation showed increased susceptibility to seizures after birth, with pathological changes of diffuse cortical dysplasia and neuronal ectopia, which was attributed to radiation-induced cell death and altered ion channels.

3.2.3 Mechanism of High-Stress Sensitivity of HPA Axis in Epilepsy

The HPA axis is a vital endocrine system that regulates the body's response to stress and is susceptible to damage during fetal development (Xiong and Zhang 2013). Studies indicate that unfavorable environmental conditions or chronic stress during intrauterine growth can lead to modifications in HPA axis programming, abnormal HPA axis activity, and stress sensitivity in offspring after birth, as well as the onset of epilepsy (He et al. 2016; Huang et al. 2015; Rogawski 2013). Austin et al. (2014) discovered that maternal glucocorticoid (GC) levels rise considerably in pregnant rats after prenatal exposure to stress, leading to altered HPA axis programming in their fetal rats and high-stress sensitivity in the offspring after birth, an alteration associated with increased susceptibility to neuropsychiatric disorders such as epilepsy and anxiety in adulthood. Previous research conducted by the authors' team revealed that adult offspring exposed to multiple exogenous substances during pregnancy have lower blood adrenocorticotrophic hormone (ACTH) and GC levels at baseline, but significantly higher levels than controls after postnatal administration of chronic stress, indicating a high-stress sensitivity of the HPA axis (Xia et al. 2014; Xu et al. 2013, 2012). Research has indicated that stress hormones related to the HPA axis, particularly hypothalamic corticotropin-releasing hormone (CRH) and GC, may contribute to epileptic seizures by disrupting the balance of excitatory and inhibitory neurotransmitters. The persistence of CRH and GC in the body due to their roles in stress response regulation could be a key factor in promoting seizures (Chen et al. 2001). CRH gene expression begins in the late gestational rat brain and persists until birth, with a significant number of CRH-expressing GABAergic interneurons present in the hippocampus of young rats between 10 and 13 days post-birth. When CRH binds to postsynaptic CRH receptors on neurons, even a small glutamate input signal can be amplified into a strong excitatory signal, leading to epileptic discharges and seizure induction. These findings suggest that altered HPA axis function is closely associated with the occurrence of epilepsy and that high-stress sensitivity programming of the HPA axis in offspring as a result of unfavorable intrauterine environments is a significant factor influencing epilepsy susceptibility.

4 Epigenetic Mechanisms of Fetal-Originated Epilepsy

Epigenetic modifications play a crucial role in regulating gene expression, inducing phenotypic alterations, and can be passed down through generations without any DNA sequence changes. These modifications are known to affect tissue-specific gene expression, particularly during fetal brain development, and have been linked to the development of various neuropsychiatric disorders, such as epilepsy (Table 18.2).

4.1 DNA Methylation and Fetal-Originated Epilepsy

DNA methylation has been shown to be associated with the silencing of transposons, which are known to contribute to epilepsy susceptibility by suppressing the expression of genes involved in brain development. Studies have demonstrated that the administration of synthetic GC to pregnant rats led to increased DNA methylation of genes related to the frontal cortex and hippocampus, as well as increased expression of the dentate gyrus DNA methyltransferase 3α (DNMT3α) in their offspring (Crudo et al. 2013). This finding is consistent with altered DNMT3α

Table 18.2 Epigenetic regulatory mechanisms in the onset of fetal-originated epilepsy

Epigenetic modifications	Genes	Species	Sex	Phenotypes	Main references
DNA methylation	BDNF BDNF	Human Rat	Male/female Male/female	Childhood epilepsy Epilepsy	Caramaschi et al. (2020) Roth et al. (2009)
Histone acetylation	$GABA_A$ CREB ACE	Rat Rat Rat	Male/female Male/female Male	Epilepsy Epilepsy Temporal lobe epilepsy	Bohnsack et al. (2017) Zhang et al. (2013) Hu et al. (2020)
miRNAs	miR-34a miR-129-5p	Rat Human	Male Male/female	Epilepsy Epilepsy	Hu et al. (2012) Wan and Yang (2020)
	miR-146a	Human	Male/female	Childhood epilepsy	Nguyen et al. (2018)

expression observed in the brains of epileptic patients. It is suggested that alterations in DNA methylation due to environmental influences during pregnancy may play a crucial role in the development of epileptic seizures. Furthermore, high choline intake during pregnancy and early postnatal development has been shown to protect the brain from neurological damage associated with epilepsy, and its protective effect was linked to the level of DNA methylation modifications regulating epilepsy-related genes in the brain (Jan et al. 2017). A genome-wide DNA methylation analysis of children with fetal alcohol spectrum disorder (FASD) due to prenatal alcohol exposure identified 658 significantly differentially expressed methylation sites and 101 differentially methylated regions containing two or more CpGs between FASD cases and controls in one study. An analysis of genes with hypermethylation has shown that differentially methylated genes are significantly enriched in signaling pathways associated with neurodevelopmental and neuropsychiatric disorders, including epilepsy, anxiety, and autism spectrum disorders. This finding suggests that differential DNA methylation sites may be potential epigenetic targets for FASD (Portales-Casamar et al. 2016). In a prospective cohort study, Caramaschi et al. (2020) examined approximately 800 children in cord blood at birth and peripheral blood during childhood and adolescence for genome-wide DNA methylation and found significant correlations between alterations in DNA methylation and seizures. Among these genes, the brain-derived neurotrophic factor (BDNF) gene was highly expressed in the brain and showed a strong correlation with brain methylation levels, suggesting that it might be an important mechanism underlying seizures. In conclusion, alterations in DNA methylation caused by adverse factors during pregnancy may persist after birth, leading to the development of fetal-originated epilepsy.

4.2 Histone Acetylation and Fetal-Originated Epilepsy

Histone acetylation is the most prevalent form of histone epigenetic modification and is influenced by external factors during growth and development. Posttranscriptional histone acetylation is a significant mechanism of chromatin remodeling that contributes to the epigenetic regulation of gene transcription and is closely associated with the development of the nervous system and normal neuronal function. Studies have shown that histone H4 acetylation of the BDNF gene was significantly upregulated in rat hippocampal CA3 neurons following 3 hours of pilocarpine-induced status epilepticus (Huang et al. 2002). Additionally, ethanol exposure during pregnancy has been linked to the mediation of epileptic seizures by regulating GABA receptor expression via histone deacetylases (Bohnsack et al. 2017). The cyclic AMP-response element-binding protein (CREB) is a transcription factor that regulates GABA receptor expression in the hippocampus. It has been observed that electroconvulsive seizures produced by repetitive electrical stimulation can lead to a decrease in histone acetylation in the promoter region of the CREB gene in the hippocampus of rats (Tsankova et al. 2004). Early maternal licking of the litter has been shown to affect the level of histone acetylation in the hippocampal CREB promoter. Conversely, early separation of the litter from the mother can lead to convulsions after birth by decreasing the level of histone acetylation in the hippocampal CREB promoter (Zhang et al. 2013). Recently, the authors' team demonstrated that prenatal exposure to dexamethasone caused a heritable susceptibility to temporal lobe epilepsy in male offspring after birth, which was associated with an increased level of H3K27ac in the hippocampal local angiotensin-converting enzyme (ACE) promoter region and activation of the renin-angiotensin system classical pathway, inducing hippocampal neuronal glutamate excitatory damage (Hu et al. 2020). These findings suggest that adverse environments during pregnancy may mediate epilepsy susceptibility in offspring after birth through alterations in histone acetylation modifications closely related to epilepsy. Histone deacetylase inhibitors have been proposed as a potential intervention strategy for human epilepsy (Citraro et al. 2017, 2020).

4.3 MiRNAs and Fetal-Originated Epilepsy

MicroRNAs (miRNAs) are a class of small noncoding RNAs that exert epigenetic regulation by modulating the translation and stability of mRNA transcripts via specific binding to the 3'UTR of target genes. The regulatory role of miRNAs in the development of the central nervous system has been increasingly recognized in recent years. Specifically, miR-34a has emerged as a key mediator of dendritic morphology, ion channel levels, neuronal migration, and glial cell function, and its upregulation has been implicated in epileptic seizures (Mollinari et al. 2015; Hu et al. 2012). Furthermore, pregnancy stress has been shown to alter the expression of several miRNAs, including miR-34a, in the hippocampus of offspring, and may thus increase the susceptibility to postnatal seizures (Rodgers et al. 2013). In addition, Ge et al. (2018) demonstrated that prenatal exposure to fluoride and aluminum induces seizures in offspring rats, which is associated with upregulation of miR-132 and miR-204 expression in the brain. Wan and Yang (2020) revealed that lncRNA NEAT1 inhibits miR-129-5p and regulates the Notch signaling pathway in epileptic models, implying that miR-129-5p could serve as an early biomarker of epilepsy. Moreover, miR-146a upregulation has been reported as the most common dysregulation event in neurodevelopmental disorders, including epilepsy, autism spectrum disorders, and intellectual disabilities, and is significantly associated with the occurrence of childhood epilepsy (Nguyen et al. 2018). Collectively, alterations in miRNA expression induced by adverse prenatal conditions can directly impact neurodevelopment and subsequently mediate epilepsy susceptibility later in life.

5 Summary and Prospect

This review provides a comprehensive analysis of the intrauterine origins of epilepsy and the consequences of unfavorable pregnancy environments on fetal neurodevelopment. The underlying mechanisms leading to fetal-originated epilepsy involve the imbalance in neuronal excitatory/inhibitory signaling, impaired cortical development, abnormal neuronal signaling, and altered programming of the HPA axis, resulting from the regulation of key gene expression through traditional genetic variants or epigenetic modifications (Fig. 18.1). Recent research has highlighted the

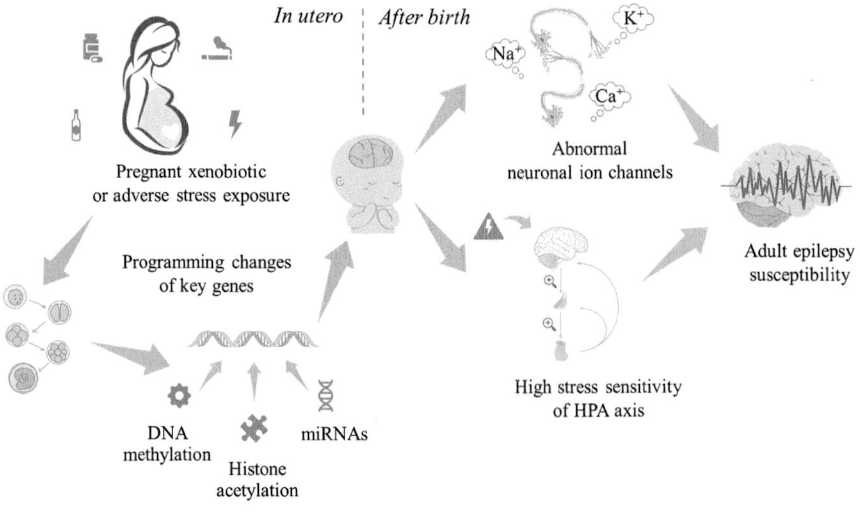

Fig. 18.1 Mechanisms leading to fetal-originated epilepsy due to adverse environment during pregnancy

crucial role of epigenetic modifications in fetal-originated disorders, and identifying key epigenetic loci will facilitate the identification of biomarkers for offspring epilepsy caused by adverse prenatal environments. Future studies should focus on identifying the developmental sensitivity window to mitigate the effects of adverse factors on the developmental programming of the nervous system. Moreover, the development of new diagnostic and prevention techniques for personalized medical treatment strategies will guide eugenics and improve population quality.

References

Ahmadzadeh R, Saboory E, Roshan-Milani S, et al. 2011. Dev Psychobiol, 53(8): 806–812.
Austin C K, Amanda D G, Tara S P, et al. 2014. Behav Brain Res, 259: 106–118.
Barbora B, Maurits W C, Anna U, et al. 2021. Eur J Paediatr Neurol, 30: 88–96.
Beck S L, Gavin D L. 1976. Science, 193(4251): 427–428.
Biervert C, Schroeder B C, Kubisch C, et al. 1998. Science, 279(5349): 403–406.
Bohnsack JP, Patel VK, Morrow AL. 2017. J Pharmacol Exp Ther, 363(1): 1–11.
Boison D, Steinhäuser C. 2018. Glia, 66(6):1235–1243.
Bosnjak J, Vukovic-Bobic M, Mejaski-Bosnjak V. 2002. Epilepsy Behav, 3(6): 502–509.
Brunton P J, Russell J A. 2010. J Neuroendocrinol, 22(4): 258–271.
Caramaschi D, Hatcher C, Mulder RH, et al. 2020. Clin Epigenetics, 12(1): 8.
Cepeda C, Andre V M, Wu N, et al. 2007. Epilepsia, 48 Suppl 5: 79–85.
Chao L, Chun Q Z, Xin C, et al. 2020. J Mol Neurosci, 70(5): 796–805.
Chen Y, Bender R A, Frotscher M, et al. 2001. J Neurosci, 21(18): 7171–7181.
Cho C H. 2013. Front Cell Neurosci, 7: 127.
Citraro R, Leo A, Santoro M, et al. 2017. Curr Pharm Des, 23(37): 5546–5562.
Citraro R, Leo A, De Caro C, et al. 2020. Mol Neurobiol, 57(1). 408–421.
Cossa A C, Lima D C, do Vale T G, et al. 2016. Metab Brain Dis, 31(4): 891–900.
Crudo A, Petropoulos S, Suderman M, et al. 2013. Endocrinology, 154(11): 4170–4181.
Daif A, Lukas RV, Issa NP, et al. 2018. Epilepsy Behav, 80:331–336.
Ding D, Zhou D, Sander J W, et al. 2021. Lancet Neurol, 20(4): 316–326.
Donnell-Luria AH, Pais LS, Faundes V, et al. 2019. Am J Hum Genet, 104(6):1210–1222.
During M J, Ryder K M, Spencer D D. 1995. Nature, 376: 174–177.
de Lanerolle N C, Lee T S, Spencer D D. 2010. Horm Behav, 42(4): 437–447.
Denis J, Villeneuve N, Cacciagli P, et al. 2019. Epilepsia, 60(5): 845–856.
Edwards HE, Dortok D, Tam J, et al. 2002. Horm Behav, 42(4):437–47.
Elenkov I J, Iezzoni D G, Daly A, et al. 2005. Neuroimmunomodulat, 12(5): 255–269.
Fiest K M, Sauro K M, Wiebe S, et al. 2017. Neurology, 88(3): 296–303.
Frye C A, Bayon L E. 1999. Dev Psychobiol, 34(3): 227–234.
Ge Q D, Tan Y, Luo Y, et al. 2018. Environ Toxicol Pharmacol, 63: 60–68.
Guo F, Sun F, Yu J L, et al. 2010. Brain Res Bull, 81(4–5): 510–516.
He Z, Zhu C, Huang H, et al. 2016. Toxicol Res, 5(2): 388–398.
Hu K, Xie Y Y, Zhang C, et al. 2012. BMC Neurosci, 13: 115.
Hu S, Yi Y, Jiang T, et al. 2020. Arch Toxicol. 94(9): 3201–3215.
Huang Y, Doherty JJ, Dingledine R. 2002. J Neurosci, 22(19):8422–8.
Huang H, He Z, Zhu C, et al. 2015. Toxicol Appl Pharmacol, 288(1): 84–94.
Hye E K, Se H K, Hoon-Chul K, et al. 2020. Seizure, 82: 50–55.
Ingrid E S, Rima N. 2019. Epilepsia, 3: S17-S24.
Jacobs K M, Graber K D, Kharazia V N, et al. 2000. Epilepsia, 6: S153–161.
Jan K B, Barbara E S, Tiffany J M. 2017. Nutrients, 9(8): 815.

Keret A, Shweiki M, Bennett-Back O, et al. 2018. Seizure, 58: 29–34.
Keun S L, Bo L L. 2020. Childs Nerv Syst, 36(3): 647–650.
Lennox AL, Hoye ML, Jiang R, et al. 2020. Neuron, 106(3):404–420.e8.
Lily C W, Thomas B, Mai-Lan H, et al. 2018. Epilepsy Res, 145: 1–17.
Li J, Vestergaard M, Obel C, et al. 2008. Epilepsy Res, 81(1): 52–57.
Lin Y, Hu S, Hao X, et al. 2021. Epilepsia, 62(11): 2640–2650.
Mao L Y, Ding J, Peng W F, et al. 2013. Epilepsia, 54(9): e142–e145.
Maria T, Joan L, Anika B, et al. 2018. Brain Pathol, 28(2): 155–171.
Maria J N, Hanka V, Laurens W, et al. 2020. Genet Med, 22(4): 797–802.
Marie-Coralie C, Tristan T S, Maria R C. 2018. Semin Fetal Neonatal Med, 23(3): 204–212.
Mollinari C, Racaniello M, Berry A, et al. 2015. Cell Death Dis, 6(1): e1622.
Nejatbakhsh M, Saboory E, Bagheri M. 2018. Int J Dev Neurosci, 68: 66–71.
Nickels K. 2019. Epilepsy Curr, 19(1): 33–35.
Nguyen LS, Fregeac J, Bole-Feysot C, et al. 2018. Mol Autism, 9: 38.
Omran A, Peng J, Zhang C, et al. 2012. Epilepsia, 53(7): 1215–1224.
Patel D C, Tewari B P, Chaunsali L, et al. 2019. Nat Rev Neurosci, 20(5): 282–297.
Pitkanen A, Lukasiuk K. 2011. Lancet Neurol, 10(2): 173–186.
Philip H I, Peter B C. 2017. Annu Rev Pathol, 12: 547–571.
Pfisterer U, Petukhov V, Demharter S, et al. 2020. Nat Commun, 11(1): 5038.
Portales-Casamar E, Lussier A, Jones M, et al. 2016. Epigenet Chromatin, 9: 25.
Rodgers A B, Morgan C P, Bronson S L, et al. 2013. J Neurosci, 33(21): 9003–9012.
Rogawski M A. 2013. Acta Neurol Scand Suppl, (197): 9–18.
Roland JL, Snyder AZ, Hacker CD, et al. 2017. Proc Natl Acad Sci USA, 114(50): 13278–13283.
Roth T L, Lubin F D, Funk A J, et al. 2009. Biol Psychiatry, 65(9): 760–769.
Rubboli G, Plazzi G, Picard F, et al. 2018. Ann Clin Transl Neurol, 25; 6(2): 386–391.
Sara B, Fabienne P, Nienke E V, et al. 2019. Genet Med, 21(2): 398–408.
Shang N X, Zou L P, Zhao J B, et al. 2010. Pediatr Neurol, 42(3): 181–186.
Struber N, Struber D, Roth G. 2014. Neurosci Biobehav Rev, 38: 17–37.
Stephen A T, Giridhar P K, Nitin T, et al. 2016. Seizure, 41: 211–216.
Terrone G, Balosso S, Pauletti A, et al. 2020. Neuropharmacology, 167: 107742.
Tizazu E, Ellis CA, Reichert J, et al. 2020. Seizure, 80: 38–41.
Tsankova N M, Kumar A, Nestler E J. 2004. J Neurosci, 24(24): 5603–5610.
Tubi M A, Lutkenhoff E, Blanco M B, et al. 2019. Neurobiol Dis, 123: 115–121.
van Campen J S, Jansen F E, de Graan P N, et al. 2014. Epilepsy Behav, 38: 160–171.
Velisek L. 2011. Epilepsy Res, 95(1–2): 9–19.
Walker A I, Stevenson D E, Robinson J, et al. 1969. Toxicol Appl Pharmacol, 15(2): 345–373.
Wan Y, Yang Z Q. 2020. Cell Cycle, 46(6): 546–563.
Xia L P, Shen L, Kou H, et al. 2014. Toxicol Lett, 226(1): 98–105.
Xiong F, Zhang L. 2013. Front Neuroendocrinol, 34(1): 27–46.
Xu D, Wu Y, Liu F, et al. 2012. Toxicol Appl Pharmacol, 264(3): 395–403.
Xu D, Xia L P, Shen L, et al. 2013. Acta Pharmacol Sin, 34(12): 1526–1534.
Young N A, Teskey G C, Henry L C, et al. 2006. Exp Neurol, 198(2): 303–312.
Zaman T, Helbig K L, Clatot J, et al. 2020. Ann Neurol, 88(2): 348–362.
Zhang T Y, Labonte B, Wen X L, et al. 2013. Neuropsychopharmacol, 38(1): 111–123.

Chapter 19
Fetal Origin of Testicular Dysplasia

Hao Xiao and Yi Liu

Abstract Testicular dysplasia is a condition characterized by testicular volume reduction, dysfunction of testosterone synthesis, and decreased sperm quality. Both clinical and basic studies have demonstrated that testicular dysplasia has a fetal origin. Exposure to adverse environments during pregnancy, such as exogenous substances and poor maternal health, can result in abnormal morphological and functional development of the fetal testicle. This, in turn, can lead to changes in the developmental programming of the hypothalamic-pituitary-testicular axis after birth. The underlying mechanisms include abnormal secretion of local factors, such as insulin-like growth factor 1, changes in intrauterine neuroendocrine metabolic programming, oxidative stress, mitochondrial damage, and abnormal epigenetic modifications, such as DNA methylation, histone modification, and regulation of noncoding RNA. Furthermore, epigenetic modifications of key genes can evade germ cell reprogramming and be passed on to subsequent generations, resulting in multigenerational inheritance effects of testicular dysplasia. This chapter provides a comprehensive review of the evidence supporting the fetal origin of testicular dysplasia and its intrauterine programming mechanisms. The findings have significant implications for guiding eugenics and improving the quality of the population.

H. Xiao (✉)
Division of Joint Surgery and Sports Medicine, Department of Orthopedic Surgery, Zhongnan Hospital of Wuhan University, Wuhan, Hubei, China
e-mail: HaoXiao@whu.edu.cn

Y. Liu
Department of Pharmacology, School of Basic Medical Science, Wuhan University, Wuhan, Hubei, China

H. Wang et al. (eds.), *Fetal Origin of Diseases*,
https://doi.org/10.1007/978-981-97-5730-5_19

1 Introduction

The testis is a critical component of the male reproductive system, responsible for spermatogenesis, androgen synthesis and secretion, and the maintenance of male secondary sex characteristics. In recent years, the decline in male reproductive function has become a global concern, with testicular dysplasia being a common cause. Testicular dysplasia is characterized by reduced testicular size, decreased spermatogenesis, and decreased androgen synthesis. The causes of testicular dysplasia can be attributed to genetic factors and exogenous factors. Genetic factors include sex chromosome aberrations, while exogenous factors include adverse environments during pregnancy. Recent studies by domestic and foreign scholars have revealed a strong correlation between adverse prenatal environments, fetal birth weight, and adult chronic disease susceptibility, leading to the development of the "Developmental Origins of Health and Disease (DOHaD)" theory. It is widely accepted that early life events have a significant impact on individual health. Adverse environments during pregnancy refer to various factors that can negatively affect the growth and development of the embryo or fetus during the intrauterine period. These factors include physical and chemical exposures, as well as maternal health conditions such as maternal diseases and chronic stress. Studies have demonstrated that exposure to adverse environmental conditions during pregnancy, such as cadmium exposure, can lead to morphological changes such as stromal widening in offspring, as well as inhibit spermatogenesis and testosterone synthesis. Furthermore, exposure to dexamethasone during pregnancy can lead to continued inhibition of testosterone synthesis in the offspring and result in transgenerational inheritance effects (Liu et al. 2022a). These findings suggest that testicular dysplasia may have an intrauterine origin. In this chapter, we summarize the latest research and evidence from epidemiological and basic studies to provide an overview of the intrauterine origin and programming mechanisms of testicular dysplasia.

2 Testicular Dysplasia

Testicular dysplasia is a condition characterized by reduced testicular size, decreased spermatogenesis, and decreased androgen synthesis. Both genetic and environmental factors are closely associated with the development of testicular dysplasia. The incidence of testicular dysplasia has been increasing in recent years. In this section, we will provide an overview of the definition of testicular dysplasia, its current incidence, and its main mechanism.

2.1 Definition and Incidence of Testicular Dysplasia

Testicular dysplasia is primarily characterized by a decrease in testosterone level and sperm quality. This concept has led to a new understanding of the origin of male fertility reduction. A cohort study of Danish newborns with hypospadias conducted by Boisen et al. (2005) suggested that fetuses with intrauterine growth retardation are closely associated with impaired testicular function in adulthood. Preiksa et al. (2005) found that cryptorchidism was closely related to low birth weight and preterm birth, and low birth weight was associated with other external genital congenital malformations such as cryptorchidism and hypospadias. Available data indicate that 2% to 9% and 0.2% of male newborns suffer from cryptorchidism and hypospadias, respectively. Epidemiological studies on cryptorchidism, hypospadias, low sperm quality, and testicular cancer have shown a significantly increasing trend in recent generations worldwide, which has severely affected male fertility (Skakkebaek 2004). These findings suggest that testicular dysplasia may have a fetal origin.

2.2 The Main Mechanism of Testicular Dysplasia

Through clinical and animal experimentation, the etiology of testicular dysplasia can be attributed to gene defects and polymorphism, lifestyle, exposure to exogenous substances, and intrauterine growth retardation (IUGR) (Fig. 19.1). Bisphenol A (BPA), an endocrine-disrupting chemical with weak estrogenic activity, is widely used in various industries, including plastics. Exposure to BPA has been found to

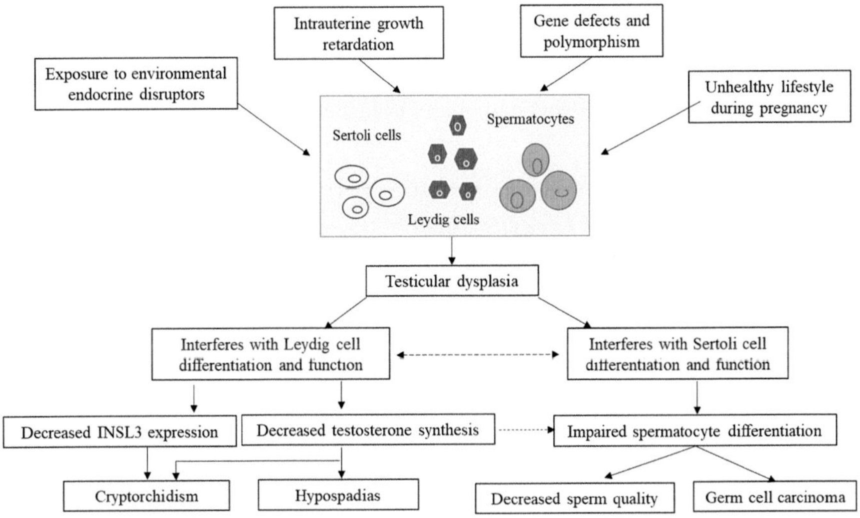

Fig. 19.1 Pathogenesis of testicular dysplasia syndrome

elevate levels of prolactin, estradiol, and sex hormone-binding globulin in men, which may contribute to male testicular dysplasia and adult infertility (Cabaton et al. 2013). Prenatal lifestyle factors, such as chronic stress and substance abuse, have also been closely associated with testicular dysplasia. Although the causal relationship between maternal smoking and testicular dysplasia in offspring remains inconclusive, smoking during pregnancy has been shown to impair fetal growth, which may increase the risk of testicular dysplasia. Main et al. (2009) reported that unhealthy maternal lifestyle habits during pregnancy, such as drinking and smoking, significantly raised the probability of testicular dysplasia in male offspring.

Testicular dysplasia is primarily induced by the disruption of androgen synthesis, secretion, and its effect (Weber et al. 2002). This disruption can result from inhibited androgen synthesis, suppressed androgen receptor expression, or the disturbance of endogenous hormone levels, ingestion of estrogen analogs, or substances with antiandrogenic activity. The factors that can cause androgen disorder include decreased Leydig cell number, abnormal expression of series androgen synthesis enzyme, inhibition of insulin-like factor 3 secretion by Leydig cells during fetal development, and negative feedback inhibiting the abnormal secretion of gonadotropins, especially follicle-stimulating hormone (FSH), which leads to the disorder of adrenal-pituitary-testicular axis. Abnormal hormone levels are the primary mechanism of testicular dysplasia, and pregnancy is a key period that can induce androgen secretion disorder. Ganmaa et al. (2001) reported that during the first pregnancy or twin pregnancy, the incidence of testicular cancer and undescended testis in offspring increased due to abnormal maternal estrogen levels.

3 Intrauterine Origin of Testicular Dysplasia

Numerous epidemiological and laboratory studies have established a significant correlation between adverse intrauterine environments and the development of testicular dysplasia postnatally, indicating an intrauterine origin of this condition.

3.1 Epidemiological Investigation

A cohort study of 343 men on semen quality revealed that higher coffee intake during pregnancy was associated with a decrease in semen volume, testosterone, and statin B concentrations in offspring (Preiksa et al. 2005). Alcohol exposure during pregnancy resulted in hypothalamic pituitary testis (HPT) axis dysfunction in male offspring, leading to reduced sperm quality in adulthood (Ramlau-Hansen et al. 2010). A study conducted in Finland, Denmark, and other European countries reported a significant 20-fold increase in the incidence of undescended testicles and hypospadias in the offspring of mothers treated with diethylstilbestrol in early pregnancy (Xing and Bai 2018). During the process of artificial insemination and in vitro fertilization-embryo transformation, the sperm, oocyte, and zygote are susceptible to exposure to abnormal hormone levels, which can induce testicular dysplasia in the offspring. The above epidemiological survey results suggest that abnormal sperm-oocyte binding and adverse pregnancy environment, including drug abuse, may be the main reasons for the occurrence of testicular dysplasia in offspring.

3.2 Clinical and Basic Laboratory Research

Numerous scholars have provided further evidence to support the intrauterine origin of testicular dysplasia through clinical and animal experiments. Currently, various animal models, including exposure to foreign substances and nutrient restriction during pregnancy, are often utilized to study and confirm the intrauterine origin of testicular dysplasia.

3.2.1 Influence of Adverse Environment During Pregnancy on Testicular Structure

The testicle is composed of distinct morphological and structural components, including the interstitial region, spermatogenic epithelium, and spermatogenic tubules. Based on the characteristics of testicular morphological and structural development, changes in hormone levels during pregnancy can lead to the emergence of undifferentiated Sertoli cells in a male testicle or even the formation of tubes composed entirely of Sertoli cells. Exposure to dibutyl phthalate during pregnancy has been reported to reduce reproductive function in male offspring, causing shortened anal centrosomes, apoptosis of spermatogenic tubule epithelial cells, and morphological damage such as shortened cross-sectional area and longest diameter, and widened interstitial region (Ma et al. 2017). Androgens have been widely reported to play an essential role in regulating the normal developmental trajectory of the reproductive system, affecting anal colonization and the normal proliferation and differentiation of spermatogenic cells (MacLeod et al. 2010). Research has indicated that exposure to dexamethasone during pregnancy can result in significant pathological changes, such as a decreased number of fetal testicular Leydig cells and cell shrinkage (Liu et al. 2018). Anal colonization is often utilized by toxicologists as a measure of fetal testicle development. Androgen deficiency during the masculinization programming window, caused by interference from certain endocrine disruptors, can lead to abnormal male reproductive organs and anus colonization, reproductive dysfunction, and an increased risk of cryptorchidism and hypospadias, making anus colonization one of the more intuitive indicators of testicular development. The normal expression of the sex-determining region of the Y chromosome (Sry) gene is crucial for the structural and functional development of the testis during gonadal development. Dexamethasone exposure during pregnancy has been found to continuously upregulate SRY expression and disrupt the expression of SRy-related high mobility group-box gene 9 (Sox9) and its downstream genes, ultimately leading to male feminization (Yun et al. 2016). Interference with the differentiation of fetal sex and phenotype is mainly manifested as the failure of normal development and formation of fetal testicular structure, such as the absence of regular spermatogenic tubules and the presence of numerous vacuoles in the spermatogenic tubules. Exposure to dexamethasone during late pregnancy has been reported to cause a decrease in the number of spermatogenic tubules per unit cross-sectional area of the testis, decreased diameter, and a thinner spermatogenic epithelium thickness (Jeje and Raji 2017). When maternal protein intake is restricted from pregnancy to postnatal 21 days, male offspring mice exhibit a shortened diameter of the semicircular tubule, decreased total embryonic cell number and Sertoli cells, and a reduction in testicular volume during early adolescence (Rodriguez-Gonzalez et al. 2012). Similarly, protein intake restriction during the second trimester leads to reduced testicular weight and decreased Sertoli cell number in offspring rats during puberty, indicating that nutritional restriction during pregnancy can result in abnormal testicular structural and functional development and weakened reproductive function in adulthood (Genovese et al. 2010).

3.2.2 Influence of Adverse Environment During Pregnancy on Testicular Function

Testicular tissue comprises various types of cells, including Sertoli cells (maintaining the homeostasis of the testicular environment), Leydig cells (synthesizing androgens), and spermatogenic cells (producing sperm), which are crucial for testicular function. Fisher analyzed Leydig cells, Sertoli cells, germ cells, and peritubular myoid cells under adverse conditions during pregnancy and found that the distribution and/or function of these four types of cells was abnormal. Fisher proposed that abnormal development of Sertoli cells during fetal life may be a key factor in testicular dysplasia, leading to a series of abnormal chain reactions of other cells and resulting in abnormal testicular function after birth. Furthermore, numerous studies have reported that exposure to glucocorticoids (GC) during pregnancy can affect androgen synthesis function in the offspring's testis (Table 19.1). For instance, when pregnant mice were treated with dexamethasone between 14 and 19 days of gestation, the testosterone synthesis ability of their offspring's testis was decreased, and the expression of testosterone synthesis series enzymes in adolescent and adult Leydig cells was inhibited, leading to reduced testosterone secretion (Page et al. 2001). Furthermore, Rhind et al. (2001) reported that prenatal stress could lead to a decrease in the blood testosterone level of male offspring rats, and it could also delay the onset of puberty and alter adult reproductive behavior. Moreover, according to recent studies conducted by the authors' team, exposure to exogenous substances, such as caffeine, ethanol, nicotine, and dexamethasone, during pregnancy, could result in testicular dysplasia in the offspring. This dysplasia could manifest as abnormal testicular morphology before and after birth, reduced testosterone synthesis function, and decreased spermatogenesis function in adulthood

Table 19.1 Reproductive programming development following glucocorticoids exposure during pregnancy

Substance	Time	Effect	Main reference
Dexamethasone	Gestational day (GD)13–23 (rat)	Delaying puberty	
	GD14–19 (rat)	Reduced blood testosterone	Page et al. (2001)
	GD15–21 (rat)	Reduces the number/diameter of vas deferens and epididymis lumen	Jeje and Raji (2017)
		Shorten the anal distance at birth	
		Reduces the sperm count and motility	
		Decreased fetal testosterone kurtosis	
Betamethasone	GD12–19 (rat)	Reduces the sperm count and motility	Borges et al. (2017)
	GD109–116 (rat)	Reduced testicular weight and spermatic cord length	Pedrana et al. (2008)
Hypothalamic pituitary adrenal axis activated	GD14–21 (rat)	Changed sexual behavior	

(Zhang et al. 2020; Pei et al. 2019; Liu et al. 2019, 2018). Additionally, exposure to heavy metals and chemicals during pregnancy could inhibit spermatogonial cell proliferation and enhance apoptosis, leading to abnormal sperm quality, including deformity and reduced number, in adulthood. The findings mentioned above indicate that the adverse environment during pregnancy could affect the three types of testicular cells, resulting in an abnormal testicular microenvironment, suppression of androgen synthesis and secretion, and weakened spermatogenic function.

4 Intrauterine Mechanism of Fetal-Originated Testicular Dysplasia

Prior research has indicated that the etiology of testicular dysplasia originating from the fetus is linked to anomalous secretion of local testicular factors, alterations in neuroendocrine and metabolic programming, oxidative stress, and damage to mitochondria.

4.1 Changes in GC-IGF1 Axis Programming

GCs play an essential role in pregnancy maintenance, delivery, and normal fetal development. However, numerous studies have demonstrated that GC exposure during pregnancy can lead to impaired fertility of the offspring before and after birth, as well as decreased testosterone synthesis ability and abnormal testicular structure, resulting in changes in male reproductive phenotype (endocrine and spermatogenic function) (Yun et al. 2016). Research has shown that the glucocorticoid receptor (GR) is widely present in testicular tissues and plays a role in regulating the growth, development, and function of various cells in the testicular tissues. In vitro studies have indicated that abnormal levels of glucocorticoids can have a direct impact on the function of several types of cells in testicular tissue. For instance, the steroidal synthase system of rat Leydig cells was significantly inhibited in a concentration-dependent manner after exposure to different concentrations of dexamethasone or corticosterone. Additionally, GC has been shown to reduce LH-induced testosterone production by inhibiting luteinizing hormone (LH) production in vivo, which is mediated by GR. In conclusion, GC exposure during pregnancy can directly or indirectly affect testicular differentiation and development and alter the offspring's reproductive phenotype in adulthood.

Insulin-like growth factor 1 (IGF1) and its receptors play a role in the regulation of proliferation, apoptosis, and testosterone synthesis in fetal testicular Leydig cells. GCs have the ability to regulate the expression of IGF1 and its receptor. The authors' team conducted studies that revealed that exposure to exogenous substances (such as caffeine, ethanol, and nicotine) during pregnancy can result in intrauterine maternal GC overexposure, which can affect the IGF1 signaling pathway and the functional development of multiple organs by altering the "intrauterine neuroendocrine metabolic programming" (Ni et al. 2015; Xu et al. 2015; Wang et al. 2014; Xia et al. 2014). The expression of IGF1 in various tissues and organs is negatively regulated by the level of blood GCs, indicating that the expression of IGF1 is inhibited at high GC levels, which appears to be a physiological rule. Therefore, based on previous systematic research, the authors' team proposed the programming theory of "GC-IGF1 axis," which suggests that excessive GC can reduce IGF1 expression in multiple organs, leading to organ dysplasia. Furthermore, a recent study indicated that exposure to dexamethasone during pregnancy can result in the inhibition of the maternal hypothalamus-pituitary-adrenal (HPA) axis, leading to a significant decrease in maternal blood GC levels, and ultimately causing the offspring to be in a low-GC environment. Moreover, in this low-GC environment, the expression of IGF1 in fetal offspring testicles is also inhibited. Furthermore, due to the continuous inhibition of the offspring's adrenal function after birth, they remain in a low-level GC environment, leading to a reduction in local IGF1 expression in the testis. These consistent changes in GC levels and IGF1 expression in the testicle before and after birth can be considered as positive programming of the "GC-IGF1" axis in the testicle, resulting in testicular dysplasia in offspring (Liu et al. 2022b). It is suggested that the inhibited steroidal synthesis function of fetal testicles caused by abnormal GC levels during pregnancy mediates low testosterone levels and testicular dysplasia in adulthood. This change in testicular development may be related to the programming of the "GC-IGF1" axis.

4.2 Changes in Neuroendocrine Programming

The term "intrauterine programming" refers to the alteration of the original developmental process of the fetus due to its exposure to an abnormal environment during development, resulting in long-lasting or permanent changes in body function. Fowden's hypothesis of "endocrine programming of intrauterine development" is widely accepted, suggesting that an adverse intrauterine environment can cause changes in fetal HPA axis development. Research on both humans and animals has confirmed that the fetal HPA axis and its hippocampal regulatory center hippocampus are vulnerable to adverse environmental influences during pregnancy, leading to increased sensitivity of the HPA axis to stress response after birth (Glover et al. 2010). The authors' team has demonstrated that fetal rats exposed to exogenous substances (such as caffeine, nicotine, and ethanol) during pregnancy were overexposed to intrauterine maternal GC and exhibited high HPA axis sensitivity to multiple chronic stresses as adults (Xia et al. 2014). Additionally, there is a close interaction between the HPT axis and the HPA axis (Fig. 19.2): excessive activation of the HPA axis can inhibit the functional development of the HPT axis, while the HPT axis can regulate the functional state of various components of the HPA axis. Therefore, the abnormal development of the HPA axis caused by an adverse environment during pregnancy (mainly manifested as abnormal GC levels) can directly

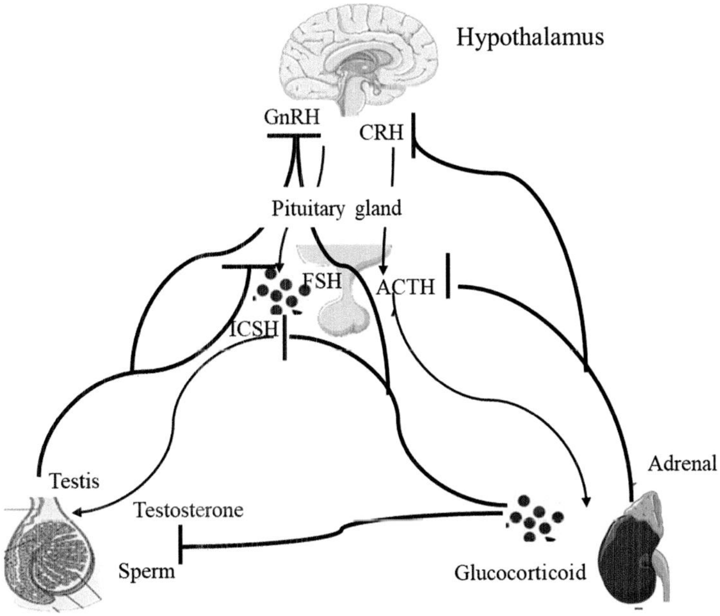

Fig. 19.2 Interaction of hypothalamic-pituitary-testicular axis and hypothalamic-pituitary-adrenal axis

lead to testicular dysplasia, or indirectly affect the activity of the HPT axis, leading to hormone secretion dysfunction, and ultimately impacting the testicular structure and functional development.

Exposure to dexamethasone during late pregnancy can have an adverse impact on the secretion of hypothalamic gonadotropin-releasing hormone (GnRH) and the migration, quantity, and distribution of neurons in the fetal brain, leading to decreased reproductive function and affecting the structural development of the hypothalamus in male offspring (Sullivan et al. 2010). In a study conducted on zebrafish, researchers treated the subjects with 2 mg/mL dexamethasone during early life and found that the expression of GnRH2 and GnRH3 in the hypothalamus increased, which was linked to the enhanced secretion of Kisspeptins polypeptides coding for tumor metastasis suppressant Gene-1 after dexamethasone treatment in the highly sensitive stage of the hypothalamic GR system. Additionally, exposure to substances that affect the proliferation and differentiation of pituitary cells during embryonic development is critical to pituitary development and function. For example, when pregnant mice are administered a high dose of dexamethasone during the late gestation period, which is the critical fetal pituitary development stage, the volume of the fetal pituitary gland becomes smaller, the number of FSH and LH cells decreases, and gonadotropin cell dysplasia results in abnormal testicular morphological and functional development. Crews and McLachlan (2006) also supported the idea that GC overexposure could affect gonadal development by reducing LH release. Therefore, the development of the HPT axis is crucial for regulating testicular development and function. In conclusion, GC exposure during pregnancy can disrupt the development of GnRH neurons and gonocytes in the embryonic brain of the offspring, leading to adverse effects on testicular development and reproductive function.

4.3 Oxidative Stress and Mitochondrial Damage

Adverse events experienced during pregnancy often result in the onset of oxidative stress, mitochondrial damage, and cell apoptosis in various tissues. The initial phase of steroid hormone synthesis takes place within the mitochondria, which is essential for the maintenance of steroid hormone synthesis function. Numerous studies have demonstrated that exogenous substances and disease states can lead to the production of various reactive oxygen species. An excessive production of reactive oxygen species or a lack of antioxidants such as glutathione can result in the destruction of the mitochondrial membrane, leading to a reduction in steroid hormone synthesis function, which is closely associated with oxidative stress and cell apoptosis. For instance, Kilic et al. (2012) found that exposure to smoke during pregnancy increased testicular Leydig cell apoptosis in newborn rats, and DNA damage was observed in the testicular epithelium and Leydig cells. In addition, Yuksel et al. (2014) reported that testicular Sertoli cells undergo oxidative stress, which results in decreased dependence on aspartic-specific cysteine proteinases, leading to increased apoptosis as well as independent DNA fragments, resulting in testicular structural damage during the neonatal period. Exposure to a phthalate mixture can also lead to oxidative stress and apoptosis of the testis and epididymis and affect spermatogenesis by damaging the spermatogonial cells or testicular Sertoli cells. Furthermore, it can also damage Leydig cells by producing reactive oxygen, resulting in increased levels of oxygen free radicals, decreased steroid system expression, and decreased testosterone synthesis and secretion.

5 The Role of Epigenetic Modification in the Development of Fetal-Originated Testicular Dysplasia

Epigenetic modification refers to heritable changes in gene expression that occur without alterations to the DNA sequence, including DNA methylation, histone modification, and noncoding RNA. These modifications are a normal part of higher eukaryotic cell development, but abnormal changes can have a significant impact on the intrauterine programming of fetal-originated testicular dysplasia.

5.1 Epigenetic Modification and Fetal-Originated Testicular Dysplasia

The researchers in this study aimed to investigate the negative effects of exogenous substance exposure during pregnancy on testicular development and its correlation with abnormal epigenetic modification. For instance, Wu et al. (2010) found that offspring mice exposed to phthalates during pregnancy exhibited reduced testosterone secretion and increased activity of DNA methyltransferases (DNMTs) in the testis before and after birth. This suggests that increased DNA methylation is a crucial mechanism of prenatal phthalate exposure leading to the developmental toxicity of offspring testis. In another study, Li et al. (2014) reported that exposure to bisphenol A (BPA) in late gestation rats resulted in the demethylation of the Gck promoter-314 site in offspring rats' sperm, decreased total sperm methylation, and sperm count. Alamdar et al. (2017) conducted cell experiments and found that H3K9 methylation in the 3β-HSD promoter region of mouse testicular Leydig cells increased following exposure to arsenic, resulting in reduced 3β-HSD expression and testosterone synthesis. Additionally, Song et al. (2014) reported that exposure to 1, 1-di-2, 2-dichloroethylene during pregnancy led to decreased spermatogenesis, reduced motility, and abnormal morphology in adult rats, ultimately resulting in decreased fertility. The possible mechanism was related to increased methylation of IGF2.

Exposure to BPA during pregnancy can reduce the synthesis function of testosterone in offspring mice due to decreased expression of steroidogenic acute regulatory protein (StAR) during testosterone synthesis, which is regulated by reduced acetylation levels of H3 histones in the StAR promoter region (Hong et al. 2016). Similarly, exposure to dexamethasone during pregnancy resulted in abnormal testicular morphology and dysplasia in adult rats due to decreased levels and expression of H3K9ac in the StAR promoter region. Exposure to enrolling during pregnancy can downregulate the expression of specific miRNAs, such as miR-23b and miR-21, in offspring mice germ cells, leading to an imbalance in the let-7/Blimp1 pathway, without significant changes in genome-wide DNA methylation (Brieño-Enríquez et al. 2015). Moreover, dexamethasone exposure during pregnancy can inhibit testosterone synthesis function in offspring rats through the miR-466b-3p/HDAC7 and miR-124-3p/HDAC5 pathways (Liu et al. 2022a, b). These findings suggest that miRNA regulation mediates the decline of offspring sperm quality. Adverse events during pregnancy can lead to abnormal epigenetic modification of the reproductive system, which can impact its development and function. Abnormal epigenetic modification is the primary intrauterine programming mechanism of reproductive system phenotype change.

5.2 Intergenerational Inheritance of Fetal-Originated Testicular Dysplasia Induced by Adverse Environment During Pregnancy

There is mounting evidence supporting the transgenerational properties of epigenetic inheritance, which can elucidate how the intrauterine environment impacts fetal development in the absence of genetic sequence changes and how this impact persists into adulthood, the next generation, and even subsequent generations (Singh et al. 2017).

Currently, the transgenerational inheritance effect of testicular dysplasia resulting from an adverse prenatal environment is mainly associated with prenatal toxic exposure. It has been discovered that prenatal exposure to ethylene sclerotia can alter the expression of DNMTs in offspring testicles, leading to the establishment of a new imprinted site in germ cells that can be passed on for multiple generations (Skinner et al. 2013). Offspring exposed to dioxins during pregnancy showed abnormal male prostate function that persisted into the F3 generation, and new DNA methylation sites were identified in F3 germ cells (Manikkam et al. 2012). Radionuclides, such as uranium, are highly deleterious, and recent studies have demonstrated that they can impact the development of the male reproductive system through epigenetic inheritance. After exposure to an acute toxic concentration of uranium (40 mg/L), male rats exhibited significant changes in DNA methylation and DNMT gene expression in testicular tissue over multiple generations. Exposure to dexamethasone during pregnancy can lead to a decrease in blood testosterone levels in male offspring (F1 and F2 generations), as well as a reduction in the number and diameter of spermatogenic tubules per unit area and sperm motility (Jeje and Raji 2017). Prenatal exposure to phthalate has been shown to result in a 30% incidence of undescended testicles in the F1 generation, a 12.5% incidence in the F2 generation, and no undescended testicles in the F3 and F4 generations. In the F1 generation, spermatogenic epithelium atrophy and a small amount of germ cell apoptosis were observed, which improved in the F2 generation and tended to be normal in the F3 and F4 generations (Chen et al. 2015). Additionally, a recent study has demonstrated that low expression of miR-466b-3p in oocytes can be programmed to cause the transgenerational inheritance of testosterone synthesis function inhibition by targeting HDAC7 in offspring testis in a rat model exposed to dexamethasone during pregnancy (Liu et al. 2022a). Therefore, testicular dysplasia in offspring resulting from adverse environmental exposure during pregnancy may persist for several generations.

6 Summary and Prospect

Reproductive health in men has recently garnered significant attention, resulting in the identification of more fetal reproductive abnormalities and a series of fundamental research. The molecular mechanisms involved in endocrine function damage are intricate and unclear, necessitating a comprehensive and detailed investigation. In recent years, the study of epigenetic modifications has emerged as a popular research topic. The analysis of epigenetic modifications offers a novel approach for comprehending the multigenerational inheritance of fetal-originated disease. These investigations play a pivotal role in the early detection and diagnosis of neonatal male reproductive dysplasia and adult reproductive diseases, in addition to having significant implications for the perinatal health care of pregnant women.

References

Alamdar A, Xi G, Huang Q, et al. 2017. Toxicol Appl Pharmacol, 326(8): 7–14.
Boisen K A, Chellakooty M, Schmidt I M, et al. 2005. J C lin E ndocrinol M etab, 90(7): 4041–4046.
Borges C D S, Pacheco T L, da Silva K P, et al. 2017. Reprod Toxicol, 71: 108–117.
Brieño-Enríquez M A, García-López J, Cárdenas D B, et al. 2015. PLoS One,10(4): e0124296.
Cabaton N J, Canlet C, Wadia P R, et al. 2013. Environ Heal Persp, 121(5): 586–593.
Chen J J, Wu S D, Wen S, et al. 2015. Plos One, 10(7): 2049–2051.
Crews D, McLachlan J A. 2006. Endocrinology, 147(6): S4–10.
Ganmaa D, Wang P Y, Qin L Q, et al. 2001. Med Hyp, 57(4): 510–514.
Genovese P, Nunez M E, Pombo C, et al. 2010. Reprod in Dom Anim, 45(2): 233–236.
Glover V, O'Connor T G, O'Donnell K. 2010. Neur Biobehav Rev, 35(1): 17–22.
Hong J, Chen F, Wang X L, et al. 2016. Mol and Cell Endocrin, 427(C): 101–111.
Jeje S O, Raji Y. 2017. J Dev Orig Health Dis, 8(1): 101–112.
Kilic S, Yuksel B, Lortlar N, et al. 2012. J Maternal-Fet Neon Med, 25(10): 1904–1908.

Li G, Chang H, Xia W, et al. 2014. Toxicol Lett, 228(3): 192–199.
Liu M, Chen B, Pei L, et al. 2018. Toxicol, 408: 1–10.
Liu M, Zhang Q, Pei L G, et al. 2019. Epigenetics, 14(3): 245–249.
Liu Y, Liu M, Liu Y, et al. 2022a. Biochem Pharmacol, 199: 115018.
Liu M, Liu Y, Pei L G, et al. 2022b. Acta Pharmacol Sin. 43(6): 1461–1472.
Ma T, Yin X Q, Han R T, et al. 2017. Int J Env Res Pub He, 14(10): 1284.
MacLeod D J, Sharpe R M, Welsh M, et al. 2010. Intern J of Androl, 33(2): 279–286.
Main K M, Skakkebaek N E, Toppari J. 2009. Endocrin Invol in Dev Syn, 14: 167–173.
Manikkam M, Tracey R, Guerrero-Bosagna C, et al. 2012. PloS One, 7(9): e46249.
Ni Q B, Wang L L, Wu Y P, et al. 2015. Toxicol Lett, 238(2): 117–125.
Page K C, Sottas C M, Hardy M P. 2001. J Androl, 22(6): 973–980.
Pedrana G, Sloboda D M, Perez W, et al. 2008. Anat Histol Embryol, 37(5): 352–358.
Pei L G, Zhang Q, Yuan C, et al. 2019. J Endocrinol, 242(1): M17-M32.
Preiksa R T, Zilaitiene B, Matulevicius V, et al. Hum Rep rod, 2005, 20(7): 1928–1932.
Ramlau-Hansen C H, Toft G, Jensen M S, et al. 2010. Hum Reprod, 25(9): 2340–2345.
Rhind S M, Rae M T, Brooks A N. 2001. Reproduction, 122(2): 205–214.
Rodriguez-Gonzalez G L, Vigueras-Villasenor R M, Millan S, et al. 2012. J Dev Orig Health Dis, 3(5): 321–326.
Singh D, Bhagat S, Raijiwala P, et al. 2017. Chemosphere, 185: 376–385.
Skakkebaek N E. 2004. Inter J Androl, 27(4): 189–191.
Skinner M K, Haque B M, Nilsson E, et al. 2013. PloS One, 8(7): e66318.
Song Y, Wu N X, Wang S M, et al. 2014. Hum Reprod, 29(11): 2512–2521.
Sullivan T M, Micke G C, Greer R M, et al. 2010. Anim Reprod Sci, 118(2–4): 131–139.
Wang L, Shen L, Ping J, et al. 2014. Toxicol Lett, 224(3): 311–318.
Weber R F A, Pierik F H, Dohle G R, et al. 2002. Bju Inter, 89(2): 143–148.
Wu S, Zhu J, Li Y, et al. 2010. Int J Toxicol, 29(2): 193–200.
Xia LP, Shen L, Kou H, et al. 2014. Toxicol Lett, 226(1): 98–105.
Xing J S, Bai Z M. 2018. Life Sci, 194: 120–129.
Xu D, Bai J, Zhang L, et al. 2015. Toxicol Res, 4(1): 112–120.
Yuksel B, Kilic S, Lortlar N, et al. 2014. ISRN Obstet Gyn, 2014: 170124.
Yun H J, Lee J Y, Hela K M. 2016. Biochem Biophys Res Commun, 471(1): 149–155.
Zhang Q, Pei L G, Liu M, et al. 2020. Food Chem Toxicol, 406: 115243.

Chapter 20
Fetal Origin of Premature Ovarian Insufficiency

Dan Xu and Jing Huang

Abstract The normal development of ovarian structure and function is crucial for maintaining female fertility. However, the prevalence of premature ovarian insufficiency (POI) in women has been increasing in recent years, which significantly affects the reproductive health of women of reproductive age and poses long-term risks, such as an increased risk of multiple diseases (e.g., bone and cardiovascular diseases). Researchers are increasingly interested in intrauterine environmental alterations and early programming of human reproductive health based on the "Developmental Origins of Health and Disease" theory. The mechanism of fetal-originated ovarian disease is associated with altered "intrauterine programming" during the fetal period. The adverse environment during pregnancy (e.g., exposure to environmental toxins, drugs, smoking, excessive alcohol, or coffee intake) affects maternal health, the structural and functional development of the offspring's ovaries, and their reproductive health in adulthood. This chapter provides a comprehensive summary of the phenomena and mechanisms of fetal-originated POI.

1 Introduction

The ovary is a crucial reproductive and endocrine organ in women, responsible for producing and discharging oocytes, secreting steroid hormones, and maintaining the balance of the endocrine system, feminine characteristics, and expected physiological metabolism of the body. However, the prevalence of premature ovarian insufficiency (POI) in women has been increasing in recent years, significantly

D. Xu (✉)
Department of Pharmacology, School of Pharmaceutical Sciences, Wuhan University, Wuhan, Hubei, China
e-mail: xuyidan70188@whu.edu.cn

J. Huang
Department of oto-rhino-laryngologie, Zhongnan Hospital of Wuhan University, Wuhan, Hubei, China

H. Wang et al. (eds.), *Fetal Origin of Diseases*,
https://doi.org/10.1007/978-981-97-5730-5_20

affecting the reproductive health of women of reproductive age and posing long-term risks, such as an increased risk of multiple diseases (e.g., bone and cardiovascular diseases). Women's ovarian dysplasia and abnormal reproductive function have an intrauterine developmental origin and are associated with altered "intrauterine programming" (Aiken et al. 2016). Epidemiological investigations and animal studies have demonstrated that adverse environmental exposures during pregnancy, such as exposure to environmental toxins, drug exposure, poor lifestyle habits, and poor nutrition, may not only affect the health of the parental generation but also adversely impact the ovarian structure, functional development, and reproductive health of the offspring in adulthood (Camlin et al. 2017; Weinberg et al. 1989), with xenobiotic exposures being a recent hot topic of research (Johansson et al. 2017). The presence of factors that interfere with oocyte apoptosis, primordial follicle pool size, and follicle depletion throughout the regulation of ovarian development can affect women's fertility and the length of the reproductive years. Adverse environments during pregnancy can affect the development of the offspring's ovaries and, consequently, their fertility in adulthood (Wang et al. 2017; Saatcioglu et al. 2016; Anderson et al. 2015).

2 Current Research Status of POI

POI is a clinical syndrome in which women experience a decline or loss of ovarian function before the age of 40. In recent years, the incidence of POI has been increasing annually, and the trend is toward a younger age. Due to current social and socioeconomic development and environmental impact, more and more women are choosing to delay their first pregnancy until after the age of 30. Consequently, POI has become a significant cause of female infertility in modern times.

2.1 Current Situation of POI

POI is characterized by ovarian hypofunction that occurs in women before the age of 40 years and is characterized by abnormal menstruation (amenorrhea, scanty or frequent menstruation), elevated gonadotropin levels [follicle-stimulating hormone (FSH) >25 U/L], and fluctuating decreases in estrogen levels (European Society for Human et al. 2016). Premature ovarian failure (POF) is the terminal stage of POI (Welt 2008). Epidemiological surveys have shown that the prevalence of POI is 0.1% in women aged 15–29 years old and 1% in women aged 30–39 years old (Goswami and Conway 2005). POI can have adverse effects on women's physical and psychological health, with short-term effects similar to spontaneous menopause and long-term effects affecting multisystem function, including abnormalities in kinetic, cardiovascular, neurological, and genitourinary functions. Therefore, the direct effect of POI is a decrease in the level of hormones secreted by the ovaries, which is an important reason for the multisystemic effects.

2.2 *The Main Mechanism of POI*

The etiology of POI is multifaceted and exhibits high heterogeneity. It has been associated with various factors, including chromosomal abnormalities, autoimmune diseases, infections, and medical, environmental, or psychological factors (De Vos et al. 2010). However, despite extensive research, the etiology of POI remains unclear in more than half of the patients. The pathogenesis of POI is characterized by disrupted follicular development and altered sex hormone levels, which can be induced by various factors such as decreased primordial folliculogenesis, hyperactivation, atresia, or increased apoptosis. These factors can lead to the development of POI, although the precise mechanisms underlying this process are not yet fully understood.

2.2.1 Depletion of the Primordial Follicle Pool

Various population surveys have indicated that the proportion of chromosomal abnormalities in patients with POI ranges from 10% to 12%, of which 94% were X chromosome abnormalities. Abnormalities in the structure or number of X chromosomes have been linked to folliculogenesis disorders, leading to POI (Jiao et al. 2017). Over 20 genetic mutations have been associated with POI, located on the X chromosome and autosomes. The expression abnormalities induced by these gene mutations can directly lead to accelerated follicular growth; promote follicular apoptosis and atresia, thereby causing an inadequate follicular pool reserve; and affect follicular granulosa cell proliferation and differentiation, ultimately inducing POI. For instance, mutations in folliculogenesis specific bHLH transcription factor (FIGLA) gene have been associated with POI (Tosh et al. 2015). In the Chinese population, FIGLA gene mutations are detected in 4% of patients with POI, who exhibit a relatively early onset of POI and an abnormal phenotype that is seen only in the ovaries (Pu et al. 2014).

2.2.2 Increased Follicle Apoptosis and Hyperactivation of Primordial Follicles

Abnormalities in the body's autoimmune system can lead to damage to ovarian tissues secondary to POI, and patients with POI frequently have two or more autoimmune diseases. The abnormal infiltration of lymphocytes in the ovary induces the production of various ovarian autoantibodies, such as anti-ovarian, anti-hyaline, and antinuclear antibodies, leading to an autoimmune response that results in apoptosis and accelerated follicular depletion, ultimately leading to POI (Dragojevic-Dikic et al. 2010). Certain medical factors, such as radiotherapy and surgery, may also cause direct damage to the ovaries. Current studies suggest that chemotherapeutic drugs acting on ovarian granulosa cells indirectly cause oocyte damage by disrupting communication between granulosa cells and oocytes, resulting in germ cell loss. Chemotherapeutic medications may also contribute to oocyte or granulosa cell apoptosis by interfering with mitochondrial DNA damage repair pathways. Additionally, chemotherapeutic drugs may induce oocyte apoptosis directly by affecting Wnt, phosphoinositide 3-kinase (PI3K), and other signaling pathways, indirectly accelerating primordial follicle activation, thus accelerating follicle depletion and inducing POI.

2.2.3 Follicular Arrest Occurs before the Sinus Follicular Stage

Environmental toxicants, such as deoxynivalenol, have been shown to induce POI by inducing massive follicular apoptosis and inhibiting the activity of the PI3K signaling pathway in follicles, as well as inhibiting the activity of the mammalian target of rapamycin complex (mTORC) signaling pathway, which affects normal follicular development (Shen et al. 2014; Keating et al. 2011). Moreover, women experiencing negative emotions, such as anxiety and depression, or significant psychological trauma may develop POI due to effects on the hypothalamic-pituitary-ovarian (HPO) axis or direct effects on ovarian function (Lv et al. 2021).

3 Evidence of Intrauterine Origin of POI

Epidemiological investigations and laboratory studies have provided evidence that offspring ovarian development is closely linked to the maternal environment during pregnancy. Folliculogenesis originates during the intrauterine period, and the typical ovarian structure is the foundation for the reproductive endocrine function of the ovary in adulthood. Consequently, POI has an intrauterine developmental origin. Any interfering factor that affects the size of the primordial follicle pool and the rate of follicle depletion can influence the ovarian reserve function and reproductive age in women. Both insufficient follicle numbers and accelerated follicular apoptotic atresia can lead to POI. Adverse environmental exposures during pregnancy can affect intrauterine ovarian development, leading to abnormal ovarian function and susceptibility to POI after birth.

3.1 Epidemiological Evidence

Epidemiological investigations have demonstrated that an adverse environment during pregnancy can result in intrauterine growth retardation (Deng et al. 2013), leading to infants who are small for their gestational age and primarily exhibit low birth weight at birth. In adulthood, these individuals may experience reduced ovarian volume, elevated FSH levels, and ovulation disturbance (Oluwole et al. 2016). These findings suggest that an adverse environment during pregnancy may lead to decreased ovarian reserve function, confirming that the intrauterine environment can affect reproductive function in offspring later in life. A study in the United States reported that among 660 girls, 19% and 39% of their mothers had prenatal smoking and prenatal smoke exposure, respectively. Prenatal smoke exposure negatively affected luteinizing hormone (LH) and inhibin B levels in their offspring, thereby affecting ovarian function (Gollenberg et al. 2015). A Danish pregnancy cohort study following the offspring of 436 women exposed to organochlorine pollutants in utero found that although menarche and the menstrual cycle were not affected, the number of ovarian follicles was significantly reduced, which could affect adult fertility (Kristensen et al. 2016).

3.2 Clinical and Basic Laboratory Research

Growing evidence suggests that maternal exposure to multiple exogenous sources, such as ethanol, caffeine, nicotine, or certain drugs, can affect the normal development of fetal ovaries. Studies have shown that phthalates disrupt ovarian function and affect ovarian reserve by interfering in different stages of follicular development, leading to susceptibility to POI and even having transgenerational genetic effects (Hannon and Flaws 2015; Zhang et al. 2015). The authors' team has confirmed that ethanol or caffeine exposure during pregnancy can cause abnormal ovarian development in female offspring, which persists into adulthood and leads to susceptibility to POI (Lv et al. 2021; Ni et al. 2019). Additionally, the authors' team has demonstrated that prenatal dexamethasone exposure has a toxic developmental effect on the ovaries of offspring, which can continue into the adult F1 generation with symptoms of POI susceptibility and further continue through oocytes to the adult F2 and F3 generations. This suggests that prenatal dexamethasone exposure has transgenerational genetic effects on inhibiting ovarian estrogen synthesis in offspring (Gong et al. 2021; Lv et al. 2018).

3.2.1 Changes in Ovarian Structure and Function before and after Birth

Numerous studies have demonstrated that an unfavorable intrauterine environment can have a significant impact on ovarian development and function (Table 20.1). Exposure to dexamethasone during pregnancy has been shown to reduce fetal ovarian volume, as well as the number of healthy primordial, primary, and secondary follicles, while also decreasing the number of apoptotic follicles (Ristic et al. 2008). This exposure may also have an effect on the reproductive capabilities of offspring in later stages. Animal experiments and in vitro studies have confirmed that maternal exposure to high doses of bisphenol A can significantly reduce the total ovarian follicle number in offspring (Zhu et al. 2018). Smoking can also negatively impact women's reproductive health, potentially through the influence of polycyclic aromatic hydrocarbons (PAH) in cigarette smoke on ovarian reserve. Fetal mouse ovaries cultured in PAH environments exhibited germ cell loss and were inhibited by selective PAH receptor antagonists. Exposure to smoke during pregnancy also significantly impacted fetal ovarian somatic cells. When rats were exposed to smoking from intrauterine to 2 weeks after birth, their offspring's ovarian granulosa cells were significantly apoptotic. The number of sinus follicles was reduced, which led to decreased ovarian reserve function in the offspring (Kilic et al. 2012).

The synthesis function of ovarian steroids during fetal development and after birth can also be negatively affected by an adverse intrauterine environment. Hormones secreted by the ovary, such as estrogen (E_2 and estrone), progesterone, and various androgens, can negatively regulate ovarian development and function. The synthesis of these hormones requires sequential catalysis by several enzymes. Metabolic disorders of these enzymes, including cholesterol carbostrepase deficiency leading to impaired steroid hormone synthesis or deficiency of key enzymes in steroid hormone synthesis leading to impaired sex hormone synthesis, can induce POI resulting from hypergonadotropemia. Studies have shown that exposure to environmental endocrine disruptors during pregnancy can alter endocrine hormone levels through fetal programming effects, ultimately leading to POI and other adverse health conditions, with transgenerational genetic effects (Rutkowska and Diamanti-Kandarakis 2016). Maternal exposure to high-dose caffeine during pregnancy has been shown to significantly reduce the ovarian quality of offspring rats, the number of primordial follicles, and the diameter of sinus follicles on days 7–120 after birth (Dorostghoal et al. 2011). In offspring mice exposed to nicotine during pregnancy, ovarian follicles and oocytes were decreased, oocyte oxidase levels were increased, and ovulation numbers were reduced (Camlin et al. 2016). Previous research by the authors' team has confirmed that prenatal exposure to dexamethasone, nicotine, and caffeine can inhibit fetal ovarian estrogen synthesis, leading to transgenerational genetic effects of abnormal ovarian development in offspring (Lv et al. 2018), ovarian dysfunction (Fan et al. 2019), and POI susceptibility (Lv et al. 2021), respectively. These findings suggest that the decrease of ovarian steroid synthesis caused by an adverse intrauterine environment may be the origin of adult ovarian-related diseases.

Table 20.1 Impacts of adverse prenatal environments on ovarian development and function

Adverse environment	Genera	Exposure chamber	Mechanism	Phenomenon	Reference
Dexamethasone	Mouse	Pregnant mice were given orally 125 g/kg· d dexamethasone until delivery	Induction of autophagy or combined with autophagy and apoptosis to enhance follicular atresia	Primordial follicle pool↓, granulosa cell layer↑, primary to secondary follicles↑, apoptotic follicles↑	Hulas-Stasiak et al. (2016)
Dexamethasone	Human being	Human fetal ovaries were cultured at 8–11 weeks after fertilization at 2,10, and 50 M dexamethasone for 14 days	The rate of apoptosis increased	Human fetal ovarian germ cell density↓	Poulain et al. (2012)
Dexamethasone	Rat	Pregnant rats subcutaneously received 0.2 mg/kg·d dexamethasone from gestational day (GD) 9–20	Insulin-like growth factor 1 (IGF1)/ steroidogenic factor 1 low-functional programming	Expression of ovarian steroid hormone enzymes↓, estradiol (E_2) genesis↓	Lv et al. (2018)
Dexamethasone	Rat	Pregnant Wistar rats subcutaneously received 0.2 mg/kg·d dexamethasone from GD 9–20	The mir-320a-3p expression was downregulated in the oocytes	Ovarian E_2 synthesis↓	Gong et al. (2021)
Bisphenol-A (BPA)	Sheep	Pregnant Suffolk ewes were treated with BPA (0.5 mg/kg·d) from GD 30–90	miRNAs regulating Sry-related high-mobility-group box (Sox) familly genes, kit ligand, and insulin-related genes	Altering the expression of steroid synthetase	Veiga-Lopez et al. (2013)
Smog	Rat	Wistar rats were exposed to smoke during pregnancy	DNA damage and apoptosis	Ovarian granulosa cells↓, number of follicles↓, birth weight in fetal mice↓	Kilic et al. (2012)

(continued)

Table 20.1 (continued)

Adverse environment	Genera	Exposure chamber	Mechanism	Phenomenon	Reference
Smog	Human being	Maternal ovaries from smoking during early pregnancy were collected	–	Number of somatic cells↓, the number of coenocytes was not changed	Lutterodt et al. (2009)
Nicotine	Rat	Wistar rats during GD 9–20 received nicotine 2.0 mg/kg·d hypodermic injection	Reduced levels of histone 3 lysine 9 acetylation (H3K9ac) and H3K27ac in the p450arom promoter region in ovarian granulosa cells	The number of oocytes↓, granulosa cell layer thickness↓, and E_2 production↓	Fan et al. (2019)
Caffeine	Rat	The pregnant Wistar rats were intragastrically administered caffeine (30 and 120 mg/kg·d) from GD 9–20	H3K27ac level of IGF1 promoter region and its expression were decreased in the ovary	Expression of ovarian steroid hormone enzymes↓, ovarian estradiologenesis↓	Lv et al. (2021)
Ethanol	Rat	The pregnant Wistar rats were intragastrically administered with 4 g/kg·d of ethanol from GD 9–20	Glucocorticoid-IGF1-axis programming abnormalities	Proportion of atretic follicles in adult rats↑, number of anti-Müllerian hormone-positive antral follicles↓	Ni et al. (2019)
Acetaminophen/ ibuprofen	Human being	In vitro culture of human fetal ovaries	Genetic changes in the germ cell surface, increased total histone H3K27me3, and altered expression of DNA methyltransferases 3A (DNMT3A) and DNMT3B	Number of germ cells↓, value-added rate↓	Hurtado-Gonzalez et al. (2018)
Acetaminophen	Rat	Acetaminophen was administered by gavage as a suspension in corn oil at a dose of 350 mg/kg	–	In F1 offspring, number of germ cells↓, ovarian size↓; in F2 offspring, ovarian size↓ and follicle number↓	Dean et al. (2016)

Adverse environment	Genera	Exposure chamber	Mechanism	Phenomenon	Reference
Di(2-ethylhexyl) phthalate (DEHP)	Mouse	Pregnant CD-1 mice were orally dosed with DEHP (20, 200 μg/kg·d and 500, 750 mg/kg·d) from GD11 until birth	–	Estrous cyclicity disrupts; E_2 levels↑ (F1 and F3), progesterone level↓ (F2), gonadotropin hormone levels were altered (F1 and F3)	Brehm et al. (2018), Rattan et al. (2018)
Perfluorononanoic acid (PFNA)	Mouse	PFNA (3 mg/kg·d) was orally administered from GD1–18	Increased fibroblast growth factor 21 production in postnatal female offspring impedes postnatal activation of suprachiasmatic nucleus-vasopressin neurons, which suppresses pubertal onset in periventricular nucleus -kisspeptin neurons and reproductive endocrine axis	Times of vaginal opening and first estrus were delayed; ovary size↓, numbers of primary↓, secondary, and antral follicles↓, corpora lutea↓	Zhang et al. (2021)
Vinclozolin/ dichloro-diphenyl-trichloroethane (DDT)	Rat	Pregnant rats were administered daily intraperitoneal injections of vinclozolin (100 mg/kg·d), DDT (25 mg/kg·d)	DNA methylation, ncRNA expression, and mRNA expression were altered in ovarian granulosa cells	–	Nilsson et al. (2018)
Chromium	Rat	Lactating dams were received 50 ppm potassium dichromate dissolved in drinking water	The oxidative stress was increased, and antioxicant enzyme levels were decreased in the ovary	Follicular atresia↑, steroidogenesis↓, and delayed puberty in F1 offspring	Stanley et al. (2014)
Maternal nutrient restriction	Rat	Pregnant rats were given a calorie-restricted (50% of control) diet during pregnancy	Phosphorylation of the phosphoinositide 3-kinase signaling pathway	Antral follicles↓, corpora lutea↓, atretic follicles↑	Chan et al. (2018)

The development and function of the ovary are crucial for female reproductive health. Adverse intrauterine environments can result in ovarian dysfunction, which can lead to reproductive dysfunction in offspring. Studies have demonstrated that exposure to di(2-ethylhexyl) phthalate (DEHP) during pregnancy can disrupt ovarian function and affect ovarian reserve by interfering with various stages of follicular development, leading to POF in offspring and transgenerational inheritance (Brehm et al. 2018; Rattan et al. 2018). In addition, maternal malnutrition during pregnancy has been shown to cause irregular estrus cycles in offspring mice, a significant reduction in follicles, luteal dysplasia, an increase in atretic follicles, and reproductive dysfunction in adulthood (Harrath et al. 2019).

3.2.2 Functional Alterations of the HPO Axis

The role of the HPO axis in female intrauterine development is not yet fully understood. However, after birth, the HPO axis function is fully established and activated, initiating puberty in females. The preoptic region of the hypothalamus expresses gonadotropin-releasing hormone (GnRH) neurons, which are central to regulating the HPO axis in mammals. Previous research by the authors' team found that in rats exposed to dexamethasone during pregnancy, although the HPO axis had a compensatory effect, hypothalamic GnRH did not promote an increase in pituitary FSH and LH release. Additionally, chronic stress significantly decreased blood estradiol (E_2) levels, leading to negative feedback promotion of FSH release at the pituitary level, while there was no change in feedback regulation at the hypothalamus level. This resulted in a dysfunction in the balanced feedback among the hypothalamus, pituitary, and ovary, ultimately affecting normal reproductive endocrine function (Lv et al. 2021).

4 Intrauterine Programming Mechanism of Fetal-Originated POI

The formation of primordial follicles is a critical process that occurs during the fetal period. Exposure to adverse factors during pregnancy can result in the inhibition of primordial follicle formation, premature activation of follicles, or oocyte apoptosis. The mechanism underlying fetal-originated POI involves abnormal signaling pathways associated with follicle development, an imbalance between oxidative defense and oxidative stress, abnormal epigenetic modification of developmental genes, and alterations in intrauterine developmental programming (Table 20.1).

4.1 Changes in the Programming of Oxidative Defense and Oxidative Stress Imbalance

Oxidative stress refers to a state of physiological stress where the body experiences damage due to internal and external factors, leading to interference with the metabolic activities of normal organs. This state arises from an imbalance between the production of free radicals and the body's antioxidant capacity. Research has indicated that excessive maternal inflammation and oxidative stress can negatively impact fetal development, as chronic inflammation in pregnant mice has been linked to growth restriction and a reduction in the proportion of follicles in fetal mice, which can seriously affect the fertility of offspring (Shalom-Paz et al. 2017). The exposure of pregnant mothers to smoke has also been shown to increase oxidative stress and reduce the number of oocytes in offspring, leading to a POI phenotype. Moreover, lactating maternal exposure to environmental pollutants such as hexavalent chromium can cause increased oxidative stress levels, increased atretic follicles, decreased steroid synthesis, and other POI symptoms in their offspring ovaries (Stanley et al. 2014). Additionally, exposure to lead and chromium during pregnancy can lead to a decrease in the activity of enzymatic antioxidants such as superoxide dismutase, catalase, and glutathione peroxidase in the ovary of offspring, while nonenzymatic antioxidants are significantly consumed. This exposure can also decrease the expression of steroid synthase and reduce the levels of estradiol and progesterone, accompanied by symptoms of ovarian dysfunction (Pillai et al. 2010). Maternal malnutrition has also been found to increase ovarian oxidative stress in offspring, which can mediate a decrease in ovarian follicle number and the expression of regulatory genes.

4.2 Alterations in Signaling Pathway Programming Related to Folliculogenesis

Multiple signaling pathways are involved in follicular development, such as the PI3K/AKT signaling pathway, mTOR signaling pathway, Bmp4/Smad signaling pathway, Notch signaling pathway, and Wnt signaling pathway.

The PI3K/AKT signaling pathway plays a crucial role in promoting primordial follicle development through AKT, and mutations in PI3K or AKT can lead to premature activation and development of follicles, ultimately leading to decreased fertility in female animals. Furthermore, abnormalities in forkhead box O transcription factor 3a (FOXO3a), P27, and ribosomal protein s6 can also cause premature follicle development and infertility (Maidarti et al. 2020). Specific inhibitors such as 4-vinylcyclohexene diepoxide can bind to and inhibit the autophosphorylation of the c-kit receptor, leading to decreased posttranscriptional regulation of PI3K downstream and affecting oocyte activity. Conversely, compounds such as cyclophosphamide and bisphenol A can increase the activation of PI3K/AKT and activate primordial follicles, resulting in the consumption of large follicles and eventual changes in ovarian reserve function (Zhou et al. 2017; Ganesan and Keating 2016). Maternal malnutrition during pregnancy has also been found to cause the phosphorylation of ovarian PI3K signaling in offspring, resulting in accelerated recruitment of primordial follicles, increased follicular atresia, and reproductive dysfunction, which can lead to POI symptoms (Chan et al. 2018).

The activation of the mTOR signaling pathway is closely linked to the reproductive process, as it regulates the formation, recruitment, growth, dominance, maturation, and ovulation of ovarian follicles in female mice (Chen et al. 2020). Research has shown that the use of an mTOR activator can significantly increase follicle diameter, the proportion of primary and secondary follicles, and the number of follicle membrane cells in ovaries cultured in vitro. Conversely, an inhibitor had the opposite effect, indicating that the mTOR signaling pathway plays a crucial role in the development of follicles and follicle membrane cells (Zhang et al. 2017).

The BMP4/Smad signaling pathway also plays a critical regulatory role in folliculogenesis. Activation of this pathway can promote the survival of primordial follicles and their transformation to primary follicles while inhibiting oocyte apoptosis, potentially due to the upregulation of the expressions of sperm-egg binding proteins basic helix protein 2 and c-kit genes. Furthermore, the Notch signaling pathway plays a crucial role in regulating the early stage of follicular development, participating in the promotion of the formation and recruitment of primordial follicles. Abnormalities in this signaling pathway can lead to abnormal ovarian development, such as the developmental inhibition of primordial follicles, the shedding of granulosa cells, and even the appearance of polyoocyte follicles. The Wnt signaling pathway is expressed in various stages of follicular development and regulates normal growth, maturation, ovulation, and luteal formation of follicles through multiple pathways (Chen et al. 2020).

4.3 Epigenetic Mechanism and Multigenerational Genetic Effect of Fetal-Originated POI

The development of POI may be regulated by epigenetic inheritance, with subsequent establishment of DNA methylation modifications during ovarian maturation. A study by Fan et al. (2019) found that prenatal nicotine exposure reduced the levels of H3K9ac and H3K27 ac in the P450arom promoter region in ovarian granulosa cells of rat offspring, leading to decreased P450arom expression and E_2 production, ultimately affecting the ovarian development of the offspring. Additionally, Lv et al. (2021) discovered that high glucocorticoid (GC) concentrations induced by prenatal caffeine exposure could also affect steroid synthesis function by altering the histone acetylation level in the promoter region of cytochrome P450 19A1 (CYP19A1) in ovarian granulosa cells (Fig. 20.1).

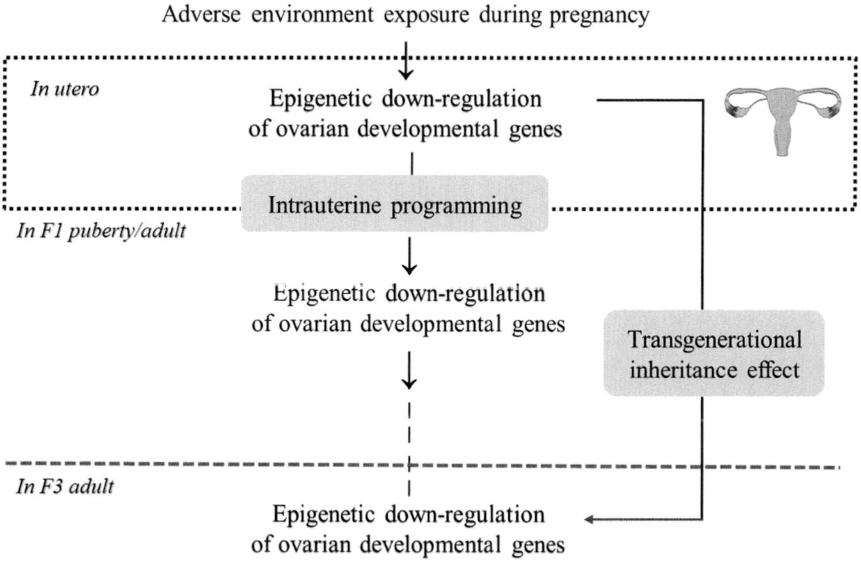

Fig. 20.1 The impact of transgenerational inheritance on fetal-originated POI

Oocytes establish relevant imprinting patterns as each follicle is recruited into the growing follicular pool. However, exposure to environmental chemicals such as bisphenol A during imprint establishment can disrupt the imprinting pattern, resulting in altered gene expression sites and developmental abnormalities in the embryo and placenta (Susiarjo et al. 2013). During germline epigenetic reprogramming, some repeats such as LINE1 lose their methylation markers and are re-acetylated with imprinted sequences, while sequences with high mutational potential such as IAP are protected from demethylation. This is believed to be the epigenetic mechanism mediating transgenerational inheritance (Fang et al. 2016; Zhao et al. 2015).

Another mechanism of epigenetic reprogramming in oocytes involves populations of small noncoding RNAs (sncRNAs), including piwi-interacting RNAs (piRNAs), microRNAs (miRNAs), nuclear sncRNAs, and endogenous small interfering RNAs (endo-siRNAs) (Nilsson et al. 2018; Sharma et al. 2018). These noncoding RNAs are associated with gamete-mediated transgenerational inheritance. Adverse environmental exposure during pregnancy can mediate the transgenerational inheritance of disease phenotypes through changes in ovarian epigenetic modifications (Fig. 20.1). Numerous studies have shown that miRNAs can regulate ovarian function by targeting different cells and factors. For example, miR-26b promoted ovarian granulosa cell apoptosis during follicle atresia by targeting ataxic telangiectasia mutation genes, while miR-23a upregulation in the plasma of POI patients decreased the X-linked inhibitor of apoptosis protein and caspase-3, leading to increased apoptosis of human granulosa cells (Lin et al. 2012; Yang et al. 2012). Exposure to environmental estrogen during fetal development reduced miRNAs targeting Sry-related high-mobility-group box (SOX) familly genes, kit ligand, and insulin-related genes in sheep, potentially leading to abnormal follicular assembly and defects in animals exposed to environmental estrogen (Veiga-Lopez et al. 2013). Moreover, prenatal dexamethasone exposure was found to decrease the expression of ovarian miR-320a-3p, which was associated with the transgenerational genetic effect of inhibiting estrogen synthesis function in the offspring's ovaries (Gong et al. 2021).

4.4 The "Two-Programming" Mechanisms of Fetal-Originated POI

The authors' team have conducted research that confirms the impact of prenatal dexamethasone exposure on miRNA-17-5p expression in ovarian granulosa cells and the inhibition of forkhead box L2 expression via the activation of local ovarian glucocorticoid receptor (GR). This results in decreased fetal ovarian reserve and impaired estrogen synthesis function. The epigenetic modifications and low ovarian function programming changes that result from this prenatal exposure may persist after birth, increasing the susceptibility to POI (Gong et al. 2021). Building upon previous studies, the authors' team propose a "two-programming" mechanism for fetal-originated POI. This mechanism suggests that adverse prenatal environments, such as xenobiotic exposure, can cause fetal overexposure to maternal or exogenous high GC environments. This overexposure can activate fetal ovarian GR and cause epigenetic modifications of important genes, including ovarian development genes and steroid synthetase, with the participation of epigenetic enzymes. As a result, estrogen synthesis inhibition or follicle development disorders may occur, leading to adult POI susceptibility. This continuous change in pre- and post-birth consistency is referred to as "first programming."

Insulin-like growth factor (IGF1) is a crucial regulator of follicle development and the expression of the steroid synthase system. Low IGF1 expression in the ovary leads to decreased expression of the steroidal synthase system and abnormal ovarian development. IGF1 promotes cell proliferation by activating the PI3K/AKT and/or MAPK/ERK signaling pathways through its receptor IGF1R. Studies have shown that IGF1/IGF1R signaling induces the proliferation of ovine ovarian granulosa cells, steroid synthetase expression, and estrogen synthesis mainly through the PI3K/AKT pathway (non-MAPK/ERK pathway) (Mani et al. 2010). In fetal tissue, high GC negatively regulates IGF1 expression by inhibiting its expression in multiple tissues through GR, and there exists a negative correlation between blood GC level and tissue IGF1 signaling pathway. The authors' team conducted a series of studies that confirmed fetal ovaries exhibit low-functional development under exposure to exogenous substances, such as caffeine, during pregnancy.

Furthermore, they found that the expression of IGF1 in fetal ovarian tissue is decreased (Lv et al. 2021). Under normal circumstances, after offspring birth, the decrease in blood GC level is coupled with an increase in ovarian IGF1 expression. However, chronic stress (high GC level) can reproduce the inhibition of ovarian IGF1 expression in offspring under intrauterine maternally high GC. This negative correlation between blood GC level and ovarian IGF1 expression suggests a negative feedback regulatory relationship between them, known as the "GC-IGF1 axis" programming. This "second programming" further inhibits ovarian development and functional differentiation in utero, resulting in reduced fetal ovarian estrogen synthesis and oocyte number. Additionally, it induces overactivation of follicles in the early postnatal period, leading to reduced adult oocyte number and POI susceptibility.

4.5 The "Two-Strike" Mechanisms of Fetal-Originated POI

The adverse environment during pregnancy is the "first strike" to the ovaries, resulting in fetal ovarian dysplasia through the "two-programming" mechanism, which continues after birth. Offspring rats in a low basal activity environment of the hypothalamic-pituitary-adrenal (HPA) axis after birth promote ovarian IGF1 expression through hypoglycemic corticosteroid levels, and low levels of estrogen act on the HPO axis through negative feedback, leading to ovarian catch-up growth and a compensatory response to estrogen synthesis function. Recent studies have suggested that adverse environments during pregnancy can increase the stress sensitivity of the HPA axis after birth, increase GC levels, and increase susceptibility to various diseases after birth. The authors' team has confirmed that prenatal caffeine exposure in offspring rats results in high-stress sensitivity of the HPA axis in adulthood, accompanied by low expression of IGF1 in ovarian tissue, decreased estrogen synthesis function and blood estradiol concentration, increased blood FSH concentration, and inhibited GnRH expression in the hypothalamus. In terms of ovarian structure, there was a significant increase in the proportion of ovarian atretic follicles, sustained thinning of the granulosa layer thickness, and decreased anti-Mullerian hormone-positive presinus follicles and sinus follicles. These changes indicate that follicle development is accelerated after birth, and more follicles enter the apoptotic state in advance, indicating adult POI susceptibility. Chronic stress during puberty, induced by high GC, constitutes a "second strike" to the progeny and further aggravates the local IGF1 signaling pathway of the ovary and the function of estrogen synthesis by inhibiting the HPO axis, ultimately inducing POI (Fig. 20.2).

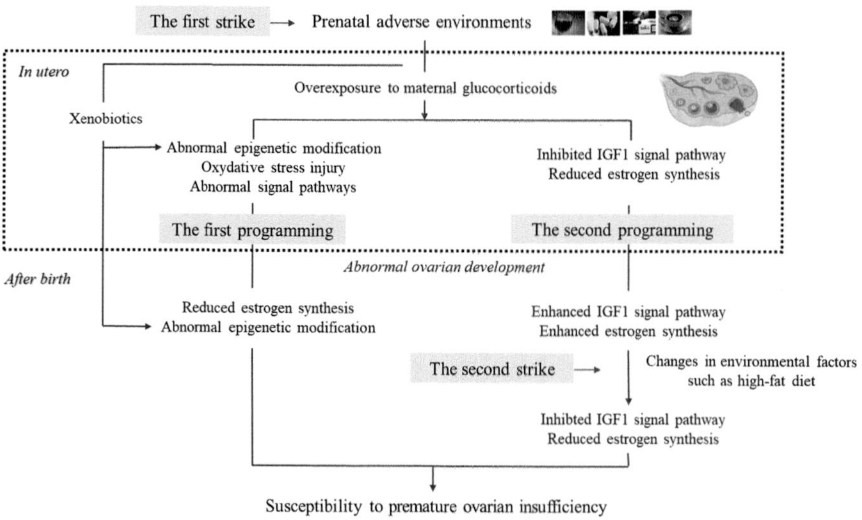

Fig. 20.2 The "two-programming and two-strike" mechanisms in the onset of fetal-originated POI

5 Summary and Prospect

Ovarian development is a complex process influenced by various factors such as heredity and environment. The adverse prenatal environment not only affects maternal health but also has irreversible effects on the development and function of the offspring's reproductive system. In this chapter, we aim to understand the intrauterine programming mechanism of fetal-originated POI by examining the effects of the adverse prenatal environment on fetal ovarian development and its pathogenesis. The mechanism of fetal-originated POI is associated with the abnormal signaling pathway of follicle development, the imbalance of oxidative stress and defense, the alteration of epigenetic modification of developmental genes, and the abnormal functional genes and organ functions caused by neuroendocrine and metabolic programming. These changes persist until after birth, resulting in ovarian dysfunction, such as impaired estrogen production, abnormal follicular development, and increased follicle consumption, leading to POI susceptibility (Fig. 20.3). However, the current research on the intrauterine programming mechanism of POI is limited. Further research on intrauterine programming changes and related epigenetic regulatory mechanisms of fetal-originated POI, as well as exploring its intervention targets, will provide new insights into the pathogenesis of fetal-originated POI and help develop early prevention and treatment strategies. This research has significant implications for guiding prepotency and enhancing people's quality of life.

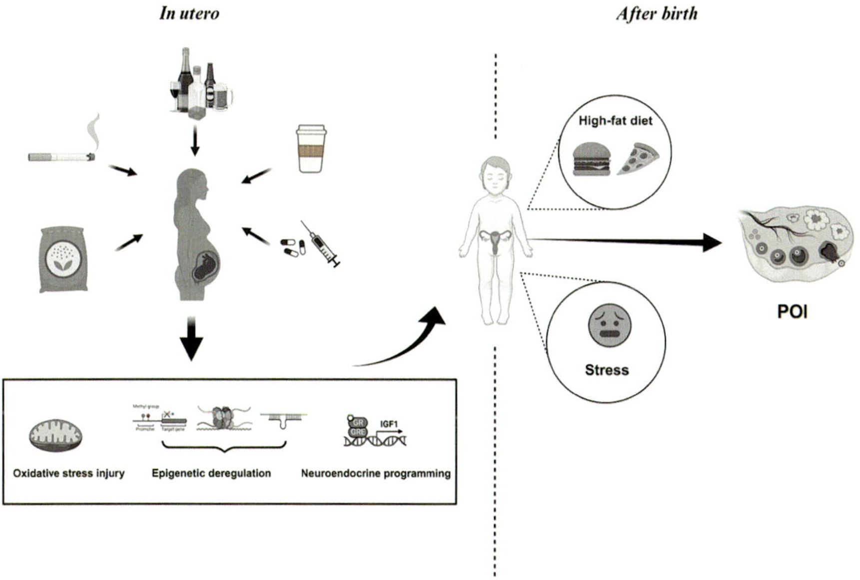

Fig. 20.3 Exploring the intrauterine developmental mechanisms in fetal-originated POI

References

Aiken C E, Tarry-Adkins J L, Ozanne S E. 2016. Mamm Genome, 27(7–8): 430–439.
Anderson R A, McIlwain L, Coutts S, et al. 2015. Mol Hum Reprod, 21(9): 753.
Brehm E, Rattan S, Gao L, et al. 2018. Endocrinology, 159(2): 795–809.
Camlin N J, Sobinoff A P, Sutherland J M, et al. 2016. Biol Reprod, 94(2): 39.
Camlin N J, Jarnicki A G, Vanders R L, et al. 2017. Hum Reprod, 32(6): 1270–1281.
Chan K A, Jazwiec P A, Gohir W, et al. 2018. Biol Reprod, 98(5): 664–682.
Chen Y, Yang W, Shi X, et al. 2020. Front Cell Dev Biol, 8: 575706.
De Vos M, Devroey P, Fauser B C. 2010. Lancet, 376(9744): 911–921.
Dean A, van den Driesche S, Wang Y, et al. 2016. Sci Rep, 6: 19789.
Deng Y, Cao H, Cu F, et al. 2013. Toxicol Appl Pharmacol, 269(1): 25–33.
Dorostghoal M, Mahabadi M K, Adham S. 2011. J Reprod Infertil, 12(1): 15–22.
Dragojevic-Dikic S, Marisavljevic D, Mitrovic A, et al. 2010. Autoimmun Rev, 9(11): 771–774.
European Society for Human R, Embryology Guideline Group on POI, Webber L, et al. 2016. Hum Reprod, 31(5): 926–937.
Fan G, Zhang Q, Wan Y, et al. 2019. Food Chem Toxicol, 128: 256–266.
Fang X, Poulsen R R, Wang-Hu J, et al. 2016. FASEB J, 30(9): 3238–3255.
Ganesan S, Keating A F. 2016. Toxicol Sci, 152(1): 169–180.
Gollenberg A L, Addo O Y, Zhang Z, et al. 2015. Horm Res Paediatr, 83(1): 36–44.
Gong X, Zhang J, Ge C, et al. 2021. Pharmacol Res, 165: 105435.
Goswami D, Conway G S. 2005. Hum Reprod Update, 11(4): 391–410.
Hannon P R, Flaws J A. 2015. Front Endocrinol (Lausanne), 6: 8.
Harrath A H, Alrezaki A, Alwasel S H, et al. 2019. J Dev Orig Health Dis, 10(5): 587–594.

Hulas-Stasiak M, Dobrowolski P and Tomaszewska E. 2016. Reprod Fertil Dev, 28(7): 1038–1048.
Hurtado-Gonzalez P, Anderson R A, Macdonald J, et al. 2018. Environ Health Perspect, 126(4): 047006.
Jiao X, Zhang H, Ke H, et al. 2017. J Clin Endocrinol Metab, 102(7): 2281–2290.
Johansson H K L, Svingen T, Fowler P A, et al. 2017. Nat Rev Endocrinol, 13(7): 400–414.
Keating A F, S M Fernandez , C J Mark-Kappeler, et al. 2011. Biol Reprod, 84(4): 743–51.
Kilic S, Yuksel B, Lortlar N, et al. 2012. J Matern Fetal Neonatal Med, 25(10): 1904–1908.
Kristensen S L, Ramlau-Hansen C H, Ernst E, et al. 2016. Environ Int, 92–93: 366–372.
Lin F, Li R, Pan Z X, et al. 2012. PLoS One, 7(6): e38640.
Lutterodt M C, Sorensen K P, Larsen K B, et al. 2009. Hum Reprod, 24(10): 2558–2566.
Lv F, Wan Y, Chen Y, et al. 2018. Endocrinology, 159(3): 1401–1415.
Lv F, Fan G, Wan Y, et al. 2021. Sci Total Environ, 789: 147691.
Maidarti M, Anderson R A, Telfer E E. 2020. Cells, 9(1).
Mani A M, Fenwick M A, Cheng Z, et al. 2010. Reproduction, 139(1): 139–151.
Ni Y, Xu D, Lv F, et al. 2019. J Endocrinol, JOE-19-0063.R0061.
Nilsson E, Klukovich R, Sadler-Riggleman I, et al. 2018. Epigenetics, 13(8): 875–895.
Oluwole O F, Salami S A, Ogunwole E, et al. 2016. J Basic Clin Physiol Pharmacol, 27(5): 483–491.
Pillai P, Pandya C, Gupta S, et al. 2010. J Biochem Mol Toxicol, 24(6): 384–394.
Poulain M, Frydman N, Duquenne C, et al. 2012. J Clin Endocrinol Metab, 97(10): e1890–e1897.
Pu D, Xing Y, Gao Y, et al. 2014. Eur J Obstet Gynecol Reprod Biol, 182: 226–237.
Rattan S, Brehm E, Gao L, et al. 2018. Toxicol Sci, 163(2): 420–429.
Ristic N, Nestorovic N, Manojlovic-Stojanoski M, et al. 2008. J Microsc, 232(3): 549–557.
Rutkowska A Z, Diamanti-Kandarakis E. 2016. Fertil Steril, 106(4): 948–958.
Saatcioglu H D, Cuevas I, Castrillon D H. 2016. PLoS Genet, 12(8): e1006215.
Shalom-Paz E, Weill S, Ginzberg Y, et al. 2017. J Endocrinol Invest, 40(10): 1125–1131.
Sharma U, Sun F, Conine C C, et al. 2018. Dev Cell, 46(4): 481–494.
Shen L, Liu Z, Gong J, et al. 2014. Toxicol Appl Pharmacol, 274(2): 263–273.
Stanley J A, Sivakumar K K, Arosh J A, et al. 2014. Biol Reprod, 91(1): 12.
Susiarjo M, Sasson I, Mesaros C, et al. 2013. PLoS Genet, 9(4): e1003401.
Tosh D, Rani H S, Murty U S, et al. 2015. Menopause, 22(5): 520–526.
Veiga-Lopez A, Luense L J, Christenson L K, et al. 2013. Endocrinology, 154(5): 1873–1884.
Wang C, Zhou B, Xia G. 2017. Cell Mol Life Sci, 74(14): 2547–2566.
Weinberg C R, Wilcox A J, Baird D D. 1989. Am J Epidemiol, 129(5): 1072–1078.
Welt C K. 2008. Clin Endocrinol (Oxf), 68(4): 499–509.
Yang X, Zhou Y, Peng S, et al. 2012. Reproduction, 144(2): 235–244.
Zhang C, Liu X R, Cao Y C, et al. 2017. Reprod Fertil Dev, 29(4): 768–777.
Zhang X F, Zhang T, Han Z, et al. 2015. Reprod Fertil Dev, 27(8): 1213–1221.
Zhang Y, Xu Y, Ding H, et al. 2021. Chemosphere, 269: 128776.
Zhao L, Zhang S, An X, et al. 2015. Cytogenet Genome Res, 147(1): 41–47.
Zhou L, Xie Y, Li S, et al. 2017. J Ovarian Res, 10(1): 56.
Zhu X, Tian G G, Yu B, et al. 2018. Arch Toxicol, 92(4): 1581–1591.

Chapter 21
Methods of in Vivo and in Vitro Study of Fetal-Originated Disease

Hui Wang, Hao Xiao, and Wen Hu

Abstract Intrauterine growth retardation (IUGR) is intimately associated with the onset of numerous chronic diseases in adulthood. Owing to the difficulties in obtaining clinical biological samples and ethical considerations in medical research, animal models have become the primary means of exploring the phenomena of IUGR-related fetal-originated disease and deciphering their underlying mechanisms. Concurrently, with the swift progress in cellular and molecular biology, developmental biology, and embryology, in vitro research methodologies have also seen substantial advancement and refinement. These in vitro methods can be generally divided into three main systems: cell culture, organ culture, and embryo culture, in addition to the widely utilized zebrafish culture system. This chapter collates and reviews the methodologies for establishing models pertinent to IUGR-associated fetal-originated disease. It clarifies the relevant in vivo and in vitro research methods, providing invaluable insights for experimental studies on diseases originating from the fetus. Moreover, this compilation lays the groundwork for future explorations into the mechanisms and potential therapeutic targets of fetal-originated disease.

H. Wang (✉)
Department of Pharmacology, School of Basic Medical Science, Wuhan University, Wuhan, Hubei, China
e-mail: wanghui19@whu.edu.cn

H. Xiao
Division of Joint Surgery and Sports Medicine, Department of Orthopedic Surgery, Zhongnan Hospital of Wuhan University, Wuhan, Hubei, China

W. Hu
Department of Pharmacy, Zhongnan Hospital of Wuhan University, Wuhan, Hubei, China

H. Wang et al. (eds.), *Fetal Origin of Diseases*,
https://doi.org/10.1007/978-981-97-5730-5_21

1 Introduction

Epidemiological studies indicate that diverse adverse environmental exposures during pregnancy can result in not only unfavorable outcomes such as intrauterine growth retardation (IUGR) but also repercussions that extend into postnatal and adult life. These consequences encompass programmed development of multiple tissues and organs, modifications in functional homeostasis, and heightened susceptibility to chronic diseases in adulthood (Doi et al. 2022). Diseases originating from the fetus refer to adult diseases that stem from detrimental stimuli encountered by the fetus during intrauterine development. This leads to impaired development of various organs, which persists into postnatal life, thereby increasing susceptibility to a variety of chronic diseases in adulthood. IUGR specifically refers to the phenomenon where fetal growth and development are impeded in utero, resulting in a lower birth weight. This condition is diagnosed when the birth weight falls below two standard deviations from the average weight of normal fetuses. The incidence of IUGR varies from 2.75% to 15.53%. Infants with IUGR not only have a higher perinatal mortality rate but also encounter difficulties in postnatal physical and intellectual development, often accompanied by an increased susceptibility to various diseases.

At present, the early prevention and treatment of diseases originating from the fetus present significant challenges due to a variety of causative factors, an extended pre-pathogenesis period, intricate mechanisms, ambiguous targets, and a dearth of advanced warning technologies. Crucial areas of focus include disease models, mechanistic theories, and intervention targets, with a particular emphasis on the development of animal models for various fetal-originated disease. Our team's prior research has confirmed that prenatal exposure to exogenous substances such as caffeine, nicotine, ethanol, and dexamethasone in rats can induce IUGR, programmed development of multiple organs, and heightened susceptibility to adult diseases. These effects display marked gender differences and multigenerational inheritance (Xiao et al. 2020b; Chen et al. 2021; He et al. 2022; Liu et al. 2023b; Lu et al. 2023). Moreover, drawing on the fetal programming theory of glucocorticoids, we have systematically proposed the "two-programming and two-strike" mechanism of diseases originating from the fetus. We have also established a suite of early warning and intervention technologies for these diseases. This chapter collates and categorizes recent in vivo and in vitro models of diseases originating from the fetus, providing theoretical and experimental foundations for further investigations into the mechanisms and early prevention of such diseases.

2 Establishment and Assessment of Animal Models for Fetal-Originated Disease

Drawing on the "two-strike" theory, it is feasible to construct in vivo and in vitro models for diseases originating from the fetus. The "first strike" refers to the incidence of IUGR brought on by adverse environmental exposures during pregnancy, which can heighten susceptibility to adult diseases. The environmental alterations post-birth, acting as inducers of adult diseases, serve as the "second strike" that directly triggers the onset of adult diseases.

2.1 *Selection of Animal Strains*

At present, the experimental animals employed to establish IUGR models encompass rodents such as rats, mice, guinea pigs, rabbits, and domesticated animals such as sheep, pigs, and cows. An optimal IUGR animal model necessitates a high degree of similarity to human anatomy and physiological characteristics, excellent repeatability, ease of detection and evaluation, and economic viability. Rodents, particularly rats and mice, are the most frequently utilized animals in scientific experiments due to their larger size, swift reproduction, robust adaptability, short gestation period, and lower modeling costs. In 2011, Fung et al. induced IUGR in mice by injecting U-46619 thromboxane A, but this method also incited alterations in the maternal mice's pregnancy-induced hypertension. While rat IUGR models have been extensively studied and utilized, models based on sheep and pigs provide certain benefits over rat models. The development and anatomical features of sheep and lambs throughout various physiological stages bear similarity to humans. The size of the individual is suitable for experimental procedures such as surgery, dissection, sampling, and blood collection, resulting in models that closely resemble humans. Some breeds exhibit traits such as high lambing rates and early sexual maturity, simplifying the acquisition of IUGR individuals and the establishment of control groups.

In China, lambs are relatively easy to raise artificially, with a broad range of feed sources and lower experimental costs. However, the process of model establishment is lengthier and more complex. For IUGR piglets, a common strategy involves selecting weak piglets born under natural conditions, which introduces a high degree of randomness. Another method is to induce IUGR samples through techniques such as restricting the diet of pregnant sows, ligating uterine arteries, passive smoking, or reducing air oxygen content. In conclusion, when selecting animal strains for IUGR modeling, factors such as the success rate, time and feeding costs, and the convenience of obtaining animals should be comprehensively considered.

2.2 Establishment and Assessment of IUGR Models

IUGR is one of the most prevalent complications during the perinatal period. The establishment of animal models of IUGR that accurately replicate human clinical pathological features is crucial for understanding its pathogenesis and exploring potential prevention and treatment strategies. The general principles for creating IUGR animal models aim to meet IUGR assessment criteria without leading to fetal abortion, stillbirth, and other adverse outcomes. These criteria typically include fetal weights that are more than two standard deviations below the average weight of the control group or below the tenth percentile of normal weight (Sharma et al. 2016). The currently recognized methods for establishing IUGR models include the following:

1. External Substance Exposure Models During Pregnancy: ① Caffeine Exposure Model: Pregnant rats are administered oral gavage of caffeine (30, 60, 120 mg/kg.d) during mid-to-late gestation. This method induces IUGR by increasing maternal glucocorticoid levels in fetal blood (Ge et al. 2021). ② Dexamethasone Exposure Model: Pregnant rats are subcutaneously injected with dexamethasone (0.1, 0.2, 0.4 mg/kg.d) during mid-to-late gestation. This method induces IUGR by directly exposing the fetus to dexamethasone (Liu et al. 2023a). Despite a relatively long administration duration compared to other models, these models are distinguished by their milder yet more persistent effect, higher IUGR incidence, and lower perinatal mortality in rodent offspring. ③ Streptozotocin D Model: Pregnant rats receive a single intraperitoneal injection of 0.1 mg/kg streptozotocin D on gestation day 10. This method induces IUGR by impairing placental function, leading to a reduced nutrient supply to the fetus (Nagai et al. 2001). In summary, IUGR animal models obtained through exposure to external substances align more closely with human IUGR occurrences in daily life, rendering them ideal for practical research.
2. Nutritional Deficiency Models During Pregnancy: ① Restricted Feeding Model: Pregnant rats in the control group have unrestricted access to food, while those in the nutritionally deficient group receive only 30%–50% of the control group's diet (or 50% of the control group's caloric intake) from gestation day 11 until birth. ② Protein Malnutrition Model: Pregnant rats in the control group receive a normal diet (with a protein content of 20%–23%), while those in the protein malnutrition group are fed a low-protein diet (with a protein content of 8%–10%) starting from gestation day 11 (Armengaud et al. 2020). IUGR animal models induced by nutritional deficiencies work by reducing the supply of nutrients to the fetus, thereby impeding its growth and development.
3. Uterine Artery Ligation Model During Pregnancy: This model involves surgical ligation of uterine artery branches in pregnant mice at gestation day 17. The underpinning principle of this model is to reduce fetal blood supply, thereby affecting fetal development. A distinct advantage of this model is the ability to maintain an internal control. Specifically, fetuses located in the anterior part of

the uterus are designated as IUGR fetuses, while those in the posterior part are classified as normal.

2.3 *Establishment and Assessment of Models for Fetal-Originated Disease*

Key attributes of well-established animal models for adult diseases: ① Similarity: A fundamental principle in creating animal disease models is to emulate human disease conditions as closely as possible. This approach ensures that the model replicates physiological and pathological processes that bear a strong resemblance to those observed in humans, thus providing a more accurate simulation of real-world clinical scenarios. ② Stability: The developed animal model should consistently mirror human diseases. Once specific treatment factors are introduced, the model should stably manifest disease-specific characteristics, functional changes, metabolic shifts, or structural modifications. Stability is a critical criterion for assessing the model's reliability, ensuring that disease manifestations remain consistent and enduring without significant variations. ③ Practicality: Animal models designed for medical experimental research should strike a balance between clinical relevance and the ease of controlling disease progression. The design of the model should allow researchers to conveniently observe and manage the disease, fostering both research progress and potential clinical applications.

However, traditional models for adult diseases, while capable of partially replicating certain disease processes, often fail to effectively encapsulate their "fetal-origin" characteristics. The key to creating successful models for diseases originating in fetal development lies in fully capturing these unique features. Such diseases stem from inadequate fetal organ development during pregnancy, with disease phenotypes only becoming apparent after exposure to environmental changes in adulthood (the so-called second strike after birth). The most defining features of diseases originating in fetal development are "in utero programming" and "heritability," which involve specific epigenetic modifications to functional genes. Several commonly used animal models for diseases originating in fetal development are detailed below.

Fetal-originated epilepsy: Epilepsy, intricately linked with early-life stress, is a condition predominantly manifesting in infancy or childhood, and the precise mechanisms of its onset remain elusive. Temporal lobe epilepsy (TLE) stands as the most prevalent intractable focal epilepsy. Eriksson's cohort study identified a heightened risk of epilepsy in male newborns exposed to prenatal glucocorticoids during early- or late-term periods among 7827 preterm cases, compared to females. Young et al., in their study conducted on gestational days 15–18 using dexamethasone or betamethasone, found that prenatal betamethasone treatment increased seizure thresholds in two models, while prenatal dexamethasone treatment increased ignition thresholds but not seizure thresholds. In a preceding study, our team successfully established a fetal-originated TLE model by administering 0.2 mg/kg.d dexamethasone subcutaneously to pregnant Wistar rats during mid- to late gestation (gestational days 9–20). The pregnant rats were anesthetized on gestational day 21 for fetal extraction, or natural delivery was allowed for obtaining offspring rats. A second hit was administered postnatally through lithium chloride injection from postnatal weeks 22–28. This model was evaluated using adult epilepsy-like EEG feature alterations and epigenetic modifications in the angiotensin-converting enzyme (ACE) promoter region, providing a means to identify the occurrence of fetal-originated TLE (Hu et al. 2020). Thus, employing the "two-strike" theory facilitates the construction of animal models for fetal-originated epilepsy.

Fetal-originated fatty liver: Metabolically associated fatty liver disease is a common liver condition marked by the accumulation of lipids in the liver. Epidemiological studies suggest that an unfavorable intrauterine environment, which includes maternal malnutrition, insufficient uterine placental function, maternal stress, or fetal overexposure to glucocorticoids, can trigger developmental programming changes in offspring, thereby increasing their susceptibility to metabolic syndrome in adulthood (Valsamakis et al. 2006). Animal studies have further confirmed that an adverse intrauterine environment leads to abnormal lipid metabolism in IUGR rats, and this alteration persists after birth, resulting in an increased susceptibility to metabolic syndrome in adulthood (Darendeliler 2019). The authors' team research has further shown that IUGR rats exposed to prenatal caffeine display increased expression of hepatic lipid synthesis genes (Hu et al. 2019). In contrast, offspring rats exposed to prenatal caffeine show increased expression of fetal liver lipid synthesis genes, accompanied by decreased expression of lipid oxidation and efflux genes (Xu et al. 2018). This leads to the accumulation of lipids in fetal livers. Subsequent exposure of these offspring to a high-fat diet after birth serves as the second strike, further increasing their susceptibility to fatty liver disease. Epigenetic modifications in the lipid synthesis gene *Srebp1c* can be used to detect the occurrence of fetal-originated fatty liver disease (Hu et al. 2019). Thus, based on the "two-strike" theory, it is feasible to construct animal models for fetal-originated fatty liver disease.

Fetal-originated osteoarthritis: Osteoarthritis, a widespread joint disease characterized by the degeneration of joint cartilage, has been linked to low birth weight and preterm birth in retrospective and cohort studies. These studies have identified these factors as significant risks for adult hip osteoarthritis, particularly resulting in the need for total hip replacement later in life (Hussain et al. 2018). Research from the authors' team suggests that prenatal caffeine exposure acts as the "first strike," affecting cartilage development in a rodent model. This vulnerability potentially lays the groundwork for susceptibility to osteoarthritis in offspring when they experience the "second strike" in adulthood, such as through excessive exercise or chronic stress. The evaluation of epigenetic modifications in the *Tgfbr1* gene provides a reliable method to identify fetal-originated osteoarthritis. Therefore, the application of the "two-strike" theory aids in constructing animal models for fetal-originated osteoarthritis.

Fetal-originated osteoporosis: Osteoporosis is a systemic bone disorder marked by decreased bone mass and microstructural deterioration. These changes result in increased bone fragility and a significantly higher risk of fractures. Research has shown a significant statistical correlation between the weight of female infants at 1 year of age and the bone mineral content (BMC) of the lumbar spine and femoral neck in adulthood, independent of adult weight and body mass index factors (Dötsch 2011). Studies investigating newborn BMC and maternal lifestyle habits have found that maternal smoking or inadequate energy intake during pregnancy can affect the BMC of newborns (Balasuriya et al. 2017). The earlier research by the authors' team discovered that prenatal dexamethasone exposure results in delayed postnatal long bone development, primarily linked to sustained overexpression of ACE, which inhibits osteogenic differentiation (Xiao et al. 2018). However, when subjected to a "second strike" in adulthood, such as postmenopausal conditions or dietary vitamin D deficiency, these factors further impact bone metabolic function, triggering and accelerating the onset of fetal-originated osteoporosis. Epigenetic modifications in ACE can be used as a distinguishing marker for fetal-originated osteoporosis (Xiao et al. 2020a). Therefore, it is feasible to develop an animal model for fetal-originated osteoporosis based on the "two-strike" theory.

3 Establishment and Assessment of in Vitro Models for Fetal-Originated Disease

In recent years, advancements in cell molecular biology, developmental biology, and embryology have contributed to the evolution and enhancement of in vitro research methodologies for fetal diseases. In vitro systems, encompassing cell culture, organ culture, embryo culture, and zebrafish culture, have become indispensable tools in investigating the mechanisms of exogenous developmental toxicity and in the screening of developmental toxins.

3.1 Establishment and Assessment of Cell Models

Cell culture, with its minimal or nonexistent animal usage, offers an advantage in terms of ease of operation and standardization. In vitro cytotoxicity assessments typically employ cells sourced directly from human tissues or those from closely related model organisms. However, a significant limitation of these systems is their relative simplicity. They only allow for studying the impact on individual mechanisms, such as cell proliferation, inhibition, or differentiation, and do not provide insights into the various regulatory mechanisms and pattern formations that are crucial for embryonic development. Toxicology studies frequently utilize cell lines such as A549, HEK-293, HepG2, C3A, HepaRG, and Huh7, among others. The use of these cell lines circumvents the issue of inter-donor variability, showcasing relatively stable phenotypes, ease of experimental manipulation, and extended life cycles. Common indicators for in vitro cytotoxicity evaluation involve methods designed to assess changes in cell toxicity characteristics based on cell quantity, morphology, structure, and physiological features.

Live cell counters are used to quantify the number of live cells in a test population. Widely used chemical colorimetric methods include 3-(4,5-dimethylthiazol-2-yl)-2,5-diphenyltetrazolium bromide (MTT), water-soluble tetrazolium salt 1 (WST-1), Cell Counting Kit-8 (CCK8), and 2,3-bis(2-methoxy-4-nitro-5-sulfophenyl)-2H-tetrazolium 5-carboxanilide (XTT). The Trypan blue exclusion assay is another method, which relies on the inability of Trypan blue to penetrate intact cell membranes, but it can stain damaged cell membranes. This allows for the calculation of the ratio of damaged cell membranes through microscopic examination. Lactate dehydrogenase (LDH) leakage assays leverage the property that cytoplasmic LDH can leak out of damaged cell membranes. The extent of cell membrane damage is assessed by detecting LDH activity in the cell culture medium. Cellular energy metabolism is most directly reflected in the rate of adenosine triphosphate (ATP) generation and consumption. The state and content of ATP within cells, indicative of cellular energy metabolism, can be determined using specific reagents or dedicated ATP detection kits. Each indicator has its unique detection method, and the choice of suitable cells and evaluation methods depends on the characteristics of the developmental toxicity under investigation.

The research team utilized Wharton's jelly mesenchymal stem cells (WJ-MSCs) derived from infants clinically diagnosed with IUGR and from normal newborns. The first step involved treating the cells with excess cortisol during differentiation, simulating intrauterine exposure to maternal glucocorticoids in IUGR fetuses. In the second step, differentiated human WJ-MSCs were treated with interleukin 1β to mimic osteoarthritis susceptibility. The team proposed that acetylation of histone H3 on lysine 9 (H3K9ac) on the TGFβRI gene in human umbilical cord cells could serve as a potential early warning biomarker predicting susceptibility to developmental cartilage defects and osteoarthritis in newborns (Qi et al. 2021). This two-step method for constructing a fetal-originated osteoarthritis cell model can be used for research on susceptibility and toxicity targets in fetal-originated osteoarthritis (Qi et al. 2021).

3.2 Establishment and Assessment of Organ Models

Organ culture is a method that involves the isolation and cultivation of a part or the entirety of a living organism. It is a more comprehensive form of tissue culture, as it involves maintaining the normal tissue structure during cultivation. This method allows for the preservation of the tissue's three-dimensional structure and the simulation of organ functions under various conditions without disrupting the normal tissue architecture. In vitro organ culture has several advantages, including speed, simplicity, cost-effectiveness, precise control over the dosage and exposure time of test substances, elimination of maternal interference factors, and excellent reproducibility of results. For example, exposing mouse embryonic palatal shelves to varying concentrations of different chemicals in vitro for a specific duration results in a dose-dependent decrease in the fusion rate of the palate. The 50% inhibitory concentration (IC50), calculated based on the fusion rate, is considered a reliable parameter for evaluating drug developmental toxicity. This can be applied in screening drugs and other environmental factors for developmental toxicity. Additionally, the use of precisely sectioned liver slices has proven beneficial in studying compound properties related to fibrogenesis and antifibrogenesis. This is due to the inclusion of all liver cell types and their metabolic capabilities within the natural structure (Palma et al. 2019).

In studies of reproductive developmental toxicity, ovarian organ culture is frequently used to evaluate female reproductive toxicity. This technique involves culturing ovaries from either large or small mouse embryos or postnatal mouse ovarian organs. Alternatively, ovaries may be cultured *via* perfusion through an intact vascular system. This method is primarily used to investigate the effects of ovarian blood flow, the initiation of quiescent follicles, follicle growth and development, ovulation, and the ability of follicles to produce steroid hormones. However, one limitation of this method is the relatively short duration of in vitro culture, which is not suitable for long-term cultivation. The research team employed extracted fetal mouse ovaries for in vitro organ culture to simulate the effects of prenatal exposure to dexamethasone on fetal ovaries. They confirmed the programmed mechanism of suppressed estrogen synthesis in the ovaries of female offspring rats (Gong et al. 2021). Therefore, ovarian organ culture can also be utilized for the toxicological evaluation of fetal-originated ovarian dysfunction.

In reproductive developmental toxicity studies, ovarian organ culture is often employed to assess female reproductive toxicity. Ovaries from large or small mouse embryos, or postnatal mouse ovarian organs, can be cultured. Alternatively, ovaries can be cultured *via* perfusion through an undamaged vascular system. This method is mainly used to study the effects of ovarian blood flow, initiation of quiescent follicles, follicle growth and development, ovulation, and the ability of follicles to produce steroid hormones. However, its limitation lies in the short duration of in vitro culture, which is not conducive to long-term cultivation. The research team utilized extracted fetal mouse ovaries for in vitro organ culture to simulate the impact of prenatal dexamethasone exposure on fetal ovaries, confirming the programmed mechanism of suppressed estrogen synthesis in the ovaries of female offspring rats (Gong et al. 2021). Therefore, ovarian organ culture can also be used for the toxicological evaluation of fetal-originated ovarian dysfunction.

3.3 Establishment and Assessment of Embryo Models

The European Centre for the Validation of Alternative Methods (ECVAM) has endorsed three distinct embryotoxicity testing methods, which include whole embryo culture, micromass culture, and the embryonic stem cell test.

Whole-embryo culture currently encompasses two primary methods: intermittent air-inflated rotating bottle pipeline culture and continuous air-inflated rotating tube culture. These methods involve observing the development of embryos and documenting parameters such as embryo survival, blastocyst and yolk sac diameter, and somite and body length. The presence of embryo survival is indicated by factors such as heartbeat and blood circulation. Indices including yolk sac diameter, crown-rump length, head length, somite number, and embryo weight are utilized to evaluate embryonic growth and development. The Brown scoring system is then employed to assess the extent to which the test substance may impact organ morphological differentiation (Zhang et al. 2013). Studies utilizing whole-embryo culture have explored the developmental toxicity of the disinfectant haloacetic acids found in tap water. These studies have revealed that high concentrations and prolonged exposure to haloacetic acids can result in embryo malformations (Hunter et al. 2006).

Embryo cell micromass culture is an in vitro experimental technique that sits between single-cell culture and organ culture. It is extensively utilized for screening the teratogenicity of chemical substances due to its benefits such as cost-effectiveness, short duration, simplicity, and high precision. This method operates on the principle of qualitative and quantitative assessment of the reduction in the number of cell aggregates during culture, which acts as an indicator of the teratogenic effects of chemical substances. The use of chick embryo hearts in micromass culture has been applied to study the cardiac toxicity of four antiepileptic drugs—valproic acid, phenytoin sodium, phenobarbital, and trimethadione—during embryonic development. The study revealed that all four drugs could inhibit the differentiation of embryonic cardiac cells (Ahir and Pratten 2014).

The embryonic stem cell test primarily involves cultivating mouse embryonic stem cells using the hanging drop method to form embryoid bodies. When these embryoid bodies are seeded on tissue culture dishes, they differentiate into contracting myocardial cell foci within 10 days. The endpoint of the test involves counting the contracting muscle foci under a microscope, with the compound's inhibition of myocardial cell differentiation serving as the measurement. To provide a comprehensive assessment of various endpoints in embryonic developmental toxicity, the test not only reflects traditional cardiac cell differentiation but also concurrently detects the inhibitory effects of chemicals on osteoblast, chondrocyte, and neuronal cell differentiation. Despite the widespread application of in vitro embryotoxicity tests in toxicological screening and mechanistic research, these tests still have limitations. For instance, in vitro embryotoxicity tests can only detect structural changes, not functional changes. The mammalian reproductive cycle is intricate, and individual in vitro tests cannot simulate the entire reproductive cycle to detect the toxicity of chemicals during embryonic reproductive development.

3.4 Establishment and Assessment of Zebrafish Models

Zebrafish, indigenous to South Asian countries such as India, Pakistan, Bangladesh, and Nepal, are named for their striped pattern reminiscent of a zebra. Zebrafish have risen to become the third most utilized vertebrate model organism, following mice and rats. In recent years, zebrafish have been extensively used in genetic studies, ecotoxicology, screening for pharmacological and bioactive substances, drug metabolism, drug toxicity screening, and other research areas. In assessing developmental and reproductive toxicity, zebrafish offer several notable experimental advantages over traditional mammalian testing. These advantages include the transparency of embryos and larvae, higher throughput, shorter testing times, lower costs, lower compound quantities required, ease of operation, and direct compound delivery. In zebrafish research, 12 primary endpoints and 8 secondary endpoints can be used to evaluate substance toxicity. The 12 primary endpoints include the heart, brain, jaw, eyes, liver, intestine, trunk, tail, atrioventricular valve, muscles, synapses, and body pigmentation, circulation, body swelling, and bleeding. The nine secondary endpoints include fins, ears, swimming, bladder, erythropoiesis, kidney cysts, pancreas, locomotion, and body length. For instance, using 10% gentamicin, acetaminophen, and tenofovir to observe the impact on zebrafish fry's kidneys, it was found that gentamicin and acetaminophen could decrease kidney clearance, while acetaminophen and tenofovir could enlarge the proximal tubules. The nephrotoxic mechanisms were associated with mitochondrial changes, including mitochondrial deformation, swelling, cristae disruption, and loss of matrix granules (Gorgulho et al. 2018). Traditional mammalian reproductive toxicity assessments are complex, expensive, and time-consuming, with limited potential for high-throughput or large-scale analyses. Contrastingly, zebrafish has emerged as a simpler and more appealing animal model for studying reproductive toxicity and teratogenicity. The Organisation for Economic Co-operation and Development (OECD) 229 recommends the use of zebrafish for assessing reproductive toxicity indicators, primarily focusing on ① gonadal growth index (gonadal weight); ② reproductive capacity, including egg production and sperm quality; ③ gonadal histology; ④ determination of vitellogenin expression; and ⑤ analysis of the hypothalamic-pituitary-gonadal (HPG) axis gene expression. The analysis of HPG axis gene expression has become a pivotal method for assessing the functionality of the endocrine system.

3.5 Establishment and Assessment of Organ Chip Models

Organoids, a type of in vitro organ model that has emerged in recent years, primarily refer to cell tissue complexes that form through the self-organization of stem cells under three-dimensional (3D) culture conditions in vitro. The advent of organoid chips largely owes to the rapid advancement of organ chip technology. By integrating the benefits of organoids and organ chips, it is possible to simulate complex tissue microenvironments in vitro. This aids in directing the growth, differentiation, and morphogenesis of stem cells, thus overcoming the limitations present in existing organoid research. Organoid chips can precisely mimic the tissue structure of target organs. They utilize microchannels as sources and distribution pathways for soluble factors, controlling the distribution of biochemical concentration gradients in the cell-extracellular matrix and inducing tissue-like regionalization. For instance, by combining microchannels with patterned culture scaffolds, stem cells and differentiated epithelial organoids can be confined to crypts and villi-like structures within the organoid chip (Wang et al. 2017). Moreover, organoid chips can enhance nutrient exchange and orchestrate the formation of tissue structure by controlling the degree of "vascularization" and neural networks. This avoids the death of central cells in organoids and allows the growth and development of organoids to millimeter diameters (Yu et al. 2019).

Pluripotent stem cells, with their ability to differentiate into target cells exhibiting primary tissue-like phenotypes and their boundless supply, have proven invaluable in research. Scientists have successfully established various organoid chip systems, including those for the brain, intestines, liver, pancreatic islets, kidneys, and retinas, among others (Tang et al. 2022). One notable example of their application is in the study of liver toxicity due to prolonged drug exposure. In this context, researchers utilized liver cells derived from human-induced pluripotent stem cells (iPSCs). These cells were exposed to hepatotoxic compounds at relevant concentrations over periods of 2, 7, and 14 days. The study found that, during extended exposure periods, iPSC-derived liver cells exhibited increased sensitivity to both amiodarone and aflatoxin B1. Furthermore, phospholipid deposition and lipid degeneration were observed in the cells (Holmgren et al. 2014).

Presently, the study of fetal-originated disease often encounters limitations as in vitro experiments struggle to effectively simulate systemic effects. Therefore, organoids can be employed to assess the developmental toxicity of drugs, examining their toxic impacts on organs such as the liver, kidneys, and heart. A multi-organoid system has been developed, comprising the heart, lung, and liver organoids, all situated within a single recirculating system that features a universal culture medium with biosensing capabilities. This system can be utilized to evaluate the pharmacological effects and toxic reactions of drugs on multiple organs (Skardal et al. 2017).

4 Summary and Prospect

In vivo animal models play a pivotal role in unraveling the pathogenesis and prevention of developmental toxicity-related fetal-originated disease. Despite their importance, in vivo experiments grapple with challenges such as lengthy timelines, high costs, and ethical considerations linked to the extensive use of animals. The 3R principle—reduce, refine, and replace—underscores the need to lessen animal usage in experiments. These challenges, encompassing cost, time, and ethical concerns, have fostered an increasing emphasis on identifying cost-effective and reliable in vitro methods as alternatives to in vivo experiments. Propelled by advances in cell biology, molecular biology, developmental biology, and molecular embryology, in vitro research is emerging as a focal point of scientific inquiry. It offers a unique and invaluable perspective for studying developmental toxicity mechanisms and screening potential developmental toxins. In vitro methods confer several advantages, such as cost-effectiveness, rapid results, and the capacity to assess a large number of compounds for developmental toxicity.

These methods adhere to the 3R principle and contribute to the reduction of animal use in experiments. However, they face challenges including metabolic differences, the lack of toxicokinetic processes, difficulties in species extrapolation, and variations in sensitivity between somatic and germ cells. In vivo and in vitro research should be viewed as complementary rather than mutually exclusive. In vitro findings can inform in vivo research, and results require validation through in vivo experiments and, ideally, confirmation through population studies. The continual development and refinement of in vitro models are anticipated to address these current limitations. Future research should focus on enhancing the reliability and accuracy of in vitro methods to increase their applicability in developmental toxicity assessment. Collaborative efforts between in vivo and in vitro researchers can pave the way toward a more comprehensive understanding of developmental toxicity mechanisms and assist in the development of effective prevention and treatment strategies for fetal-originated disease.

References

Ahir B K, Pratten M K. 2014. Toxicol In Vitro, 28(5): 948–960.
Armengaud J B, Dennebouy Z, Labes D, et al. 2020. Br J Nutr, 123(8): 892–900.
Balasuriya C N D, Evensen KA I, Mosti M P, et al. 2017. J Clin Endocrinol Metab, 102(7): 2491–500.
Chen Y, Xu D, Xia X, et al. 2021. Pharmacol Res, 174:105942.
Darendeliler F. 2019. Best Pract Res Clin Endocrinol Metab, 33(3): 101260.
Doi M, Usui N, Shimada S. 2022. Front Endocrinol (Lausanne), 13: 860110.
Dötsch J. 2011. Dermatoendocrinol, 3(4): 240–242.
Ge C, Xu D, Yu P, et al. 2021. BMC Med, 19(1): 311.
Gong X H, Zhang J Z, Ge CY, et al. 2021. Pharmacol Res, 165: 105435.
Gorgulho R, Jacinto R, Lopes S S, et al. 2018. Arch Toxicol, 92(1): 411–423.
He B, Zhang Q, Guo Y, et al. 2022. Food Chem Toxicol, 168: 113384.
Holmgren G, Sjögren A K, Barragan I, et al. 2014. Drug Metab Dispos, 42(9): 1401–1406.
Hunter E S, 3rd, Blanton M R, Rogers E H, et al. 2006. Reprod Toxicol, 22(3): 443–448.
Hu S, Yi Y, Jiang T, et al. 2020. Arch Toxicol, 94(9): 3201–3215.
Hu S, Xia L, Luo H, et al. 2019. Toxicology, 417: 23–34.
Hussain S M, Ackerman I N, Wang Y Y, et al. 2018. Arthritis Res Ther, 20(1): 121.
Liu L, Wen Y, Ni Q, et al. 2023b. Biol Res, 56(1): 61.
Liu Y, Liu Y, Chen G, et al. 2023a. Toxicol Sci, kfad016.
Lu Z, Guo Y, Xu D, et al. 2023. Acta Pharm Sin B, 13(2): 460–477.

Nagai K, Ikenoue T, Mori N. 2001. J Matern Fetal Med, 10(2): 136–140.
Palma E, Doornebal E J, Chokshi S. 2019. Hepatol Int, 13(1): 51-57.
Qi Y, Li B, Wen Y, et al. 2021. Stem Cell Res Ther, 12(1): 163.
Sharma D, Shastri S, Farahbakhsh N, et al. 2016. J Matern Fetal Neonatal Med, 29(24): 3977–3987.
Skardal A, Murphy S V, Devarasetty M, et al. 2017. Sci Rep, 7(1): 8837.
Tang X Y, Wu S, Wang D, et al. 2022. Signal Transduct Target Ther, 7(1): 168.
Valsamakis G, Kanaka-Gantenbein C, Malamitsi-Puchner A, et al. 2006. Ann N Y Acad Sci, 1092: 138–147.
Wang Y, Gunasekara D B, Reed M I, et al. 2017. Biomaterials, 128: 44-55.
Xiao H, Wen YX, Pan ZQ, et al. 2018. Cell Death Dis, 9(6): 638.
Xiao H, Wu Z, Li B, et al. 2020b. Br J Pharmacol, 177(20): 4683–4700.
Xiao H, Wen Y, Wu Z, et al. 2020a. Bone, 141: 115578.
Xu D, Luo H W, Hu W, et al. 2018. FASEB J. fj201701557R.
Yu F, Hunziker W, Choudhury D. 2019. Micromachines (Basel), 10(3): 165.
Zhang C, Panzica-Kelly J, Augustine-Rauch K. 2013. Methods Mol Biol, 947: 423–450.

Chapter 22
Targets for Early Warning and Drug Intervention for Fetal-Originated Disease

Hui Wang, Xiaoqian Lu, and Yu Peng

Abstract As bioinformatics and molecular genetic technologies continue to advance, coupled with the establishment of animal models simulating pregnancies, an increasing body of research has confirmed that adverse environmental exposure during pregnancy can result in abnormal development of multiple organs in offspring and heighten susceptibility to adult diseases. These include metabolic disorders, neuropsychiatric conditions, and diseases linked to the reproductive system. This chapter offers a comprehensive overview of the mechanisms and potential biomarkers of developmental programming changes in multiple organs in offspring, as elicited by an unfavorable prenatal environment. These insights are drawn from various animal models of pregnancy and in vitro studies. Furthermore, this chapter encapsulates the early warning signs and intervention targets for a range of fetal-originated diseases, thereby laying a foundation for the exploration of preventative and therapeutic strategies for such conditions.

1 Introduction

The "Developmental Origins of Health and Disease (DOHaD)" theory asserts that detrimental prenatal environments constitute significant triggers for multi-organ developmental abnormalities and heightened disease susceptibility in offspring. The concept of fetal developmental programming is essential for understanding healthful developmental pathways across the lifespan. In recent years, the spotlight has increasingly been cast on fetal-originated diseases, which are driven by rising environmental pollution and the increasing number of older mothers. These diseases have emerged as a critical research frontier internationally. Studying fetal-originated diseases associated with developmental toxicity presents a considerable challenge

H. Wang (✉) · X. Lu · Y. Peng
Department of Pharmacology, School of Basic Medical Science, Wuhan University, Wuhan, Hubei, China
e-mail: wanghui19@whu.edu.cn

H. Wang et al. (eds.), *Fetal Origin of Diseases*,
https://doi.org/10.1007/978-981-97-5730-5_22

due to their diverse causes, extended premorbid periods, intricate mechanisms, and elusive targets. These complexities contribute to a dearth of early warning technology and complicate early clinical prevention and treatment strategies. Consequently, conducting fundamental and translational research on fetal-originated diseases related to developmental toxicology holds substantial theoretical and clinical importance.

Biomarkers, characterized by their immediacy and high repeatability, hold promising potential in early disease diagnosis, prognosis determination, and complication risk assessment. For fetal-originated disease, there is an urgent need for precise biomarkers to identify at-risk newborns early and to offer targets for intervention. This chapter delves into the developmental programming changes in multiple organs (such as the liver, hippocampus, adrenal glands, and bones) and the susceptibility to various diseases in offspring due to unfavorable prenatal environments. It consolidates early warning signs and intervention targets for a range of fetal-originated diseases, thereby providing a theoretical and empirical basis for accurately understanding the onset and progression of these diseases. Furthermore, it lays a theoretical foundation for prenatal healthcare, as well as the early warning, comprehensive prevention, and treatment of fetal-originated diseases.

2 Exposure to Adverse Prenatal Environments and the Origination of Fetal Diseases

The exogenous environment primarily involves exposure to external substances. Substances demonstrated to exhibit developmental toxicity include (i) environmental toxins such as organic volatiles, heavy metals, smoke, and air pollutants (Forand et al. 2012; Walfisch et al. 2013; Dessì et al. 2018); (ii) medications such as dexamethasone, prednisone, caffeine, antibiotics, and others (Xu et al. 2011; Dobson et al. 2014; Wang et al. 2018); and (iii) adverse dietary factors, including alcohol, coffee, and tea (Nykjaer et al. 2014).

The maternal environment during pregnancy refers to maternal stress levels, nutritional status, and disease state. Research has shown that maternal anxiety and depression or imbalanced nutrient intake during pregnancy can adversely affect fetal growth and development (Painter et al. 2008; Song et al. 2020). Additionally, conditions such as gestational diabetes and preeclampsia can negatively impact the mother and lead to fetal dysplasia (Crispi et al. 2009; Jenner et al. 2018). Therefore, both exogenous and maternal factors need to be carefully managed to ensure the optimal development of the fetus.

The adverse environment during pregnancy can lead to fetal developmental toxicity, causing both short-term and long-term harm. Short-term harm often presents as adverse pregnancy outcomes, while long-term harm can increase susceptibility to various diseases post-birth, including metabolic diseases, neuropsychiatric disorders, and reproductive system diseases. Studies have shown that excessive maternal exposure to harmful chemicals (such as arsenic, lead, and chloroprene) and physical factors (such as radiation, noise, and high temperatures) during pregnancy can significantly increase the risk of miscarriage (Patelarou and Kelly 2014). Exposure to diethylstilbestrol during pregnancy may lead to an increased risk of depression in offspring during adulthood (Titus et al. 2019).

Research by the authors' team has found that rats with intrauterine growth retardation (IUGR) caused by various adverse environments (such as xenobiotic exposure and nutritional deficiencies) during pregnancy are more susceptible to metabolic syndrome in adulthood (He et al. 2015; Kou et al. 2017). Factors such as alcohol, nicotine, caffeine exposure, and dietary restriction during pregnancy can cause abnormal development of the hypothalamic-pituitary-adrenal (HPA) axis in offspring (Zhang et al. 2013; He et al. 2017; Lu et al. 2018; Xu et al. 2018), thereby increasing susceptibility to neuropsychiatric diseases (Dong et al. 2018; Jiang et al. 2022). Additionally, caffeine exposure during pregnancy can also lead to abnormal testicular development in offspring and insufficient sperm count in adulthood (Pei et al. 2019).

3 Developmental Programming of Multiple Organs and Associated Biomarkers

Maternal exposure to adverse environmental factors during pregnancy can lead to developmental toxicity in offspring, primarily evident as growth retardation, and functional defects or mortality in developing organisms. The underlying mechanism is the alteration in multi-organ developmental programming, which can ultimately heighten susceptibility to diseases across multiple organs. Drawing from research findings, our team systematically proposes the mechanisms of "neuroendocrine-metabolic programming" and "multi-organ developmental programming alteration" triggered by exogenous substances. We suggest that in various animal models exposed to exogenous substances during pregnancy, fetuses experience maternal glucocorticoid (GC) overexposure, inhibition of hypothalamic-pituitary-adrenal (HPA) axis functional development, and alterations in peripheral tissue glucose and lipid metabolism. Utilizing metabolomics technology, we have identified features of metabolic spectrum changes in fetal blood and amniotic fluid. These findings corroborate that exogenous substances can induce changes in intrauterine neuroendocrine metabolic programming related to the HPA axis. In adult offspring rat models, where disease susceptibility is due to xenobiotic exposure during pregnancy, we have confirmed that these offspring are susceptible to metabolic syndrome and related metabolic diseases such as diabetes mellitus and nonalcoholic fatty liver disease (NAFLD). This was determined by observing the effects of a high-fat diet and chronic stimulation on peripheral glucose and lipid metabolism. From the perspective of exogenous substance metabolism activation, the fetal adrenal gland possesses a robust exogenous substance metabolism enzyme system, encompassing phase I, phase II, and antioxidant enzymes, among others. This validates the significant role of the fetal adrenal gland in exogenous substance metabolism and proposes the mechanism of in situ metabolic damage caused by exposure to exogenous substances during pregnancy. From the perspective of the epigenetic regulation of embryonic development, exogenous substances can induce epigenetic modifications of a series of functional genes in the fetal adrenal glands, hippocampus, and liver. This suggests an epigenetic programming mechanism due to exogenous substance exposure during pregnancy.

The liver is the largest and most vital metabolic organ in the body. Hormones such as GC, insulin-like growth factor 1 (IGF-1), leptin, and adiponectin can bind to their respective receptors on the hepatocyte cell membrane, thereby regulating glucose and lipid metabolism (Zabeau et al. 2004; Netchine et al. 2011). Our team proposes the "two-programming and two-strike" mechanism of liver developmental toxicity caused by an adverse environment during pregnancy. The "first strike" is intrauterine maternal GC overexposure due to prenatal exogenous substance exposure. This overexposure has a dual effect: firstly, it increases de novo synthesis of fatty acid in the fetal liver tissue, with these changes persisting after birth and potentially into adulthood (the "first programming"); secondly, it inhibits the hepatic IGF1 signaling pathway and lipid metabolism. Under the condition of good nutritional status after birth, the increased expression of the IGF1 signaling pathway in the liver of IUGR offspring leads to enhanced compensatory function of lipid metabolism. This results in catch-up growth and increased susceptibility to metabolic diseases (the "second programming") (Liu et al. 2012; Xu et al. 2012; Wang et al. 2014; Xia et al. 2014). Postnatal environmental changes, such as high mental stress and overnutrition, serve as a "second strike." This can further exacerbate liver lipid metabolism disorders, culminating in metabolic syndrome and related metabolic disease susceptibility.

The hippocampus serves as a critical regulatory hub of the HPA axis, ensuring its normal functionality. Research has demonstrated that IUGR offspring exhibit elevated GC concentrations in adulthood due to factors such as alcohol exposure, dietary restriction, and maternal malnutrition during pregnancy. This increase manifests as an overreaction in the hypothalamus and an enhanced negative feedback inhibition in the hippocampus under stress conditions (Vieau et al. 2007). The authors' team discovered that prenatal caffeine exposure (PCE) disrupts the hippocampal regulation of the fetal HPA axis. This disruption is achieved by inducing an overexpression of glutamate decarboxylase 67 (GAD67) in the hippocampus of offspring rats, leading to heightened stress sensitivity of the HPA axis in adulthood (Xu et al. 2018). Prenatal dexamethasone exposure (PDE) can stimulate the glucocorticoid receptor (GR) in offspring rats, promoting the conversion of glutamate to γ-aminobutyric acid (GABA). This process induces an imbalance in the paraventricular nucleus of the hypothalamus, resulting in reduced basal activity of the HPA axis both prenatally and postnatally (Lu et al. 2020). This combination of low basal activity and heightened stress sensitivity of the HPA axis is a key mechanism underlying the increased susceptibility to multiple diseases in offspring. Moreover, epigenetic modifications induced by adverse prenatal environments play a crucial role in altering hippocampal developmental programming. For instance, prenatal maternal stress can inhibit the maturation of newborn neurons in the dentate gyrus of the mouse hippocampus. This inhibition is achieved by inducing hypermethylation of the GAD67 promoter region in the offspring, which subsequently leads to depressive-like behavior (Zhong et al. 2020).

As the terminal effector of the HPA axis, the adrenal gland is the primary organ responsible for the synthesis and secretion of GC (Busada and Cidlowski 2017). Synthetic GC treatment during pregnancy has been observed to significantly reduce fetal adrenal volume and delay fetal growth, as detected by two-dimensional ultrasound (Karsli et al. 2019). The authors' team discovered that an elevated level of fetal blood corticosterone, induced by exposure to exogenous substances such as caffeine and alcohol during pregnancy, is associated with inhibition of the IGF1 signaling pathway in rat models. This inhibition ultimately leads to adrenal hypoplasia. As blood GC concentration decreases during the early stages of postnatal life, the IGF1 signaling pathway progressively strengthens. By adulthood, the IGF1 signaling in these individuals gradually reaches levels observed in normal rats. This suggests that changes in the programming of the "GC-IGF1 axis" mediate both intrauterine adrenal dysplasia and postnatal catch-up growth in offspring (Huang et al. 2015; He et al. 2019). Epigenetic imprinting plays a crucial role in the early development of the adrenal gland. The authors' team discovered a "contemporary programming" effect of maternal GCs on adrenal steroid synthesis in PCE offspring. This mechanism is related to the regulatory interaction of maternal GC with the methylation level of the imprinted gene H19 promoter region and the expression of microRNA let-7 (He et al. 2021).

The testes and ovaries are integral parts of the male and female reproductive systems, respectively. The authors' team discovered that alterations in testicular and ovarian development were associated with the programming of the "GC-IGF1 axis." Excessive GC can lead to decreased IGF1 and dysplasia of the testes, ovaries, and other organs. Recent research conducted by the team indicates that the HPA axis of PDE mothers is in an inhibitory state, causing a significant decrease in maternal blood GC levels. Consequently, the fetus exists in a low physiological GC environment within the uterus, and the expression of IGF1 in the testes is consistently suppressed both before and after birth. This results in the testes undergoing "GC-IGF1" co-programming, leading to abnormal testicular development in offspring (Liu et al. 2022). It is suggested that the low steroidal synthetic function of the fetal testes, due to abnormal GC levels during pregnancy, mediates the occurrence of low testosterone levels and testicular hypoplasia in adult life. The hypothalamic-pituitary-gonadal (HPG) axis is another significant endocrine axis involved in the programming changes of testicular and ovarian development. Adverse environments during pregnancy can impact the HPG axis at various stages, leading to offspring gonadal dysplasia and abnormal sex hormone secretion. This can further result in related dysfunctions such as impaired gender differentiation and reduced reproductive function (Sullivan et al. 2010; Lv et al. 2021).

In 1993, the World Health Organization (WHO) formally defined a biomarker in its report on the effectiveness of biomarkers in environmental risk assessment.

According to the WHO, a biomarker is an indicator that reflects the interaction of a potential physical or chemical hazard (including chemical, physical, or biological hazards) with an organism. This interaction can be detected at multiple levels, including physiology, cellular biochemistry, function, or molecular interaction.

Biomarkers, based on the strategies of measurement, can serve as indicators of the exposure itself (exposure biomarkers), the resultant effects of the exposure (effect biomarkers), and/or the inherent or acquired susceptibility (susceptibility biomarkers) that could predict the severity of the effect. Certain studies also categorize biomarkers based on the effector organs or systems they represent, such as biomarkers indicative of liver toxicity, kidney toxicity, and neurotoxicity, among others. It is crucial to understand that the classification of biomarkers is not rigid. A specific purpose might require the use of multiple biomarkers, and a single biomarker could fall under different categories depending on its varied applications in diverse scenarios.

Studies have shown that epigenetic modification mechanisms serve as mediators linking adverse environmental exposure during pregnancy with programmed changes in gene expression during offspring growth and development (Goyal et al. 2019; Mandala et al. 2020). Increasing studies suggest that prenatal adverse environment can affect fetal morphological and functional development through "multi-pathway, multi-organ, multi-cell, and multi-target" (Lu et al. 2023), in which the expression of specific genes and their epigenetic modifications can be used as potential biomarkers of changes in the programming of multi-organ development under prenatal adverse environment (Table 22.1).

4 Early Detection and Intervention Targets for Fetal-Originated Disease

Embryonic development (fetal stage) is incredibly sensitive to external environmental changes during the critical early developmental period, making fetuses the most susceptible population to alterations in environmental factors. Adverse environmental conditions during pregnancy instigate gene-environment interactions, leading to epigenetic modifications of specific functional genes in exposed individuals. These epigenetic alterations can persist postnatally, inducing developmental toxicity in multiple systems and organs and ultimately predisposing the individual to various diseases of fetal origin. Hence, investigating early warning signals and therapeutic intervention strategies for diseases originating in fetal development holds significant importance.

Table 22.1 Potential biomarkers indicating programmatic changes in multi-organ development due to adverse prenatal environmental exposures

Organ	Prenatal adverse environments	Changes in organ development/related diseases	Potential biomarkers	Main references
Adrenal	Dexamethasone	Low steroid synthesis	Male/female: 11β-HSD2-H3K27ac Male: IGF1-H3K27ac, SF1-H3K27ac	Chen et al. (2021a, b, 2022)
	Caffeine	Low steroid synthesis	Female: H19 methylation, miR let-7	Men et al. (2017), He et al. (2021)
Hippocampus	Dexamethasone	Impaired hippocampal function	Male/female: BDNF IV-H3K14ac	Huang et al. (2019)
		Epilepsy	Male: ACE/HMGB1	Hu et al. (2020)
		Depression	Female: miR-134-5p/Sox2	Jiang et al. (2022)
		Learning and memory disorders	Male: CDK5	Luo et al. (2023)
		Convulsion	Male: KCC2-H3K14ac	Xie et al. (2022)
	Bisphenol A	Autism	Male/female: BDNF DNA methylation	Kundakovic et al. (2015)
Liver	Dexamethasone	Nonalcoholic fatty liver disease	Male: miR122/YY1/ACE2-MAS1	Liu et al. (2021)
		Hypercholesterolemia	Female: miR148a/LDLR, HMGCR-H3K27ac Male: LDLR-H3K27ac	Li et al. (2020) Li et al. (2022) Data to be published
		Cholestasis liver injury	Female: CYP27A1-H3K14a	Data to be published
	Bisphenol A	Abnormal glucose metabolism	Male: GCK methylation	Ma et al. (2013), Li et al. (2014)
Pancreas	Dexamethasone	Pancreatic development inhibition	Female: AT2R-H3K14ac/H3K27ac	Kou et al. (2020)

(continued)

Table 22.1 (continued)

Organ	Prenatal adverse environments	Changes in organ development/related diseases	Potential biomarkers	Main references
			Male: ACE2	Dai et al. (2022)
Kidney	Dexamethasone	Fetal kidney dysplasia	Male/female: AT2R-H3K9ac	Li et al. (2019)
Testicular	Dexamethasone	Inhibition of testosterone synthesis	Male: miR-466b-3p, Star-H3K9ac	Liu et al. (2018), Liu et al. (2022)
Ovaries	Dexamethasone	Inhibition of estradiol synthesis	Female: miR-320a-3p, RUNX2/P450arom	Gong et al. (2023)
		Suppression of follicle reserve function	Female: miR-17-5p/FOXL2/CDKN1B	Gong et al. (2021)
Cartilage	Dexamethasone	Growth plate development change	Male/female: UGDH-H3K9ac	Wen et al. (2020)
		Osteoarthritis	Male/female: TGFβRI-H3K9ac	Qi et al. (2021)
Long bone	Dexamethasone	H-type vascular development changes	Female: miR34c/PDGFRβ	Shangguan et al. (2021)
		Osteoporosis	Male: ACE-H3K27ac	Xiao et al. (2018), Xiao et al. (2020a)
	Caffeine	H-type vascular development changes	Female: miR375/CTGF	He et al. (2021)
		Osteoporosis	Male: 11β-HSD2-H3K9ac	Xiao et al. (2020b)

4.1 Early Warning for Fetal-Originated Disease

Early warning involves diagnosing a disease before clinical symptoms manifest and monitoring relevant patient indicators to effectively prevent the disease's onset and progression. Once maternal-fetal circulation is established, maternal-cord-fetal blood can serve as a test sample for biomarkers of fetal development and disease susceptibility (Birch et al. 2005). IUGR is the most common manifestation of fetal dysplasia caused by adverse environments during pregnancy, and its associated early warning biomarkers have been extensively studied. Clinical studies suggest that in the plasma of normal pregnant women and those with pregnancy complications such as preeclampsia or hydatidiform mole, the methylation pattern of the signal transducer and activator of transcription 5A promoter in free DNA mirrors that in placental villi. This suggests that it could serve as a novel fetal DNA epigenetic marker (Rahat et al. 2016). The level of the delta-like homologue 1 gene (DLK1) in maternal circulation can predict the embryonic quality in mice and distinguish healthy young gestational-age infants from pathological young infants in a human cohort. Therefore, measuring DLK1 in maternal blood may be a valuable method for diagnosing diseases associated with impaired DLK1 expression in humans and predicting IUGR and pregnancy complications (Cleaton et al. 2016).

Adverse environmental conditions during pregnancy can negatively impact fetal liver development, leading to susceptibility to various metabolic diseases such as NAFLD, hypercholesterolemia, and diabetes. Some researchers have proposed that miR-122 is a promising early warning marker for fetal-originated NAFLD (Oses et al. 2019). Blood miR-122 levels in children with NAFLD have been found to positively correlate with liver injury markers and NAFLD grade (Brandt et al. 2018). Liu et al. (2021) demonstrated that changes in miR-122 expression could induce NAFLD in male offspring rats with PDE. Liver glucokinase (GCK) dysfunction can lead to different types of diabetes. Clinical studies suggest that blood GCK hypermethylation could predict the risk of type 2 diabetes in Chinese men (Tang et al. 2014). Male offspring exposed to bisphenol A during pregnancy exhibited insulin resistance, hypermethylation of the GCK promoter region, and low GCK expression (Ma et al. 2013). The F2 generation males displayed consistent changes with F1 generation males (Li et al. 2014), suggesting that the level of GCK methylation is a potential early warning marker of diabetes susceptibility in male offspring. The authors' team discovered that PDE offspring rats exhibited a predisposition to hypercholesterolemia. In these offspring, males consistently demonstrated a reduction in the expression of the low-density lipoprotein receptor (LDLR) and a decrease in the level of histone H3 lysine 27 acetylation (H3K27ac) in the LDLR promoter region, both in the liver and peripheral blood mononuclear cells (PBMC). These findings were corroborated by blood samples obtained from PDE male newborns, which were consistent with the overall results at the animal level (data to be published). Conversely, the expression of hydroxymethylglutaryl CoA reductase (HMGCR) in the liver of PDE female offspring remained consistently high, both prenatally and postnatally. Increased cholesterol levels and heightened HMGCR expression in PBMC were identified in blood samples from newborns exposed to dexamethasone during pregnancy (Li et al. 2022). Therefore, a decrease in the H3K27ac level of the LDLR promoter region and an elevated expression of HMGCR in cord blood PBMC may serve as early warning markers of susceptibility to fetal-originated hypercholesterolemia in PDE males and females, respectively.

Disruptions in early hippocampal development can precipitate cognitive impairments, including learning and memory disorders, depression, autism, and increased susceptibility to various neuropsychiatric diseases (Kim et al. 2015). A prospective case-control study leveraged the mRNA expression profile of PBMC to identify potential biomarkers for early diagnosis of autism spectrum disorder (ASD) (Glatt et al. 2012). Moreover, a metabolic index based on levels of succinic acid, glycerol, 3-hydroxybutyric acid, and O-phosphate choline in early cord blood holds promise for predicting nervous system dysplasia due to perinatal hypoxia in newborns (Boruczkowski et al. 2019). Nested birth cohort studies of ASD, focusing on bisphenol A exposure during pregnancy, have confirmed that the methylation level of the brain-derived neurotrophic factor (BDNF) promoter region and its expression in PBMC at birth could serve as predictors of BDNF DNA methylation and gene expression in the brain. This suggests its potential as a biomarker of ASD in children exposed to bisphenol A during gestation (Kundakovic et al. 2015). The authors' team discovered that the expression of cyclin-dependent kinase 5 (CDK5) in the hippocampus of PDE male offspring remained consistently elevated, leading to the inhibition of excitatory synaptic transmission and, ultimately, significant impairments in the recognition of novel objects and spatial learning and memory in adult male offspring rats. Blood samples collected from male offspring exposed to prenatal dexamethasone therapy revealed increased CDK5 expression in PBMC, suggesting that CDK5 in the peripheral blood of offspring may serve as an early warning marker of fetal-originated learning and memory impairment in PDE males (Luo et al. 2023).

Osteoarthritis (OA) is a chronic joint disease typified by degenerative pathology of the articular cartilage. It has been established that OA has fetal origins and is intimately associated with IUGR (Aigner and Richter 2012). The transforming growth factor β (TGFβ) signaling pathway plays a crucial role in cartilage formation, growth, development, maturation, and phenotypic maintenance. The authors' team has shown that hypofunctional programming of the articular cartilage TGFβ signaling pathway mediates poor quality of articular cartilage and susceptibility to OA both prenatally and postnatally in offspring rats exposed to various conditions during pregnancy, including caffeine and dexamethasone (Li et al. 2020; Zhao et al. 2020; Tie et al. 2022). Interestingly, the authors' team employed a "two-step strategy" to develop a three-dimensional chondrogenic differentiation model based on Wharton's jelly-derived mesenchymal stem cells (WJ-MSCs). It was confirmed that WJ-MSCs from IUGR individuals exhibited diminished chondrogenic differentiation capacity and increased susceptibility to an osteoarthritis-like phenotype. This was attributed to the decreased level of H3K9ac at the TGFβ receptor I (TGFβ RI) (Qi et al. 2021). In conclusion, the H3K9ac level of TGFβ RI in the human umbilical cord could serve as a potential early warning marker for predicting neonatal cartilage dysplasia and susceptibility to OA.

The adrenal gland is a vital endocrine organ in the human body, with steroidogenic factor 1 (SF1) serving as the most reliable biomarker to confirm the origin of the adrenal cortex. The authors' team has shown that the level and expression of H3K27ac in the SF1 promoter region of PBMC in male PDE offspring rats decreased both before and after birth. These changes were found to correlate positively with similar indicators in the adrenal gland. Simultaneously, a clinical study revealed that prenatal dexamethasone treatment led to a reduction in the level of H3K27ac in the SF1 promoter region and its expression in male neonatal PBMC. This suggests that a decreased level of H3K27ac in the SF1 promoter region and reduced SF1 expression in PBMC could serve as early warning indicators of adrenal insufficiency in male offspring resulting from prenatal dexamethasone treatment (Chen et al. 2022).

4.2 Potential Intervention Targets for Fetal-Originated Disease

Interventional treatment refers to proactive preventative measures taken to address the causes and exacerbating factors of a disease. The peroxisome proliferator-activated receptor (PPAR) governs many metabolic processes within the body. Research has indicated that a maternal high-fat diet during pregnancy can lead to obesity in offspring, and treatment with the PPAR agonist bezafibrate can effectively reduce obesity in these offspring (Magliano et al. 2013). The authors' team discovered that exposure to ethanol during pregnancy can inhibit the expression of Sirtuin 1 (SIRT1) in the liver of offspring, thereby leading to susceptibility to NAFLD. However, intervention with the histone deacetylase SIRT1 agonist resveratrol can partially reverse this phenomenon. This suggests that SIRT1 could be a target for interventional therapy to reduce susceptibility to NAFLD under prenatal ethanol exposure (Shen et al. 2021). Another significant study by the authors' team confirmed that in female PDE offspring, a high level of H3K14ac in the promoter region of cholesterol-27alpha-hydroxylase (CYP27A1) and its expression could persist until after birth. This is the intrauterine programming mechanism for the occurrence of cholestatic liver injury in PDE female adult offspring. Early administration of nilvadipine, an active inhibitor of CYP27A1, after birth could effectively reverse the liver injury and have long-term efficacy. This suggests that CYP27A1 could serve as an early intervention target for cholestatic liver injury in PDE female offspring (data to be published).

BDNF is known to regulate neuronal sensitivity to glutamate and neuronal plasticity. The authors' team discovered that PDE could reduce the level of H3K14ac at the BDNF IV site in the hippocampus, leading to neurobehavioral changes and an imbalance in the hippocampal glutamate system. These changes resulted in spatial memory deficits and anxiety-like behaviors in female offspring rats. It was further hypothesized that supplementing BDNF and enhancing the glutamatergic system could be potential therapeutic strategies for fetal-origin neurobehavioral abnormalities (Huang et al. 2019). In addition, adult offspring of PDE demonstrated bipolar disorder (BD)-like behavior and hyperexcitability of hippocampal neurons. This was associated with damage to the offspring's hippocampal parvalbumin interneurons due to programming changes in the activation of the neuregulin 1 (NRG1)-Erb-b2 receptor tyrosine kinase 4 (ErbB4) signaling pathway. Intervention in this signaling pathway could help mitigate the dexamethasone-induced damage to parvalbumin interneurons, suggesting that the hippocampal NRG1-ErbB4 pathway could serve as a therapeutic target for fetal-originated BD (Zhang et al. 2021).

The renin-angiotensin system is known to directly participate in bone metabolism and mediate the onset of diseases such as osteoporosis. The authors' team discovered that the peak bone mass of male PDE offspring was reduced, and the level of H3K27ac in the promoter region of the angiotensin-converting enzyme (ACE) was persistently elevated (Xiao et al. 2018). Intraosseous injection of ACE small interfering RNA (siRNA) could effectively reverse this phenomenon (Xiao et al. 2020a), suggesting that ACE is a potential therapeutic target for PDE-induced low peak bone mass in male offspring. 11β-Hydroxysteroid dehydrogenase type II (11β-HSD2) is an enzyme that inactivates GC. PDE led to a persistent inhibition of the local GC barrier function in bone and accumulation of active GC by decreasing the level of H3K9ac in the promoter region of 11β-HSD2 and its expression. This resulted in offspring rats being susceptible to osteoporosis. This phenomenon could be partially reversed after the injection of 11β-HSD2 overexpression lentivirus into the bone marrow (Xiao et al. 2020b), suggesting that 11β-HSD2 could be a target for intervention in the susceptibility to fetal-originated osteoporosis.

PDE can lead to abnormalities in pancreatic function and glucose metabolism in adult offspring. The mechanism in female offspring rats is related to a decrease in the level of H3K27ac in the angiotensin II receptor 2 (AT2R) promoter region and inhibited AT2R expression in pancreatic β cells (Kou et al. 2020). In male offspring, the mechanism is related to a persistent reduction in acetylation levels (H3K9ac, H3K14ac, and H3K27ac) in the angiotensin-converting enzyme 2 (ACE2) promoter region (Dai et al. 2022). These findings suggest that AT2R and ACE2 can be potential intervention targets for impaired pancreatic development in PDE female and male offspring rats, respectively. Prenatal glucocorticoid therapy can also affect kidney function in offspring, leading to conditions such as glomerulosclerosis, ultrastructural damage to podocytes, and reduced expression of marker genes in adult PDE offspring. This is associated with decreased levels of H3K9ac in the AT2R promoter region and AT2R expression in the fetal kidney. Cellular experiments showed that overexpression of AT2R could reverse dexamethasone-induced suppression of podocyte marker gene expression, suggesting that AT2R is a potential intervention target for PDE-induced renal development damage (Li et al. 2019).

5 Summary and Prospect

Adverse environments during pregnancy can have profound impacts on the growth, development, and health of offspring, leading to IUGR, multi-organ developmental programming changes, and increased susceptibility to multiple chronic diseases in adulthood. The underlying mechanisms are primarily related to changes in multi-organ developmental programming brought about by abnormal epigenetic modifications of functional genes in offspring, and the shift in endocrine axis developmental programming, leading to increased susceptibility to multiple diseases. This chapter provides a systematic review of the research progress on the toxicity of multiple organ development in offspring caused by adverse pregnancy environments. It proposes early warning signs and potential intervention targets for various fetal-originated disease, providing guidance for the prevention and treatment of such diseases. As research into the intrauterine programming mechanisms of fetal-originated disease deepens, the field of translational medicine is also continually promoting the transformation of primary research findings into clinical practice or applications.

References

Aigner T, Richter W. 2012. Nat Rev Rheumatol, 8(2): 70–72.
Birch L, English C A, O'Donoghue K, et al. 2005. Clin Chem, 51(2): 312–320.
Boruczkowski D, Pujal J M, Zdolińska-Malinowska I. 2019. Int J Mol Sci, 20(10).
Brandt S, Roos J, Inzaghi E, et al. 2018. Pediatric obesity, 13(3): 175–182.
Busada J T, Cidlowski J A. 2017. Curr Top Dev Biol, 125: 147–170.
Chen G, Ai C, Duan F, et al. 2022. Cell Biol Toxicol, 39(5): 2051–2067.
Chen Y, Xia X, Fang M, et al. 2021a. Sci Total Environ, 797: 149084.
Chen Y, Xu D, Xia X, et al. 2021b. Pharmacol Res, 174: 105942.
Cleaton M A, Dent C L, Howard M, et al. 2016. Nat Genet, 48(12): 1473–1480.
Crispi F, Comas M, Hernández-Andrade E, et al. 2009. Ultrasound Obstet Gynecol, 34(6): 660–665.
Dai Y, Kou H, Gui S, et al. 2022. Sci Total Environ, 826: 154095.
Dessì A, Corona L, Pintus R, et al. 2018. Expert Rev Proteomics, 15(8): 647–656.
Dobson N R, Patel R M, Smith P B, et al. 2014. J Pediatr, 164(5): 992–998.e993.
Dong W, Xu D, Hu Z, et al. 2018. Toxicol Lett, 283: 1–12.
Forand S P, Lewis-Michl E L, Gomez M I. 2012. Environ Health Perspect, 120(4): 616–621.
Glatt S J, Tsuang M T, Winn M, et al. 2012. J Am Acad Child Adolesc Psychiatry, 51(9): 934–944.e932.
Gong X, Dai S, Wang T, et al. 2023. Cell Biol Toxicol, 39(3):867–883.
Gong X, Zhang J, Ge C, et al. 2021. Pharmacol Res, 165: 105435.
Goyal D, Limesand S W, Goyal R. 2019. J Endocrinol, 242(1): T105-t119.
He H, Luo H, Liu L, et al. 2021. Faseb j, 35(2): e21370.
He X, Lu J, Dong W, et al. 2017. Arch Toxicol, 91(12): 3927–3943.
He Z, Li J, Luo H, et al. 2015. Sci Rep, 5: 17679.
He Z, Zhang J, Huang H, et al. 2019. Toxicol Lett, 302: 7–17.
He Z, Zhang J, Chen G, et al. 2021. Sci Total Environ, 792: 148440.
Hu S, Yi Y, Jiang T, et al. 2020. Arch Toxicol, 94(9): 3201–3215.
Huang H, He Z, Zhu C, et al. 2015. Toxicol Appl Pharmacol, 288(1): 84–94.
Huang S, Dong W, Jiao Z, et al. 2019. Toxicol Sci, 171(2): 369–384.
Jenner Z B, O'Neil Dudley A E, Mendez-Figueroa H, et al. 2018. Am J Perinatol, 35(5): 515–520.
Jiang T, Hu S, Dai S, et al. 2022. Cell Biol Toxicol, 38(1): 69–86.
Karsli T, Strickland D, Livingston J, et al. 2019. J Matern Fetal Neonatal Med, 32(3): 377–383.
Kim D R, Bale T L, Epperson C N. 2015. Curr Psychiatry Rep, 17(2): 5.
Kou H, Shen L, Luo H W, et al. 2017. Reprod Toxicol, 74: 85–93.
Kou H, Gui S, Dai Y, et al. 2020. Toxicol Appl Pharmacol, 404: 115187.
Kundakovic M, Gudsnuk K, Herbstman J B, et al. 2015. Proc Natl Acad Sci U S A, 112(22): 6807–6813.
Li B, Zhu Y, Chen H, et al. 2019. Toxicology, 411: 32–42.
Li G, Chang H, Xia W, et al. 2014. Toxicol Lett, 228(3): 192–199.
Li J, Xiao H, Luo H, et al. 2020. Food Chem Toxicol, 140: 111279.
Li L, Hu W, Liu K, et al. 2020. Toxicol Appl Pharmacol, 395: 114979.
Li X, Hu W, Li L, et al. 2022. Biochem Pharmacol, 206: 115306.
Liu H, He B, Hu W, et al. 2021. Biochemical Pharmacology, 185(1): 114420.
Liu L, Liu F, Kou H, et al. 2012. Toxicol Lett, 214(3): 307–313.
Liu M, Chen B, Pei L, et al. 2018. Toxicology, 408: 1–10.
Liu M, Liu Y, Pei L G, et al. 2022. Acta Pharmacol Sin, 43(6): 1461–1472.
Liu Y, Liu M, Liu Y, et al. 2022. Biochem Pharmacol, 199: 115018.
Lu J, Jiao Z, Yu Y, et al. 2018. Cell Death Dis, 9(6): 659.
Lu J, Li Q, Xu D, et al. 2020. Toxicol Lett, 331: 33–41.
Lu Z, Guo Y, Xu D, et al. 2023. Acta Pharm Sin B, 13(2): 460–477.
Luo M, Yi Y, Huang S, et al. 2023. Acta Pharmaceutica Sinica B, 13(9):3708–3727.

Lv F, Fan G, Wan Y, et al. 2021. Sci Total Environ, 789: 147691.
Ma Y, Xia W, Wang D Q, et al. 2013. Diabetologia, 56(9): 2059–2067.
Magliano D C, Bargut T C, de Carvalho S N, et al. 2013. PLoS One, 8(5): e64258.
Mandala A, Janssen R C, Palle S, et al. 2020. Nutrients, 12(10).
Men Y, Fan Y, Shen Y, et al. 2017. Endocrinology, 158(2): 402–409.
Netchine I, Azzi S, Le Bouc Y, et al. 2011. Best Pract Res Clin Endocrinol Metab, 25(1): 181–190.
Nykjaer C, Alwan N A, Greenwood DC, et al. 2014. J Epidemiol Community Health, 68(6): 542–549.
Oses M, Margareto Sanchez J, Portillo M P, et al. 2019. Nutrients, 11(12):2890.
Painter R C, Osmond C, Gluckman P, et al. 2008. Bjog, 115(10): 1243–1249.
Patelarou E, Kelly F J. 2014. Int J Environ Res Public Health, 11(6): 5904–5933.
Pei L G, Zhang Q, Yuan C, et al. 2019. J Endocrinol, 242(1): M17-M32.
Qi Y, Li B, Wen Y, et al. 2021. Stem Cell Res Ther, 12(1): 163.
Rahat B, Thakur S, Bagga R, et al. 2016. Placenta, 44: 46–53.
Shangguan Y, Wu Z, Xie X, et al. 2021. Biochem Pharmacol, 185: 114414.
Shen L, Xie L, Chen L, et al. 2021. J Toxicol Sci, 46(9): 413–423.
Song L, Yan J, Wang N, et al. 2020. Biol Reprod, 102(3): 705–716.
Sullivan T M, Micke G C, Greer R M, et al. 2010. Anim Reprod Sci, 118(2–4): 131–139.
Tang L, Ye H, Hong Q, et al. 2014. Gene, 547(2): 329–333.
Tie K, Zhao Z, Wu Z, et al. 2022. Biochem Pharmacol, 203: 115196.
Titus L, Hatch E E, Palmer J R, et al. 2019. Epidemiology, 30(5): 679–686.
Vieau D, Sebaai N, Léonhardt M, et al. 2007. Psychoneuroendocrinology, 32 Suppl 1: S16–20.
Walfisch A, Nikolovski S, Talevska B, et al. 2013. Arch Gynecol Obstet, 287(6): 1131–1136.
Wang G, He B, Hu W, et al. 2018. Toxicology, 408: 70–79.
Wang L, Shen L, Ping J, et al. 2014. Toxicol Lett, 224(3): 311–318.
Wen Y, Shi H, Wu Z, et al. 2020. Faseb j, 34(9): 12834–12846.
Xia L P, Shen L, Kou H, et al. 2014. Toxicol Lett, 226(1): 98–105.
Xiao H, Wen Y, Pan Z, et al. 2018. Cell Death Dis, 9(6): 638.
Xiao H, Wen Y, Wu Z, et al. 2020a. Bone, 141: 115578.
Xiao H, Wu Z, Li B, et al. 2020b. Br J Pharmacol, 177(20): 4683–4700.
Xie L, Jiao Z, Zhang H, et al. 2022. Toxicol Lett, 364: 12–23.
Xu D, Chen M, Pan X L, et al. 2011. Environ Toxicol Pharmacol, 32(3): 356–363.
Xu D, Liang G, Yan Y E, et al. 2012. Toxicol Lett, 209(3): 282–290.
Xu D, Zhang C, He X, et al. 2018. Toxicol Lett, 283: 39–51.
Zabeau L, Defeau D, Van der Heyden J, et al. 2004. Mol Endocrinol, 18(1): 150–161.
Zhang L, Xu D, Zhang B, et al. 2013. Arch Med Res, 44(5): 335–345.
Zhang S, Hu S, Dong W, et al. 2021. Cell Biol Toxicol.
Zhao Z, Qin J, Pei L, et al. 2020. Toxicology, 442: 152533.
Zhong H, Rong J, Zhu C, et al. 2020. Int J Neuropsychopharmacol, 23(4): 274–285.

GPSR Compliance

The European Union's (EU) General Product Safety Regulation (GPSR) is a set of rules that requires consumer products to be safe and our obligations to ensure this.

If you have any concerns about our products, you can contact us on ProductSafety@springernature.com

In case Publisher is established outside the EU, the EU authorized representative is:

Springer Nature Customer Service Center GmbH
Europaplatz 3
69115 Heidelberg, Germany

Batch number: 10756376

Printed by Printforce, the Netherlands